T0181920

Sexual Selection and the Origins of Human Mating Systems

Sexual Selection and the Origins of Human Mating Systems

Alan F. Dixson
School of Biological Sciences, Victoria University of Wellington

This book has been printed digitally and produced in a standard specification in order to ensure its continuing availability

OXFORD
UNIVERSITY PRESS

Great Clarendon Street, Oxford OX2 6DP
United Kingdom

Oxford University Press is a department of the University of Oxford.
It furthers the University's objective of excellence in research, scholarship, and education by publishing worldwide.
Oxford is a registered trade mark of Oxford University Press in the UK and in certain other countries

Reprinted 2013

British Library Cataloguing in Publication Data
Data available

Library of Congress Cataloging in Publication Data
Data available

ISBN 978-0-19-955943-5

Where are they?
I call but there is no sound.
The tide ebbs
Silently away.
Memories rise in the still air
Like smoke from many fires.
Is this the same place,
This place of ashes?

Rarawa Kerehoma

Preface

Publication of this book, in 2009, was timed to coincide with the 200th anniversary of the birth of Charles Darwin, and the 150th anniversary of the publication of *The Origin of Species*. Darwin's work changed the entire course of modern biology, and his *Descent of Man and Selection in Relation to Sex* (1871) was also, in its way, a pioneering work on sexology and evolution. Thus, Darwin attempted to apply his insights concerning evolution by sexual selection in animals to problems of human origins, sex differences in morphology, and behaviour.

The current volume addresses many of the issues first raised by Darwin (as well as some new ones) and examines them in the light of modern ideas concerning human sexuality, reproduction, and evolutionary biology. The approach taken is a comparative one, incorporating information from many fields of biology, in order to produce a credible picture of the origins of human mating systems, patterns of mate choice, and copulatory behaviour.

Many people have helped me, either directly or indirectly during the three years I spent working on this book. My wife Amanda helped with the manuscript, and in particular with typing and checking the bibliography. Our son Barnaby drew many of the graphs and histograms for me using a computer (as I still draw everything by hand and am technologically challenged). We also conducted collaborative, cross-cultural research together on various aspects of human morphology and mate choice. Senior colleagues at the Victoria University of Wellington, and especially Professor Charles Daugherty, Professor David Bibby, and Professor Paul Teesdale-Spittle, have provided every encouragement and a positive intellectual environment in which to work. At Oxford University Press, Ian Sherman and Helen Eaton advised and encouraged me during all stages of the publication process. Professor R. D. Martin (Field Museum of Natural History, Chicago) very kindly read and criticized Chapter 1, which deals with the fossil evidence of hominid evolution. As this is not my research field, I am most grateful for his input. Any errors which remain are, I hasten to add, my responsibility.

I should also like to thank the organizations and individuals listed below for permission to reproduce figures and tables, either in their original form or as redrawn and modified versions.

American Society for Reproductive Medicine: Table 3.3; **Blackwell Publishing**: Figures 3.4B, 3.13, 3.23, 4.6, and Table 3.1; **Cambridge University Press**: Figures 2.1, 2.4, 3.2, 4.5, and 8.1; **Elsevier Inc.**: Figures 3.1, 3.11, 3.14, 7.9, 8.2, and Table 2.5; **Professor F. Grine and the South African Journal of Science**: Figure 1.5; **Johns Hopkins University Press**: Figure 3.6; **Karger AG Ltd.**: Figures 3.3, 6.3, and 8.3; **Nature Publishing Group**: Figures 1.14, 3.4A, and 3.18; **New York Academy of Sciences**: Table 5.3; **Norton and Co.**: Figure 1.13; **Orion Books**: Figure 8.5; **Oxford University Press**: Figures 1.3, 1.12, 3.16, 3.20, 3.21, 4.2, 5.3, 5.4, 5.8, 5.9, 6.6, 6.7, 6.8, 6.9, 6.10, 7.1, 7.8, 7.18, 8.4, 8.7, 8.8, and Tables 1.1, 1.2, 4.1, 4.2, 5.1, 5.2, 5.4, and 8.1; **Plenum Press**: Figure 5.2; **Random House Inc., and Alfred A. Knopf**: Figure 7.13; **Raven**

Press: Figure 3.9; **The Royal Society**: Figures 7.14, 7.16, and 7.17; **Sinauer**: Figure 5.1; **Professor Devendra Singh**: Figure 7.8; **Society for the Scientific Study of Sex**: Figure 7.19; **Society for Reproduction and Fertility:** Figures 5.6, 6.1, and 6.2; **Society for the Study of Reproduction**: Figure 3.10; **University Science Books**: Figure 3.15; **Professor Frans de Waal**: Figure 5.5; **University of Chicago Press**: Table 7.3; **Wiley**: Figures 1.10, 3.5, 6.4, and 7.4.

Professor Alan Dixson
School of Biological Sciences
Victoria University of Wellington
New Zealand
alan.dixson@vuw.ac.nz

Contents

List of Abbreviations

BMI	body mass index
DSP	daily sperm production
EPC	extra-pair copulation
MP	multiple partners
SP	single partners
WHR	waist-to-hip ratio

CHAPTER 1

A Glance at the Terrain

'There is a gift of being able to see at a glance
What prospects are offered by the terrain.'

Napoleon

'We see nothing till we truly understand it.'

John Constable

The purpose of these essays is to explore the evolutionary origins of human mating systems, the physical determinants of sexual attractiveness, mate choice, and patterns of copulatory behaviour. These are ambitious goals, given that sexual behaviour does not leave a fossil record.

We are all the products of an unbroken chain of sexual and reproductive events, extending back through time, to connect present generations with the ancestors of *Homo sapiens* and with human origins in Africa, more than 195,000 years ago. From such remote beginnings, and from small numbers, our populations have expanded across the globe, to exceed 6 billion people by the start of the present millennium. The most worrying graph ever drawn in biology may be that which depicts the growth of human numbers. I say this because, during my own lifetime, the population of the world has more than doubled. The overall biodiversity of the earth has, in consequence, declined, due to human activities which are now seriously fuelling global warming and climate change. It has been predicted that this catastrophic growth in human numbers may slow by 2050, to plateau at approximately 9.5 billion people. Even assuming that such demographic projections are accurate, and there is no assurance that they are, many of our descendents will find that life upon this population 'plateau' is far from pleasant. For, lacking access to clean water, adequate food, and medical care, let alone the benefits of a decent education, existence for billions of people will continue to be, in the words of Hobbs, 'nasty, brutish and short'.

The great geneticist Theodosius Dobzhansky famously commented that 'nothing in biology makes sense except in the light of evolution.' Thus, apart from its intrinsic interest, some attempt to understand the origins and nature of human sexuality is important, as it may provide a useful basis for consideration of current problems associated with human reproduction. Darwin's (1859; 1871) discoveries, concerning the laws of natural and sexual selection, by which evolution operates, paved the way for advances in every field of biology. His studies of intra-sexual and inter-sexual selection in animals led him to propose that these same processes had also shaped the evolution of various sexually dimorphic traits in human beings. He was intrigued by differences in the physique and facial and secondary sexual traits displayed by different human populations, and speculated as to whether sexual selection had influenced these variations. Both Charles Darwin and Thomas Henry Huxley (1863) also deduced that the great apes represent our closest relatives among the extant primates. For Darwin, the African apes were especially important in this context. He reasoned that humans had originated in Africa. These were remarkable insights since, at that time, there was no fossil evidence of human evolution in Africa and information about primate behaviour was largely anecdotal in nature. The only fossil hominids known to Darwin were a few Neanderthal specimens. Contrary to popular belief, fossil evidence played very little part in the pioneering work of Darwin and Huxley on human evolution. No formal science of genetics existed except for Mendel's pioneering work using pea plants, which remained largely unknown to his contemporaries. The scientific study of reproductive physiology and human sexuality was, likewise, still in its infancy. Indeed, Darwin's (1871) contributions regarding sexual selection and human evolution were largely neglected until many years after his death.

This book, in large measure, pays homage to Charles Darwin's early insights into the nature of sexual selection and its impact upon human evolution. However, for modern biologists who seek to understand the origins of human sexual behaviour, the intellectual 'terrain' now offers many advantages and a depth of understanding which was not available to Darwin.

1. A much more substantial fossil record of human evolution now exists, both as regards early hominids in Africa (such as the australopithecines) and the subsequent evolution of the genus *Homo*, in Africa and beyond. The fossil evidence of human evolution will be addressed below. Although these fossils cannot convey direct information concerning sexual behaviour, they provide an essential basis for understanding the origins of distinctive human traits such as bipedalism and large brain size. As we shall see in a moment, information about the degree of sexual dimorphism in fossil hominids is also potentially valuable for understanding their mating systems. For example, marked body size dimorphism is associated with effects of sexual selection via inter-male competition, and is characteristic of polygynous mammals, such as the gorilla. A controversial question concerns the existence and degree of such sexual dimorphism among the various fossil hominids.

2. Genetic studies have confirmed the close phylogenetic relationship between *H. sapiens* and the African apes. The pioneering work of Allan Wilson and his colleagues, carried out at Berkeley during the 1960s (Sarich and Wilson 1967; Wilson and Sarich 1969), employed calculations of DNA mutation rates to calibrate a 'molecular clock', which was used to time the evolutionary divergence between human and chimpanzee ancestors at some point between 5 and 4 million years ago (MYa). Forms which were ancestral to the gorilla diverged somewhat earlier. Modern research places the time of divergence between chimpanzee and human ancestors at around 8 million years, whilst fossil evidence of a possible ancestor of the gorilla has been discovered in Ethiopia, and dated at 10 million years (Suwa et al. 2007). Human population genetics has also clarified that anatomically modern *H. sapiens* most likely originated in Africa, and that all of us are the descendents of African ancestors, rather than representing the products of 'multiregional' evolution from *H. erectus* populations in different parts of the world (Stringer 2002; Manica et al. 2007; Behar et al. 2008).

3. A huge amount is now known about the sexual and social lives of the nonhuman primates (Smuts et al. 1987; Dixson 1998*a*; Jones 2003; Kappeler and Van Schaik 2004; Campbell et al. 2007), so that a rich source of comparative information is available to help place human sexual behaviour and mating systems in an evolutionary perspective.

4. Darwin considered that sexual selection operates primarily at the pre-copulatory level, via intermale competition, for greater access to females, and via inter-sexual selection to enhance attractiveness to the opposite sex. However, it is now appreciated that sexual selection also occurs at copulatory and post-copulatory levels. Females of many species mate with multiple partners, resulting in the potential for sperm competition (Parker 1970) and cryptic female choice (Eberhard 1985, 1996) to profoundly influence reproductive success. Sexual selection has affected the structure and functions of the genitalia in both sexes. Comparative studies of the anatomy and physiology of the reproductive organs as well as detailed analyses of copulatory behaviour have much to tell us about the evolution of human reproduction.

5. To this array of evolutionary tools, we may add insights gained from cross-cultural, anthropological studies of sexual behaviour and reproduction (Ford and Beach 1951; Betzig, Borgerhoff Mulder, and Turke 1988; Betzig 1997; Ellison 2001), from research in evolutionary psychology, as it relates to human

sexuality (Symons 1979; Daly and Wilson 1983; Buss 1994; 2005; Barrett, Dunbar, and Lycett 2002; Kauth 2006), and from the vast medical literature on human sexuality and reproduction (Johnson and Everitt 1988; Bancroft 1989; Knobil and Neill 1994; Le Vay and Valente 2002; Piñón 2002).

Given all these potential sources of information, and the huge scope offered by the subject of human sexuality, it will be helpful to define not only the goals, but also the limits of the current exercise. As Darwin correctly deduced, humans evolved from non-human primate ancestors, ape-like creatures which Foley (1995) has aptly called 'Humans before Humanity'. Darwin (1871) noted that, despite all his exalted powers 'Man still bears in his bodily frame the indelible stamp of his lowly origin.' Therefore, any serious attempt to understand the 'origins' of human sexuality must take account of traits that derive from earlier periods of evolution, as well as those traits which are more distinctively 'human'. Thus, in what follows, the goal is to examine how in ancestral hominids, lacking language and having smaller brains and less developed intellectual capacities than ourselves, the conditions were created for the emergence of human patterns of mate choice and copulatory behaviour. In order to attain this goal, I shall rely heavily upon insights gained from comparative studies of the anatomy and reproductive biology of extant primates. Effects of sexual selection upon the evolution of the primary genitalia and patterns of copulatory behaviour in both sexes are explored in Chapters 2–5. The evolution of sexual behaviour in relation to the menstrual cycle is considered in Chapter 6. Chapters 7 and 8 deal with the origins of sexually dimorphic traits in *H. sapiens*, and the role played by sexual dimorphism in bodily and facial cues in relation to sexual attractiveness and mate choice. The origins of human kinship systems, incest avoidance, and pair bonding have recently been analysed by Chapais (2008). His findings are incorporated into Chapter 9, which presents an overview, and conclusions, based upon the previous eight chapters.

Before entering upon this task, however, some introductory remarks are essential, concerning hominid palaeontology and the fossil evidence of human evolution.

Always Something New Out of Africa (*Ex Africa Semper Aliquid Novi*)

Pliny, The Elder

It was once thought that human evolution had progressed in a linear and stepwise fashion, from simpler to more complex forms, culminating in *H. sapiens*. However, as the fossil record has grown, it has emerged that a diverse assemblage of hominids existed in the past, some of which were ancestral to humans, whilst others occupied separate branches of an extensive evolutionary tree. The relationships between these fossil forms have occasioned much debate among palaeontologists and will, doubtless, continue to do so. The arrangement shown in Figure 1.1 should be regarded only as a summary of current knowledge, as the hominid tree will certainly expand and change in the future.

Firstly, at the base of the tree, whose African roots include the common ancestors of humans and chimpanzees, are several fossil ape genera, known only from fragmentary remains: *Sahelanthropus*, *Orrorin*, and *Ardipithecus* (Stringer and Andrews 2005; Sawyer and Deak 2007).

A single cranium of *Sahelanthropus tchadensis* has been discovered; it exhibits hominid features such as a flat vertical face, small canines, and a foramen magnum situated on its undersurface, possibly indicating that the head was held in an upright position. However, no parts of the postcranial skeleton are available, and it is not known if this animal walked upright. It existed in the Saharan region, in what is now Chad, about 7 MYa. Its brain was quite small (360 cc), and certain features of its nasal anatomy indicate that it may have pre-dated the common ancestors of chimpanzees and humans (Brunet et al. 2002; Lebatard et al. 2008). Hominid affinities of the skull have been reinforced by the results of studies using CT scans (Zollikofer et al. 2005).

Even less well known is *Orrorin tugenensis* which occurred in Kenya about 6 MYa. Only some jaw and limb fragments of this ape have been described, and although there has been speculation that it may have walked upright, there is insufficient information to reconstruct its locomotor patterns. *Ardipithecus* is likewise only known from fragments of its jaws and postcranial skeleton. Two species have been described: *A. ramidus* (4.4 MYa) and *A. kadabba*

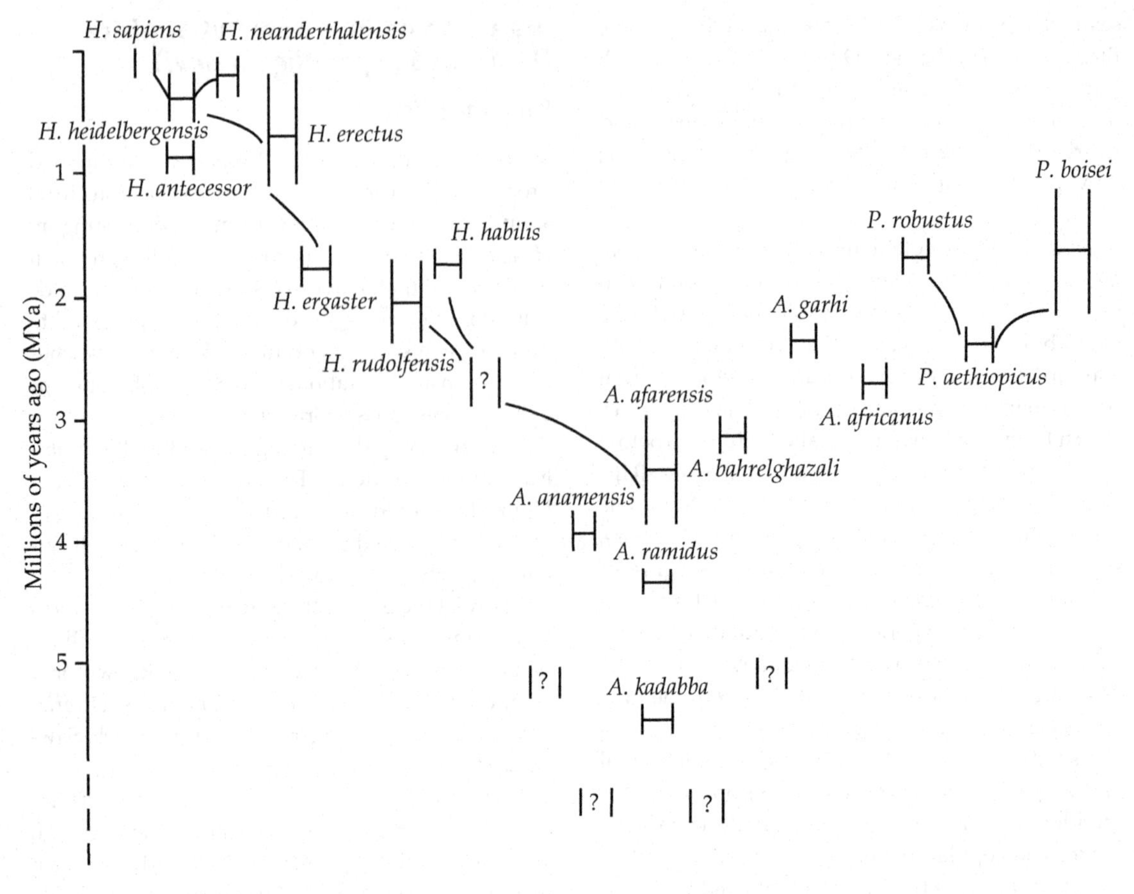

Figure 1.1 Approximate time spans and evolutionary relationships of the main hominid taxa. Please note that the abbreviation "*A*" is used for the genus *Ardipithecus* (ie *A. kadabba* and *A. ramidus*) and for members of the genus *Australopithecus*. Further details are given in the text. *Source*: Based upon information in Tattersall and Schwartz (2001); Wood (2001); Collard (2002), and Stringer and Andrews (2005).

(5.5 MYa). *Ardipithecus*, *Sahelanthropus*, and *Orrorin* occupy the base of the hominid lineage (Figure 1.1) from which the subfamily Australopithecinae and, ultimately, the genus *Homo* arose.

Five species of *Australopithecus* (literally 'southern ape') have been discovered, at widely separated sites in Southern, Eastern, and Central Africa (*Australopithecus anamensis, A. afarensis, A. bahrelghazali, A. africanus, and A. garhi*). These were relatively small-brained creatures (brain volume 400–500 cc), but with notably reduced canine teeth compared to those of modern apes. *Australopithecus* was capable of terrestrial, bipedal locomotion as well as being adept at climbing trees. Best known is *A. afarensis* (4.2–3.2 MYa), the most complete skeleton being that of 'Lucy' from Hadar, in Ethiopia (Johanson and White 1980; Johanson and Edey 1981). It displays an intriguing mixture of traits, such as long, curved fingers and toes, a funnel-shaped rib cage typical of apes, and a broad, short pelvis with some similarities to that of human beings (Figure 1.2).

Lucy may have weighed approximately 30 kg, but it has been argued that adult males of this species, and of some other australopithecines, were considerably larger than adult females. Table 1.1 includes estimates of body weight for adults of both sexes of various fossil hominids, as well as for modern *H. sapiens*. At 44.6 kg, males of *A. afarensis* were 1.52 times larger than adult females. If correct, this sex difference in body size would be consistent with effects of sexual selection, via intermale competition, such as occurs among extant

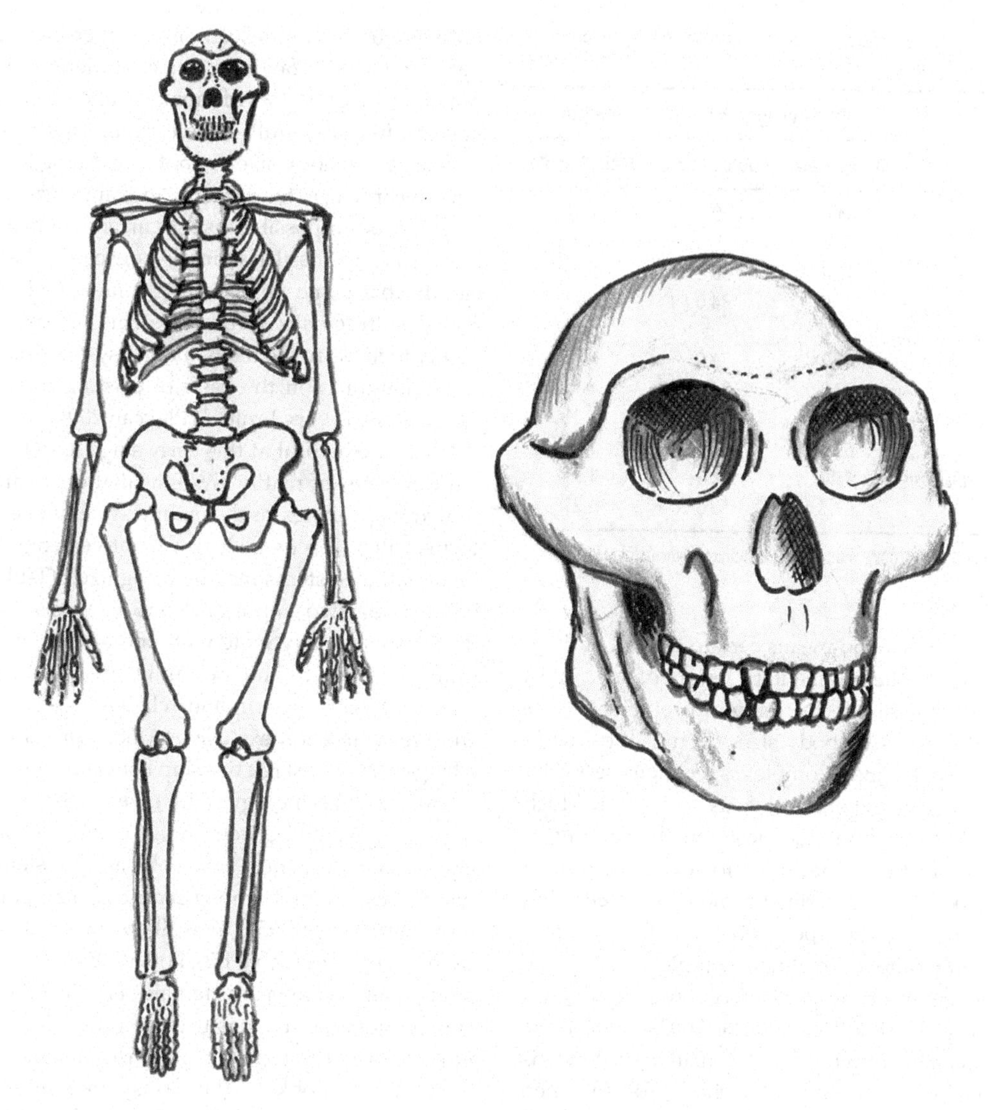

Figure 1.2 Reconstruction of the skeleton of *Australopithecus afarensis* (Lucy).
Source: Author's drawings, after Johanson and Edey (1981).

monkeys and apes which have polygynous mating systems. Intra-sexual selection has favoured the evolution of body size sexual dimorphism in a variety of Old World monkeys, such as geladas (adult male: female body weight ratio = 1.6), hamadryas baboons (1.7), and proboscis monkeys (2.0) as well as gorillas (2.3). However, a major difference between all these species and the australopithecines is that intra-sexual selection has also resulted in the evolution of markedly larger canine teeth, as well as greater body size in males of the extant forms. Why the australopithecines and all later fossil hominids have relatively small canines has not been explained. However, it is possible that some overriding selective process, perhaps connected with feeding ecology, might account for the reductions in canine size, and lack of dental sexual dimorphism in the australopithecines.

Estimates of body weights and of sexual dimorphism in the fossil hominids listed in Table 1.1 must be treated with considerable caution. Although such estimates are often cited in the anthropological

Table 1.1 Estimates of body size and sexual dimorphism in fossil hominids, and in modern *Homo sapiens*

	Body weight (kg)		Ratio
Species	**Adult male**	**Adult female**	**Male/Female**
Australopithecus afarensis	44.6	29.3	1.52
A. africanus	40.8	30.2	1.35
Paranthropus boisei	48.6	34.0	1.43
P. robustus	40.2	31.9	1.26
Homo habilis	37.0	31.5	1.17
H. ergaster	63.0	52.0	1.21
H. erectus	63.0	52.5	1.20
H. neanderthalensis	73.7	56.1	1.31
H. sapiens	63.5	52.3	1.21

Source: After Collard (2002), and based upon multiple sources.

literature, it is by no means certain that all of them are accurate. The fragmentary nature of specimens of various fossil species renders problematic any attempt to measure body size. Even in the case of Lucy, one of the most complete specimens, less than 40 per cent of the skeleton was recovered. Much less is known about the postcranial anatomy of the other australopithecines. However, Alemseged et al. (2006) have described a more complete skeleton of an immature specimen of *A. afarensis*, from Dikika in Ethiopia. An almost complete skeleton of *A. africanus* has been partly recovered from Sterkfontein in South Africa (Clarke 2002). Then there is the question of whether a particular specimen is from a male or a female, not a trivial problem when only small portions of the skull or skeleton may be available. Thus, in the case of *A. afarensis*, there are major differences of opinion among scholars concerning sexual dimorphism in body size. Reno et al. (2003) conducted a careful analysis of skeletal dimensions in this species, and reached the conclusion that it was not markedly sexually dimorphic. These authors therefore suggested that sex differences in body weight in *A. afarensis* were within the range displayed by modern humans and were consistent with a monogamous mating system, rather than with the occurrence of polygyny.

Although sex differences in body size are most pronounced in extant anthropoids which are polygynous, sexual selection, via inter-male competition, has also favoured the evolution of dimorphism among Old World monkeys and apes which have multi-male/multi-female mating systems. Sex differences in body size are not so extreme in these circumstances, as for example in chimpanzees, but it would be impossible using current estimates of body sizes in fossil hominids to exclude the possibility that some of them might have had multi-male/multi-female mating systems. Clearly, our ability to identify the likely mating systems of these early hominids on the basis of existing fossil evidence alone is very limited (Plavcan 2001).

It may be helpful at this early stage to define the various types of mating systems that occur among the extant primates, as the mating systems of extinct forms will be discussed in comparative perspective. Five mating systems may be recognized (Table 1.2) based upon two important considerations. Firstly, do females usually mate with one partner, or with multiple partners and, secondly, are these sexual relationships long term and relatively exclusive or short term and non-exclusive? Monogamy involves a long-term sexual relationship between a male and a female, whilst polygyny involves long-term sexual relationships between multiple females and a single male. Both these kinds of mating systems are widely represented among recent human cultures (Ford and Beach 1951) as well as among the non-human primates (Table 1.2). Polyandry (a long-term relationship between one female and several males) is exceptional, but it has been recorded for people in parts of northern India, and also in some of the New World monkeys (tamarins). In multi-male/multi-female mating systems, by contrast, females mate with a number of partners in a more labile fashion, and the same is true for males, given the constraints of social rank and other limiting factors within groups. Sexual relationships are short term in such systems which occur in macaques, many baboons, and chimpanzees. Human beings do not exhibit this type of a mating system. Nor is the dispersed mating system (Table 1.2) found in human populations, as this is typical of non-gregarious nocturnal primates, such as mouse lemurs (*Microcebus*) which live in individual home ranges and engage in short-term relationships and multiple-partner copulations.

Table 1.2 A classification scheme for primate mating systems

Number of males mating per female cycle	Type of sexual relationship	Mating system	Examples
One	Long term, exclusive	Monogamy	*Indri, Aotus, Callicebus, Hylobates, Homo sapiens*
One	Long term, exclusive[†] but other females also mate with the resident male	Polygyny	*Theropithecus, Papio hamadryas, Nasalis, Gorilla, Homo sapiens*
Two or more	Long term, exclusive	Polyandry	*Saguinus fuscicollis**, *Callithrix humeralifer**
Two or more	Short term, not exclusive, gregarious	Multi-male/multi-female	*Propithecus, Macaca*, most *Papio spp., Cercocebus Saimiri, Lagothrix, Pan*
Two or more	Short term, not exclusive, non-gregarious	Dispersed	*Microcebus, Daubentonia*, most *Galago spp., Perodicticus*

[†]The single male mates with other females in the 'harem' unit.
*Monogamy also occurs in these callitrichid species, and is probably the primary mating system in many cases.
Source: Modified from Dixson (1998*a*).

Returning to the australopithecines, attempts have been made to reconstruct what the various species might have looked like in the living state. As an example, Figure 1.3 shows a reconstruction of *A. africanus*, the first of the 'southern apes' to be discovered (Dart 1925). This species occurred in Southern Africa, approximately 3.0–2.3 MYa and, like *A. afarensis*, it is thought to have been proficient at climbing as well as walking upright on the ground. It had relatively longer arms and shorter legs than *Homo*. In Figure 1.3, a male and female *A. africanus* are shown walking upright, in much the same way as human beings. There is no assurance that they moved in precisely the same way as we do, however, as they possessed a combination of locomotor traits not found in humans or in any existing primate species. This explains why so much controversy has surrounded attempts to interpret the locomotor and postural capabilities of the australopithecines; indeed some earlier scholars, such as Zuckerman, refused to accept that they might be ancestral to humans. As the anatomical evidence has accumulated over the years, it reinforces the conclusion that these creatures stood and walked upright, as well as climbing and using their arms for arboreal suspension and locomotion. At Laetoli, in Tanzania, well-defined tracks of australopithecine footprints have even been discovered, dated to about 3.5 MYa. These are thought to have been made by *A. afarensis* as they walked upright through an area covered in moist volcanic ash (Leakey and Hay 1979).

The external appearance of *A. africanus*, as depicted in Figure 1.3, is inevitably subject to much uncertainty. Surface morphology is heavily defined by traits which do not fossilize, such as the hair and skin. Extant African apes have dark skin due to deposits of epidermal melanin. Likewise African populations of *H. sapiens* are, typically, dark skinned, and it is thought that this would have been the case in the earliest human populations in Africa. Melanic pigmentation offers protection against exposure to high levels of UV radiation in the tropics (Jablonski and Chaplin 2000; Jablonski 2006). In Figure 1.3, *A. africanus* is shown as having much less hair on its body and limbs than today's chimpanzees or gorillas. There is no assurance that this is correct. As apes which lived in arboreal environments, as well as in more open, grassland conditions, it is possible that the australopithecines were more hirsute than is shown in Figure 1.3. Loss of hair may only have become pronounced in later hominids, in association with increased density of cutaneous sweat glands and other specializations to improve thermoregulation in large-bodied and

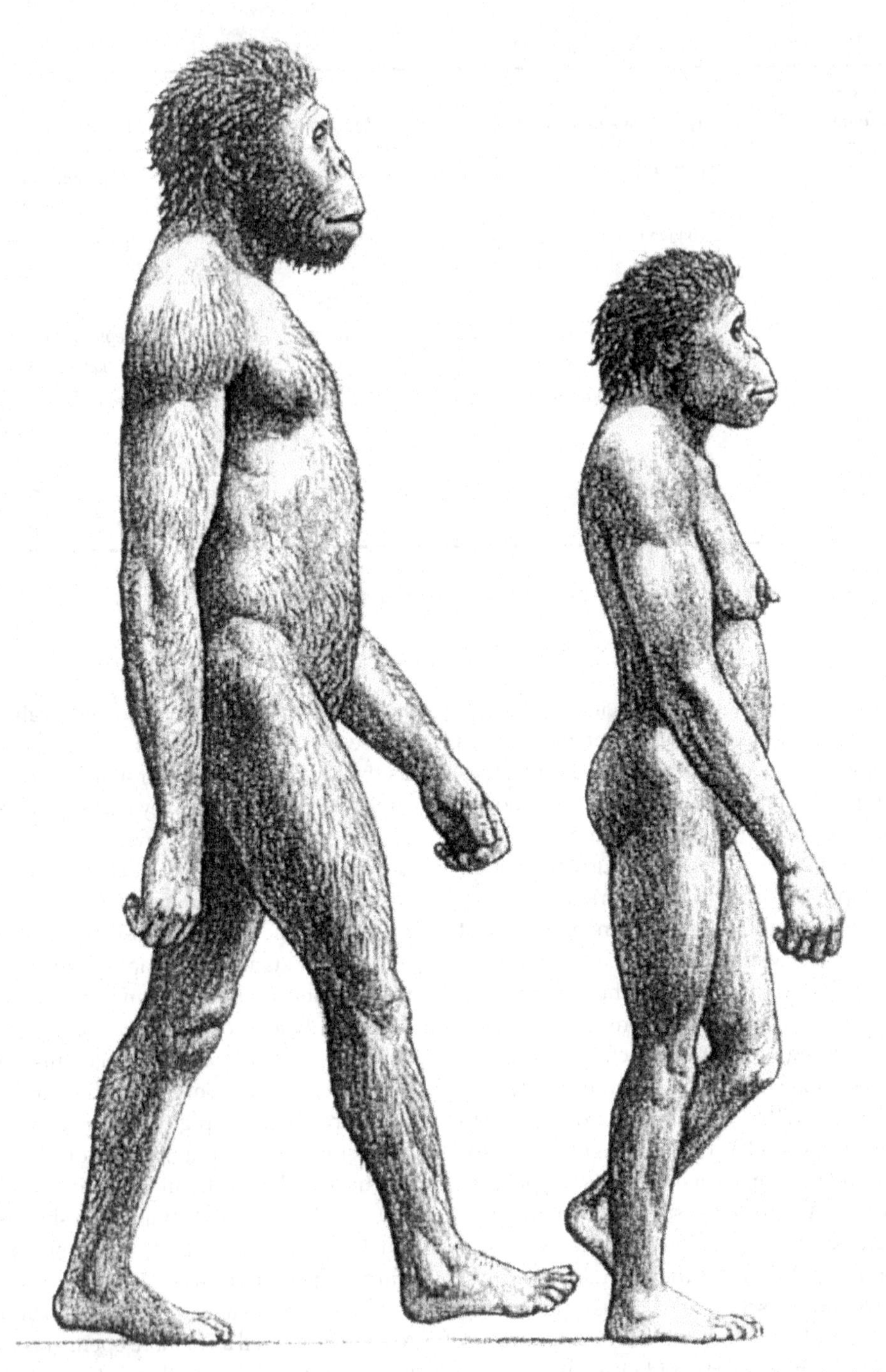

Figure 1.3 What did the australopithecines look like? Reconstruction of *Australopithecus africanus*. The adult male is shown on the left.
Source: After Tudge (2000).

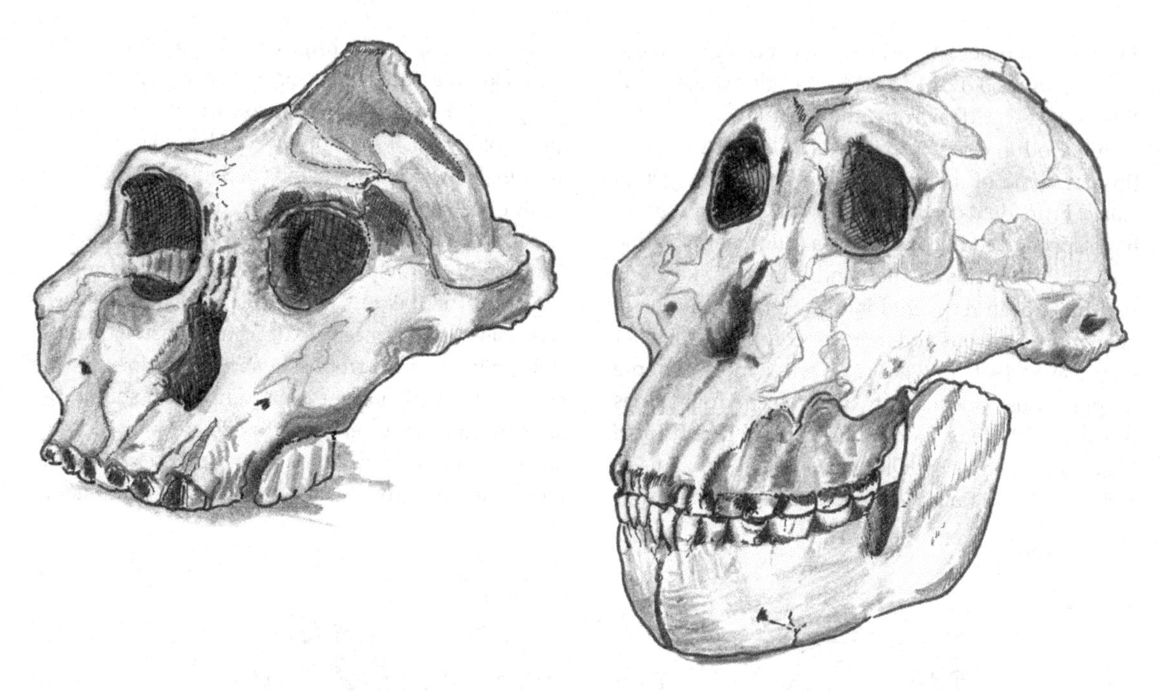

Figure 1.4 Reconstructions of the skulls of robust australopithecines. Left: *Paranthropus aethiopicus* (the 'black skull'). Right: *P. boisei* ('Zinj', or 'nutcracker man'). Author's drawings from photographs.

entirely bipedal terrestrial hunter–gatherers such as *H. ergaster* and *H. erectus*.

Although the genitalia are not visible in Figure 1.3, the larger individual on the left is clearly intended to represent an adult male. As we have seen, however, there is currently no certainty that sexual dimorphism in body size was this pronounced in the australopithecines. If indeed these creatures were markedly sexually dimorphic and polygynous, then the males in particular may also have possessed striking cutaneous secondary sexual adornments, such as capes and crests of hair, or fleshy facial elaborations, as these are often present in males of extant anthropoid species which have polygynous mating systems (Dixson, Dixson, and Anderson 2005). The smaller female has a protruding breast, with an areola area surrounding the nipple, as in *H. sapiens*. Again, inclusion of this detail owes something to artistic license. Breast enlargement due to fat deposition and the visually prominent pigmented areola area surrounding the nipple may have arisen much later in hominid evolution. As we shall see, in Chapter 7, fat deposition occurs at puberty in the breasts, buttocks, and thighs of women, and sex differences in fat distribution have been affected both by natural selection and sexual selection during human evolution. These traits are much more likely to have emerged in early *Homo* (e.g. in *H. ergaster and H. erectus*) in association with changes in body shape, loss of hair, and other features connected with a hunter–gatherer existence in Africa.

Closely related to the genus *Australopithecus*, but currently regarded as occupying a distinct side-branch of the hominid evolutionary tree, are the robust australopithecines (genus *Paranthropus*) which occurred in Eastern and Southern Africa during the late Pliocene and early Pleistocene. They are known primarily from cranial and dental remains, and the term 'robust' relates to the powerful development of their jaws and large molar and premolar teeth. Their body sizes are thought to be similar to those of *Australopithecus* (Table 1.1).

Oldest of the three currently described species is *Paranthropus aethiopicus* (2.7–2.3 MYa) from Kenya and Ethiopia. Figure 1.4 shows the most complete skull recovered (the so-called 'black skull'), with its prominent sagittal crest (for attachment of the jaw muscles) surmounting a small cranium (brain size:

410cc). Two other species have also been described: *P. robustus* (2.0–1.0MYa, from South Africa) and *P. boisei* (2.2–1.2 MYa, from Ethiopia, Kenya, and Tanzania). Both of these species were alive at the same time as some of the earliest members of the genus *Homo*. From dental evidence, *P. robustus* is thought to have specialized in consuming a tough herbivorous diet. Such remains of the hand and foot skeletons that have been recovered indicate that it may have been more dextrous and more bipedal than *A. afarensis* (Susman 1988; 1994). *P. boisei* was a more extreme 'hyper-robust' form, with molar teeth four times the size of those of modern humans. When *P. boisei* was discovered, at the Olduvai Gorge in Tanzania in 1959, it was originally assigned to a separate genus (*Zinjanthropus*) and its huge chewing teeth earned it the nickname 'nutcracker man' (Figure 1.4).

Figure 1.5 shows Grine's reconstruction of what a group of *P. robustus* might have looked like, as it foraged in the wooded grasslands of Southern Africa. An adult male *Paranthropus* standing upright is depicted as markedly larger than the females in the social group. The face of the male is more massive, and there is a suggestion of a beard. This is in accord with published accounts of pronounced sexual dimorphism and possible polygyny in the robust australopithecines. However, to counterbalance this assumption it should be kept in mind that very few fossils are known with any certainty to represent females of these apes. One female in the group is shown nursing her infant,

Figure 1.5 Grine's representation of a group of *Paranthropus robustus*, as it might have appeared in life, at Swartkrans in South Africa
Source: Professor Fred Grine and the South African Journal of Science.

whilst next to her a second female is digging for tubers with a stick. Tool use of this kind is unconfirmed for the robust australopithecines, but is likely to have occurred. The hand is thought to have been adapted for a range of movements; In Figure 1.5, the female employs a 'power grip' to wield her digging stick, whilst in the foreground another member of the group is picking berries using a 'precision grip' of the thumb and index finger. Chimpanzees are known to use a variety of simple tools (McGrew 1992), and the occasional use of sticks as tools has also been documented in the gorilla and the orangutan. *P. robustus* is thought to have resembled the chimpanzee in body weight yet, at 515 cc, its brain was approximately 100 cc larger than that of a chimpanzee. It is generally agreed that *Australopithecus* and *Paranthropus* both had larger brains, relative to body mass, than is the case for the extant great apes (Martin 1983, 1990). Thus it is reasonable to infer that *P. robustus* might have used simple tools, although there is no fossil evidence of this. Not until the advent of *H. habilis* is there evidence of an association between hominid fossil remains and more durable (stone) tools.

The genus *Homo* is not thought to have arisen directly from any of the robust australopithecines described here. However, there are indications that *Homo* is closely related to *Paranthropus* (Strait, Grine, and Moniz 1997), so an as yet undiscovered stem form may be ancestral to the human genus. Both *A. africanus* and *A. afarensis* have also been proposed as possible human ancestors at various times, but further discoveries or reinterpretations of existing material will be required to resolve this question (Fleagle 1999).

Uncertainties concerning the precise origins of the genus *Homo* are linked to uncertainties about relationships between the earliest members of the genus, several species of which are known to have occurred in Africa at around 2 MYa. *H. habilis* lived in Southern and Eastern Africa, from approximately 2.3 to 1.6 MYa. The initial discoveries of these fossils were made at the Olduvai gorge in Tanzania during the 1960s, and were assigned by Louis Leakey and his colleagues to the genus *Homo* (Leakey, Tobias, and Napier 1964). The relatively small brain size (590–690 cc) of *H. habilis*, its ape-like limb proportions and resemblance to australopithecines have led to considerable debate about its taxonomic affinities (Wood 1992; Walker and Shipman 1996). Recently, for example, it has been reported that *H. habilis* and *H. erectus* were sympatric in East Africa, approximately 1.55 MYa, making it unlikely that *H. habilis* represents a direct ancestor of modern humans (Spoor et al. 2007). Palaeontologists have found it difficult to determine the affinities of some fossils which may, or may not, belong to *H. habilis*. Oldowan tools (hammer stones and simple flakes) have been found associated with the fossilized remains of this species, hence its Latin name *H. habilis*, which means 'handy man'. It has smaller molar and premolar teeth than the australopithecines, and a more advanced foot anatomy, as well as a robust hand with curved, chimpanzee-like fingers and a large thumb.

The marked size differences exhibited by fossils of *H. habilis* have led to speculation that it might have been sexually dimorphic. However, it is unknown in many cases whether individual fossils belong to males or females, or even whether they represent members of different species. With so little evidence, speculation concerning the mating system of *H. habilis* is of little value. Smith (1984), for example, in discussing the evolutionary origins of human sexual behaviour and sperm competition, argued that *H. habilis* engaged in multi-partner matings, and that sperm competition would have led to the evolution of large testes and a large penis in males of this species. At the time when he advanced these ideas, *H. habilis* was the major candidate for the earliest species of the genus *Homo*; thus Smith attempted to identify traits consistent with sperm competition in what was then thought to be the direct ancestor of *H. sapiens*. However, as we shall see in Chapters 3 and 4, the assumption that sperm competition has played a significant role in the evolution of the genus *Homo* is not supported by extensive comparative studies of the reproductive anatomy and behaviour of extant primates.

The realization that some of the fossils originally designated as *H. habilis* were too divergent to represent a single species led to the recognition that another species of early *Homo* existed in East Africa during the same time period. This was *H. rudolfensis*, named by Alexeev (1986) in honour of its discovery site at Lake Rudolf (now Lake Turkana), in Kenya. It is best known on the basis of the skull (KNM-ER 1470), described by Leakey

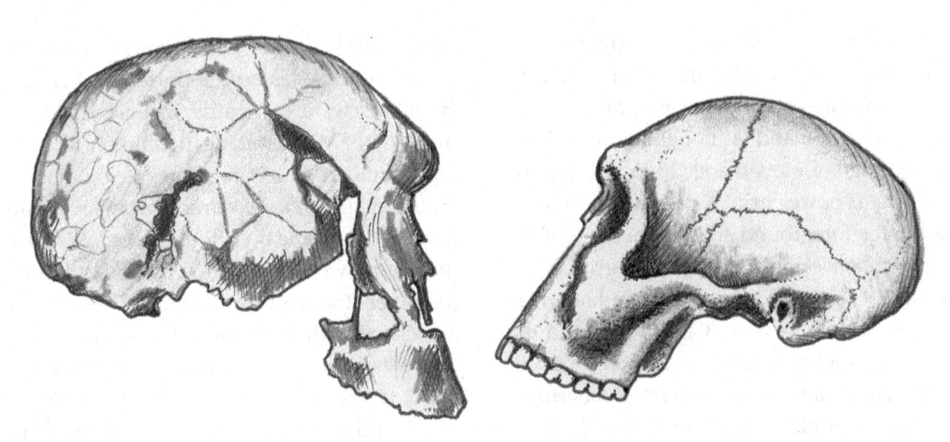

Figure 1.6 Two possible reconstructions of the KNM-ER 1470 skull (*Homo rudolfensis*). On the left, the face has a more vertical orientation, as is typical of members of the genus *Homo* (Leakey 1973). On the right, a more sloping, australopithecine facial morphology is shown (Walker and Shipman 1996). Author's drawings from photographs.

(1973), of a larger-brained (752 cc) hominid which lived approximately 1.9 MYa (Figure 1.6). Very little can be said concerning its post-cranial anatomy however, largely because of uncertainties about whether fossils currently assigned to *H. habilis* or *Paranthropus* might belong to *H. rudolfensis*.

It is important to reiterate that the taxonomic status of *H. habilis* and *H. rudolfensis* has been much debated by palaeontologists. The strong affinities shown by both these hominids with the australopithecines have led some authorities to recommend that they should not be included in the genus *Homo* (Wood and Collard 1999). However, retention of both taxa in the genus *Homo* is supported by cladistic analyses of hominoid craniofacial traits (González-José et al. 2008). On the other hand, it is interesting that Alan Walker, who has played a key role in reconstructing these fossils, believes that the face of KNM-ER 1470 was more projecting and australopithecine in its morphology than the reconstruction presented by Richard Leakey (both interpretations are shown in Figure 1.6). Thus, Walker and Shipman (1996) point out that '1470 might have a big braincase, but morphologically it was just an australopithecine.... Ignoring the cranial capacity, the overall shape of the specimen and that huge face grafted onto the braincase were undeniably australopithecine.'

Much more extensive fossilized remains of a third type of hominid, *H. ergaster*, have also been found in Africa, spanning the period between 1.9 and 1.5 MYa. By far the most complete specimen, and one of the most spectacular finds ever made in hominid palaeontology is the so-called 'Turkana (or Nariokotome) boy' (Figure 1.7). Eighty per cent of the skeleton of this individual, an immature male estimated to have been about 12 years old, was recovered from Nariokotome, to the west of Lake Turkana. As a result, a huge amount has been learned about the functional anatomy and the likely capacities of *H. ergaster* (Walker and Leakey 1993; Walker and Shipman 1996). Although not fully grown, it is likely that the Turkana boy would have reached 6 ft in height as an adult, much taller than any of the earlier hominids. The limb proportions were, indeed, similar to those of modern humans but more robust and with a longer shin bone; *H. ergaster* was probably well adapted for long-distance walking and running. The jaw was reduced in size, and the cheek teeth were smaller than in the australopithecines or *H. habilis*. The Turkana boy had a broad, flat face with a large, projecting nose, but the cranial vault was relatively small. Thus, at 800–900 cc the brain of *H. ergaster* was much smaller than in *H. sapiens*.

Figure 1.8 shows a reconstruction of what this young individual might have looked like in life. The overall proportions of the head, body, and limbs can be represented with greater confidence than in the australopithecine reconstructions described earlier. Features of surface morphology are always

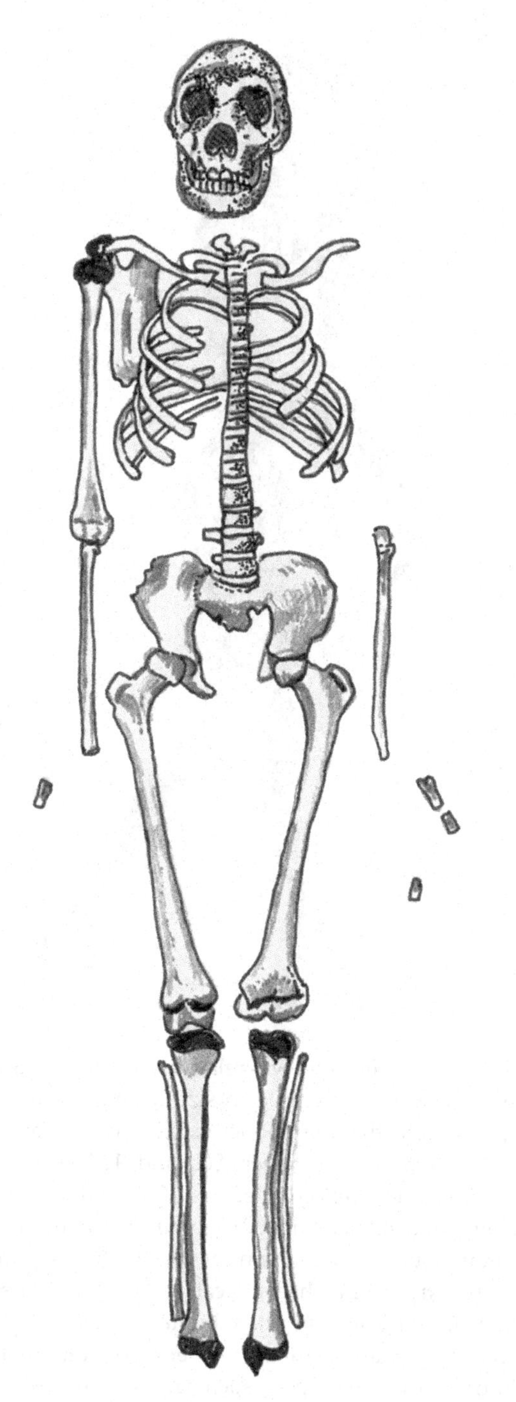

Figure 1.7 Skeleton of *Homo ergaster* based upon an immature male specimen (the 'Turkana' or 'Nariokotome boy'), dated at 1.6 MYa.

Source: Author's drawing, from photographs in Fleagle (1999) and Stringer and Andrews (2005).

problematic, but some carefully considered deductions are included in this reconstruction. Thus, at least from the neck down, *H. ergaster* probably looked very human. It is surmised that by this stage of hominid evolution loss of body hair, increased sweat gland activity, and other adaptations for long-distance, bipedal locomotion would have been present. The skin was probably black, as befits an African hominid skilled at traversing open country and exposed to high levels of UV radiation. Studies of variation at the melanocortin receptor (MC1R) locus in extant human populations have also provided evidence for increased skin pigmentation, as an accompaniment to hair loss in human ancestors about 1.2 MYa (Rogers, Iltis and Wooding 2004).

One interesting detail included in Figure 1.8 is that the eyes of the Turkana boy are represented as having a prominent white sclera surrounding the dark iris and pupil. Comparative studies of the eyes of extant primates have shown that forms which live in more open, savannah conditions have eyes that are (horizontally) wider and (vertically) narrower. Humans belong to this group. However, the white sclera of the eye is an additional trait, highly developed in humans (Kobayashi and Kohshima 2001), but present also as an occasional anomaly in apes (Jane Goodall (1986), for example, records that one of her chimps at Gombe exhibited this trait). The white of the eye probably evolved in association with non-verbal communication, as subtle facets of facial communication often involve making, or breaking, eye contact. The white sclera is especially striking when the eye is viewed against the dark skin of the face.

H. ergaster, with its relatively small brain, is not thought to have used a spoken language, but its capacity for non-verbal (especially visual) communication was likely to have been highly developed in social and sexual contexts. Alan Walker expresses the view that the eyes of *H. ergaster* probably reflected 'that deadly unknowing I have seen in a lion's blank yellow eyes. He may have been our ancestor but there was no human consciousness within that human body. He was not one of us.' (Walker and Shipman 1996). However, in Chapter 7, it will be argued that many of the effects of sexual selection upon the evolution of human body shape, skin tone, secondary sexual traits and facial sex

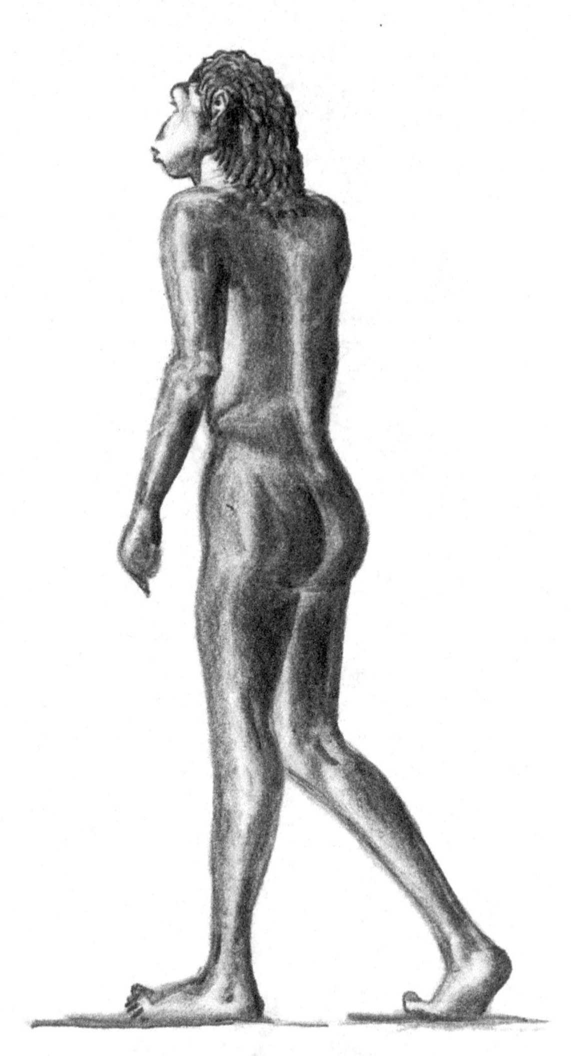

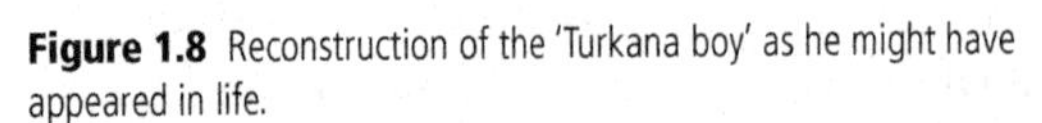

Figure 1.8 Reconstruction of the 'Turkana boy' as he might have appeared in life.

Source: Author's drawings, based upon a reconstruction exhibited at The Museum of Man, San Diego, USA.

differences derive from forms such as *H. ergaster*. As regards non-verbal communication and patterns of mate choice, *H. ergaster* may have been very much 'one of us'.

H. ergaster was very closely related to another hominid, *H. erectus*, which is best known from sites far beyond Africa, and especially from Java and China. Some authorities consider that *H. ergaster* and *H. erectus* should be placed in the same species (Rightmire 1990). Originating in Africa, *H. erectus* spread to Asia between 1.8 and 1.5MYa, where it survived until comparatively recently (some specimens from the Solo River in Java may be only 27,000 years old).

The skull of *H. erectus* had a long, low cranial vault, surmounted by a sagittal keel and with pronounced brow ridges above the eye sockets (Figure 1.9). The brain was relatively small in earlier specimens, but increased in size during the huge span of time that this species existed. Thus, cranial volumes ranging between 800 and 1250cc have been recorded, reflecting effects of individual variability, and also a tendency for brain size to increase throughout the evolution of *H. erectus* (Figure 1.10). Extensive archaeological evidence indicates that *H. erectus* was an active hunter, and perhaps a scavenger, manufacturing (Acheulian) stone tools in order to butcher prey such as antelope, horses, and elephants. Whether this extraordinary hominid used fire is debatable (Balter 1995). It is, of course, probable that *H. erectus* was a gatherer as well as a hunter, relying heavily upon plants, and not just upon animal protein, for survival.

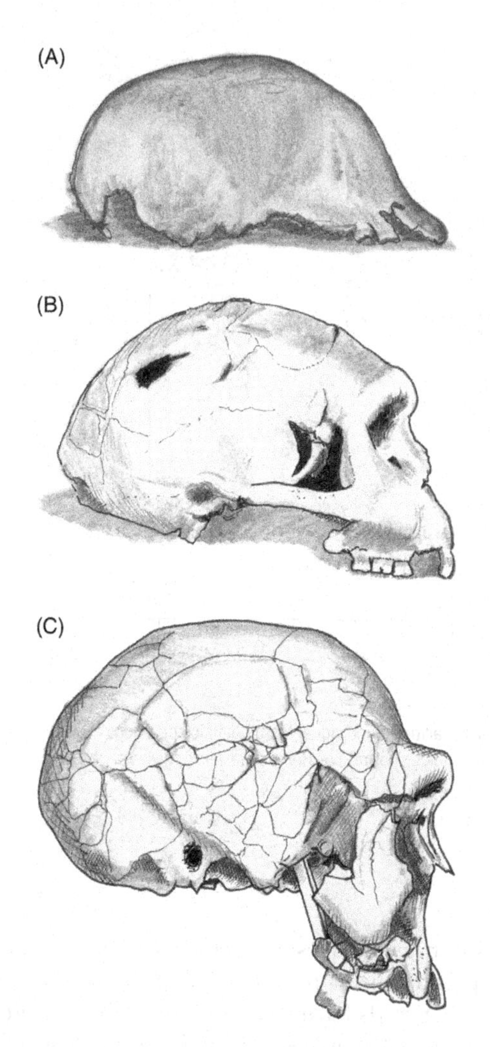

Figure 1.9 Lateral views of skulls attributed to *Homo erectus*. (A) The original skull cap, discovered at Trinil in Java, by Dubois in 1891. (B) The Sangiran 17 skull, from Java (dated at 700,000 years old). (C) The KNM-ER 3733 skull, from Koobi Fora in northern Kenya (circa 1.6 MYa). Pronounced differences between skulls (B) and (C) are interpreted by some authorities as evidence that they belong to separate species. Author's drawings, from photographs.

Fossils from Georgia (Dmanisi), including five skulls with cranial volumes ranging from 610–775 cc and dated at approximately 1.7 million years, may also belong to *H. erectus* or *H. ergaster*. They have been assigned to a separate species, *H. georgicus* (Gabunia et al. 2000), although few authorities have followed this classification. Much more problematic is a diminutive fossil hominid from the Island of Flores in southeast Asia which has been mooted as a late surviving, pygmy descendent of *H. erectus*. A brief diversion from the mainstream of human evolution is justified here, to consider this recent discovery, as it has caused considerable controversy.

Only a single skull and partial skeleton of the Flores hominid has been described, as well as a second mandible and fragmentary postcranial remains from a number of other individuals (Brown et al. 2004; Morwood et al. 2004; Morwood and van Oosterzee 2007). Known only from one location, the Liang Bua cave, the remains of this putative new hominid (*H. floresiensis*) are associated with stone tools; a hearth; and fossil remains of the *Stegodon* (pigmy elephant), Komodo dragon, and other species, probably eaten by the cave's occupants as recently as 18,000 years ago. However, claims that *H. floresiensis* was responsible for these traces of technology and hunting activity are highly contentious. The very small brain of *H. floresiensis* (400 cc) is most unusual for such a recent hominid (Figure 1.10) and its size would normally be considered as incompatible with production of the artefacts recovered from Liang Bua. This has given rise to much discussion as to whether the fossils truly represent a new species of *Homo*, or whether they should be placed in a separate genus (*Sundanthropus* was the generic classification originally considered). More than one authority has provided evidence that the skull is morphologically abnormal, and presents features found in some modern human microcephalics (Weber, Czarnetzki and Pusch 2005; Martin et al. 2006*a*; 2006*b*; see also Richards 2006; Hershkovitz, Kornreich, and Laron 2007). Rebuttals of this hypothesis have occasioned a lively debate (Falk et al. 2005*a*; 2005*b*; 2006). Most recently, Obendorf, Oxnard, and Kefford (2008) have added another twist to these debates by suggesting that the Flores hominids were members of *H. sapiens* whose stunted growth was caused by chronic thyroid deficiency (cretinism).

As more evidence comes to light, controversies surrounding *H. floresiensis* and associated finds at Liang Bua may serve as cautionary reminders of how ill-advised it is to make far-reaching interpretations based upon limited or conflicting fossil evidence. Errors of this kind have occurred in the past. As an example, Dart (1957) incorrectly interpreted

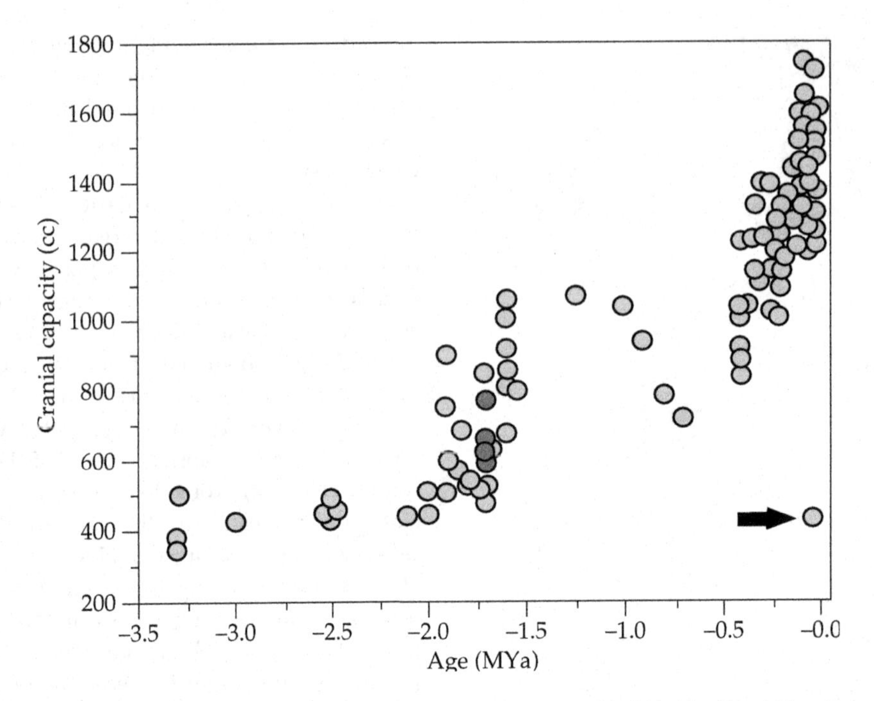

Figure 1.10 Progressive increases in brain size during hominid evolution. Cranial capacities of individual fossil hominid specimens have been plotted against time, beginning at almost 3.5 MYa. The arrow indicates the anomalous position of *Homo floresiensis*, given its recent date and small cranial capacity.

Source: After Martin et al. (2006*b*), and based upon data in Stanyon et al. (1993).

material from the limestone caves where *Australopithecus* was discovered as evidence that it had an 'osteodontokeratic culture', and used tools made of bone, teeth, or horn to kill and prepare prey for cooking. One of the discoverers of *H. floresiensis*, Mike Morwood, says that he himself let his mind

> drift into the past to try to capture the emotions and feelings of these tiny humans who had once been alive, sheltering in Liang Bua, bringing in hunted game and vegetables, or bundles of firewood to be carefully used for cooking, warmth and light. (Morwood and van Oosterzee 2007: p. 110).

However, critical reasoning indicates that, equipped with a brain smaller than that of most australopithecines, it is not at all likely that *H. floresiensis* could have made the complex stone tools at Liang Bua, or that it had mastered the use of fire, or the ability to cook. The hypothesis that its small brain was somehow 're-wired' to be exceptionally efficient for its size begs the question of why this should have happened only on the tiny island of Flores and why other hominids have failed to achieve such neurological adaptations, especially as to do so would have conveyed huge physiological advantages.

Alternative and much less problematic explanations exist which may explain the co-occurrence of *H. floresiensis* with stone tools and other evidences of human activity at Liang Bua. There is ample evidence that modern *H. sapiens* occurred on the island of Flores and would have overlapped for an extensive period of time with *H. floresiensis*. Thus, a more parsimonious interpretation of the finds at Liang Bua might be that the artefacts and animal remains are due to the activities of more advanced hominids whose remains may yet come to light in those areas of the cave that have still to be excavated.

Let us now return from this brief foray into a side channel of hominid evolution, and rejoin the mainstream of evidence provided by *H. erectus*. The variability of fossils discovered in China, Java, and Africa has led authorities to recognize that they may represent more than one species. In China, for example, fossils recovered from a cave at Zhoukoudian (near Beijing) and at several other sites are sometimes ascribed to *H. pekinensis*, but are acknowledged as sharing many features with *H. erectus* fossils from Java and Africa. It is generally argued that the sexes were probably less dimorphic in body size than in earlier hominids such as the australopithecines. Thus, in Table 1.1, the estimated ratio of adult male and female body weights for *H. erectus* is 1.2, and is the same as for modern humans. However, many uncertainties remain regarding the degree of sexual dimorphism in various populations currently assigned to *H. erectus*. Spoor et al. (2007) have described an unusually small *H. erectus* skull (endocranial capacity 691 cc) from Ileret, in Kenya. They argue that the large variations in size displayed by *H. erectus* may indicate that it was much more sexually dimorphic than modern humans. However, it is impossible on the basis of current fossil evidence to resolve this question, or to decide whether the evolution of such sexual dimorphism was due to inter-male competition within polygynous or multi-male/multi-female mating systems.

The fossil record of human evolution in Africa during the period spanning 1.2 million and 600,000 years ago is relatively poor. However, skulls of a large-brained (1225–1325 cc), and strongly built hominid species have been described from three sites (Kabwe, in Zambia; Saldanha in South Africa, and Bodo, in Ethiopia) and these are sometimes classified as *H. rhodesiensis*. *H. rhodesiensis* closely resembles another hominid, *H. heidelbergensis*, which has been described on the basis of finds at multiple sites throughout Europe, dated at between 600,000 and 250,000 years ago (Rightmire 1998). The original specimen (a jaw) was discovered in 1907 at Mauer, near Heidelberg in Germany. Still older fossils of a possible precursor of *H. heidelbergensis* have been found at Gran Dolina in northern Spain. These remains, of multiple individuals and of stone tools, are estimated to be at least 780,000 years old, and were placed in a separate species (*H. antecessor*) by their discoverers (Bermúdez de Castro et al. 1997). Most recently, the remains of a mandible, dated at 1.2–1.1 million years, have been discovered at Sima del Elefante close to the site at Gran Dolina. It seems likely that *H. antecessor*, or its forerunners, were present in Southern Europe very early in the Pleistocene (Carbonell et al. 2008).

H. heidelbergensis and the closely related forms from Africa and northern Spain occupy a pivotal role in discussions of the origins of modern humans and of Neanderthal man. *H. heidelbergensis* was a heavily built hominid, with robust limbs and a large brain. Many features of its skull align it very closely with *H. neanderthalensis* (Figure 1.11), and, indeed, it is thought to have given rise to the Neanderthals in Europe. Neanderthals were a shorter, stockier human species, distinct from *H. sapiens*, and well-adapted to life in colder conditions, although they also inhabited warmer climates in Eastern and Southern Europe, Central Asia, and the Middle East from approximately 200,000 to 30,000 years ago. A superb reconstruction of the complete Neanderthal skeleton was undertaken at the American Museum of Natural History by Sawyer and Maley (2005). Comparing this skeleton side-by-side with a modern human strikingly emphasizes the many differences. Their skulls display distinctive traits, including large brow ridges, a low forehead, receding chin, and a very large projecting nasal region. Their brains (ranging from 1200 to 1740 cc) were, in absolute terms, larger than those of *H. sapiens*, but their body size was also greater and probably exceeded 80 kg in some cases. These must have been very muscular and physically formidable humans, and the adult males were significantly larger than the females (Figure 1.12). The Neanderthals had mastered the use of fire, and they made distinctive (Mousterian) stone tools and also fashioned wooden spears and other implements (Stringer and Gamble 1993). A tangible hallmark of their humanity bequeathed to us across the millennia is the evidence that they sometimes buried their dead, a circumstance which has facilitated the preservation of their fossilized remains.

Much more is known about the Neanderthals than about their contemporaries in Africa. Both

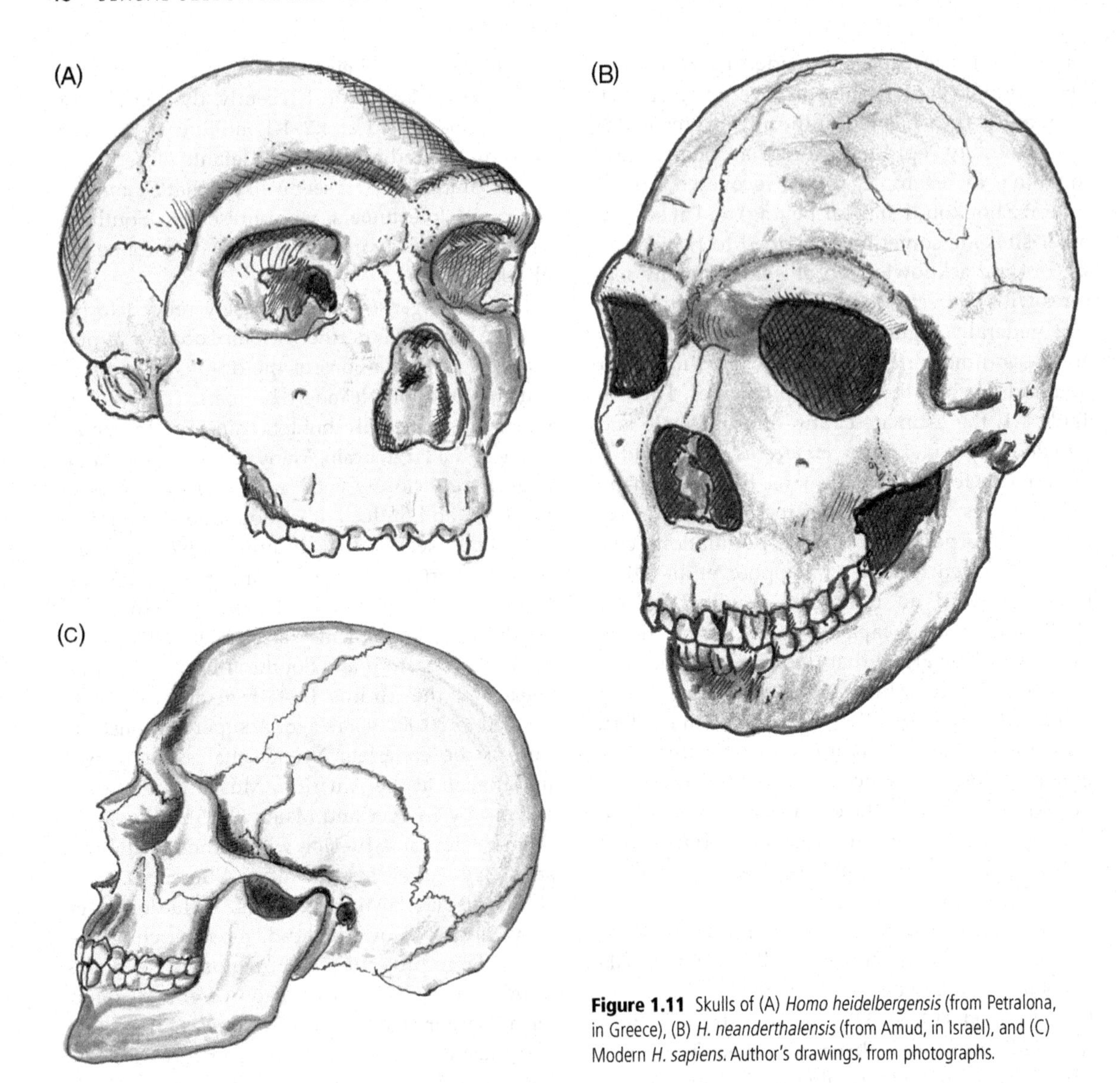

Figure 1.11 Skulls of (A) *Homo heidelbergensis* (from Petralona, in Greece), (B) *H. neanderthalensis* (from Amud, in Israel), and (C) Modern *H. sapiens*. Author's drawings, from photographs.

Neanderthals and modern humans are documented to about 100,000 years ago in the Middle East, although they never occurred together. Curiously, the stone tools used by *H. sapiens* were essentially Mousterian initially, only becoming distinctive about 40,000 years ago. However, it is now clear that modern *H. sapiens* arose in Africa, quite separately from the Neanderthals but probably from common stock, consisting either of *H. heidelbergensis* (Rightmire 1998) or of *H. ergaster* (Klein 1999), depending upon which phylogeny is adopted to account for the timing and spread of hominid species during the Pleistocene (see Figure 1.13 for alternative schemata).

The earliest fossil evidence of anatomically modern *H. sapiens* has been found in Africa (in Ethiopia, Sudan, and South Africa) and it is likely that modern humans originated there more than 195,000 years ago. It is impossible to give exact dates because times are gradually pushed back as more fossils are discovered and accurately dated. The modern human skull has a distinctive high forehead, brain size is large (1000–2000 cc, average 1350 cc), and the superciliary (brow) ridges above the eyes are much reduced but, when present, they are largest in adult males. The face is reduced in size, and the jaw is slender, but with a projecting chin which, again, is larger in the male sex (Figures 1.11 and 1.12). In 2003, White et al. reported in the journal *Nature* their discoveries of hominid crania that are morphologically intermediate between those of archaic and modern *H. sapiens*. Dated at between 160,000 and 154,000 years, these fossils from the middle Awash Valley of Ethiopia are believed to represent 'the probable immediate ancestors of anatomically modern humans' (White et al. 2003). Subsequent work on dating of anatomically modern human remains at Omo Kibish, in Ethiopia, indicates that they are of a still more ancient origin, and may be 195,000 years old.

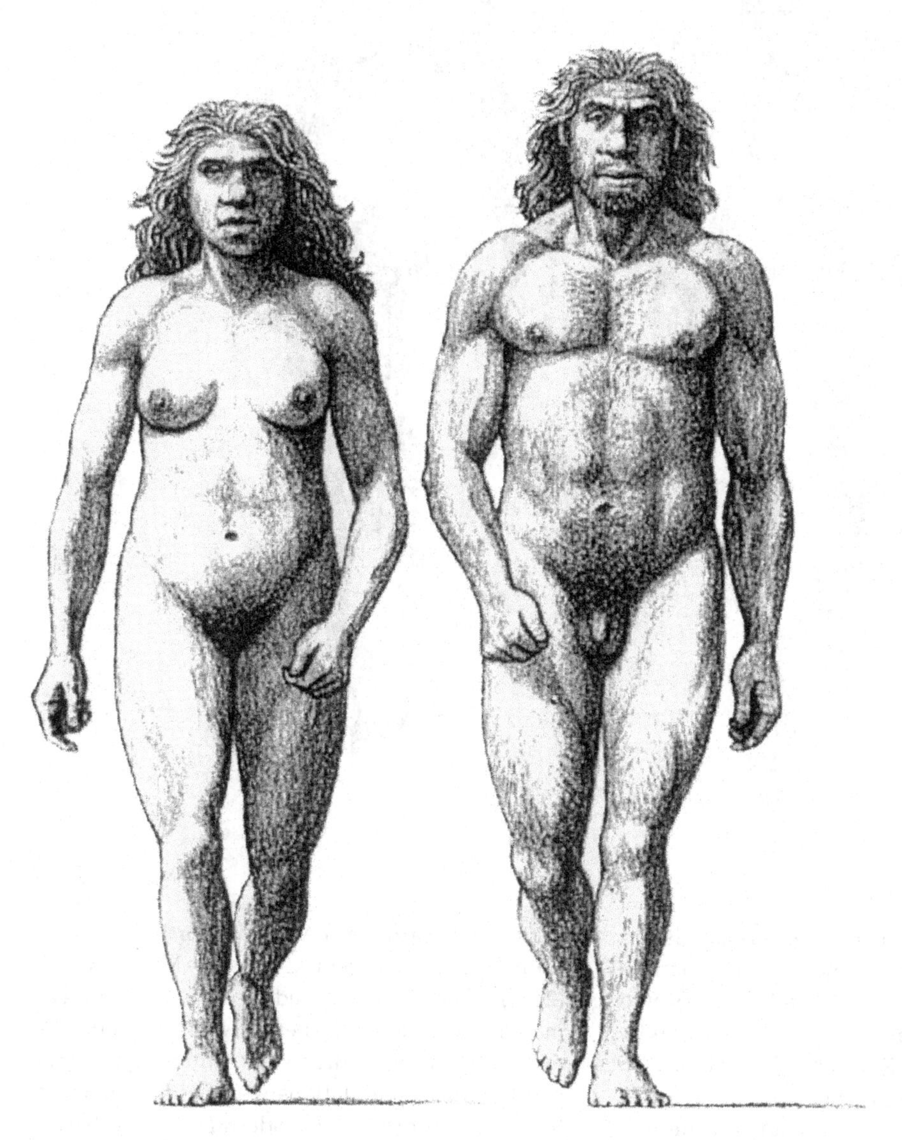

Figure 1.12 Reconstructions of (left) *Homo neanderthalensis*, as compared to (right) modern *Homo sapiens*. *Source*: After Tudge (2000).

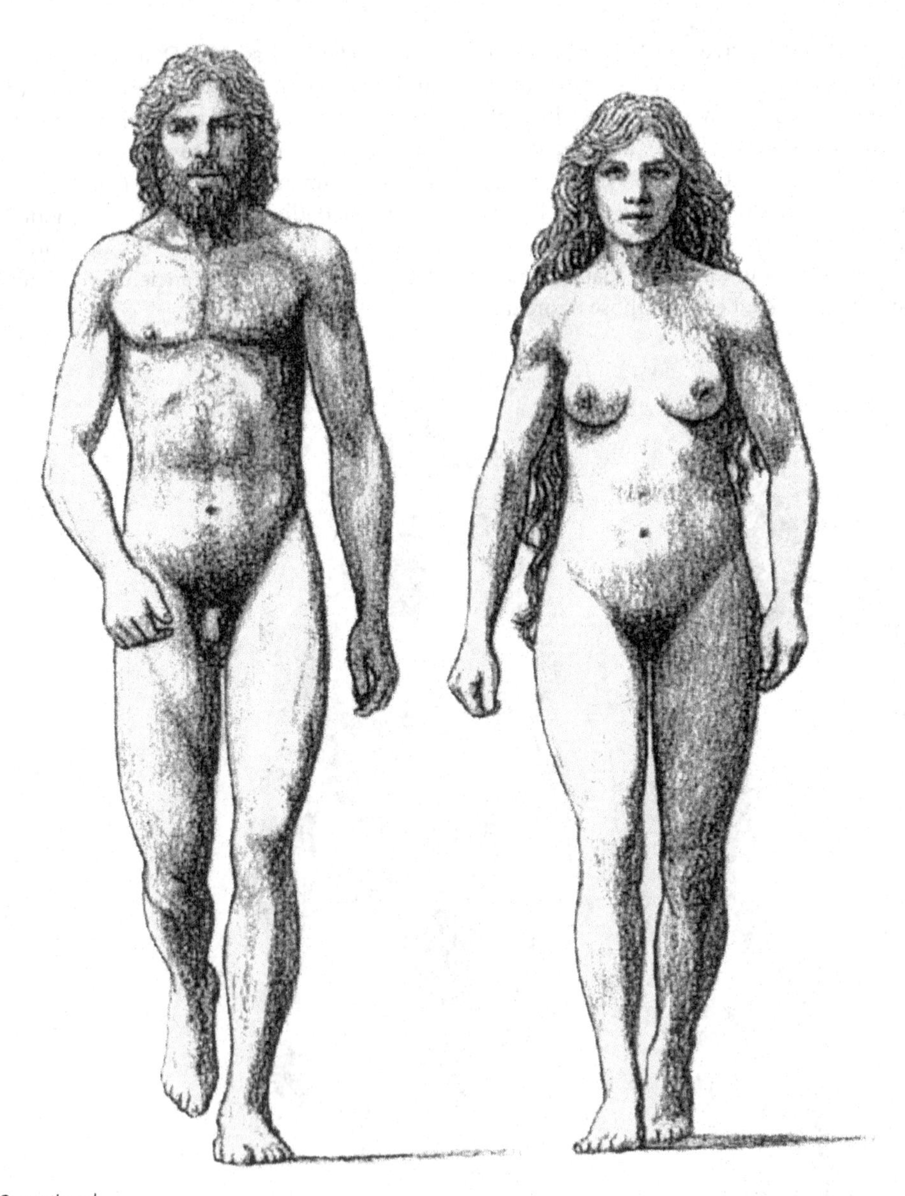

Figure 1.12 *continued*

The African origin of modern humans is also borne out by a number of genetic studies, of mitochondrial and nuclear DNA, and of Y chromosome mutations in human populations across the world (Ke et al. 2001; Ramachandran et al. 2005; Liu et al. 2006). The greatest genetic diversity exists in African populations, and it is thought that genetic 'bottlenecks' occurred as small founder populations of humans emigrated from Africa, eastwards into Asia and northwards into Europe. Indeed, recent studies by Manica et al. (2007) have demonstrated a remarkable agreement between loss of genetic diversity and reductions in phenotypic diversity (skull morphology) in 105 human populations worldwide. The most likely area of origin for modern *H. sapiens*, based upon both genetic and phenotypic studies, includes Southern, and Central/Western Africa,

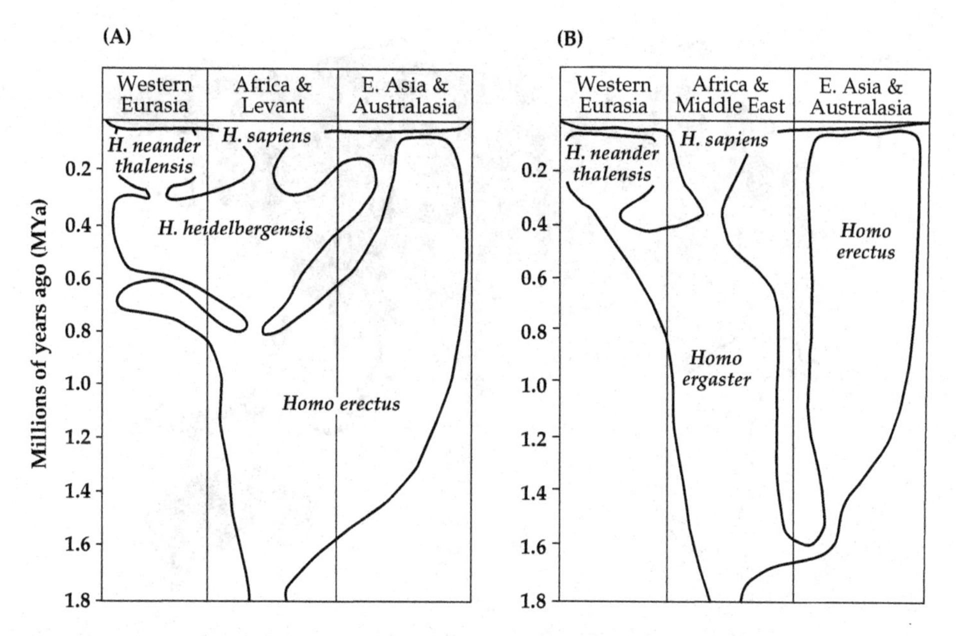

Figure 1.13 Two alternative phylogenies which have been applied to hominid evolution during the Pleistocene. (A) Rightmire classifies African and Asian *Homo erectus* as a single species. In Africa, this gave rise to *H. heidelbergensis*, and subsequently to *H. neanderthalensis* (in Europe) and *H. sapiens* (in Africa). (B) Klein's scheme places *H. ergaster* in Africa, at 1.8 MYa, and spreading to Asia, where it gave rise to *H. erectus*. *H. ergaster* also reached Europe, where it gave rise to *H. neanderthalensis*, while in Africa it was also the forerunner of *H. sapiens*.

Source: Redrawn from Boyd and Silk (2000).

as shown in Figure 1.14. Studies of the genetic diversity of *Helicobacter pylori*, a bacterium which occurs in the stomach of more than 50 per cent of all humans, also confirm the likely African origin of *H. sapiens*. There is increasing loss of genetic diversity of *Helicobacter pylori* sampled from human populations at progressive distances from Africa (Linz et al. 2007).

Modern research thus provides strong support for the *out of Africa* hypothesis of modern human origins. This has largely replaced the older *multiregional* model of human evolution. This model posited that *H. sapiens* arose in different parts of the world, from founder populations of *H. erectus* with, perhaps, some degree of genetic exchange between such populations (Thorne and Wolpoff 1981). Coon (1962), in his *Origin of Races*, for example, attempted to derive the Australian aborigines from founder populations of *H. erectus* in southeast Asia. He did not regard them as springing from the same genetic stock as geographically distant populations of *H. sapiens*, such as those in Europe. Indeed Coon considered 'that the Australian aborigines are still in the act of sloughing off some of the genetic traits which distinguish *H. erectus* from *H. sapiens*.' Only a short step would separate such misguided views from the promotion of racial superiority as being pre-ordained by biology.

In reality, extant populations of Australian aborigines, in common with all other human populations, represent the result of emigration of anatomically modern *H. sapiens* from Africa. The ancestors of modern aborigines reached Australia at least 60,000–50,000 years ago. As well as moving eastwards to occupy Asia, modern humans also populated Europe, beginning at about 45,000–40,000 years ago (Figure 1.15). In the process, they would have encountered the resident populations of Neanderthals. However, there is no evidence that *H. sapiens* and *H. neanderthalensis* interbred extensively. Analyses of ancient DNA extracted from Neanderthal fossils do not support the view that any significant

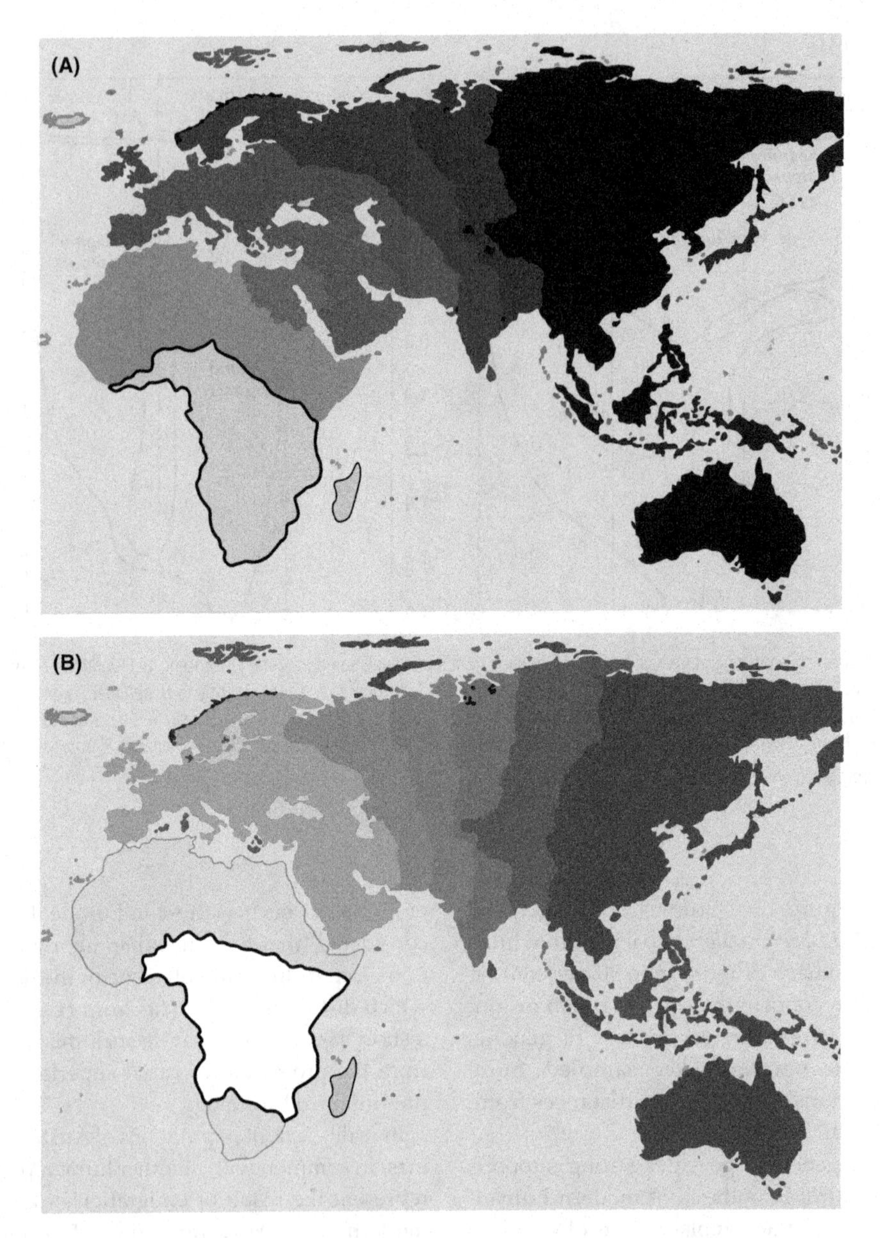

Figure 1.14 The African origin of *Homo sapiens*, and decreases in (A) phenotypic diversity (skull morphology) and (B) genetic diversity in extant human populations as indicated by darker shading at progressively greater distances from Africa. The bold line on each map encloses the probable area of origin of anatomically modern humans in Africa.

Source: After Manica et al. (2007).

exchange of genetic material took place between these two species (Krings et al. 1997; Stringer 2002). Rather, it seems probable that modern humans 'out-competed' Neanderthals (in ecological terms), so that they gradually withdrew into less favoured areas and dwindled in numbers. The Neanderthals are thought to have become extinct at around 30,000 years ago, although small numbers of them may have persisted until more recent periods (Mellars 2004).

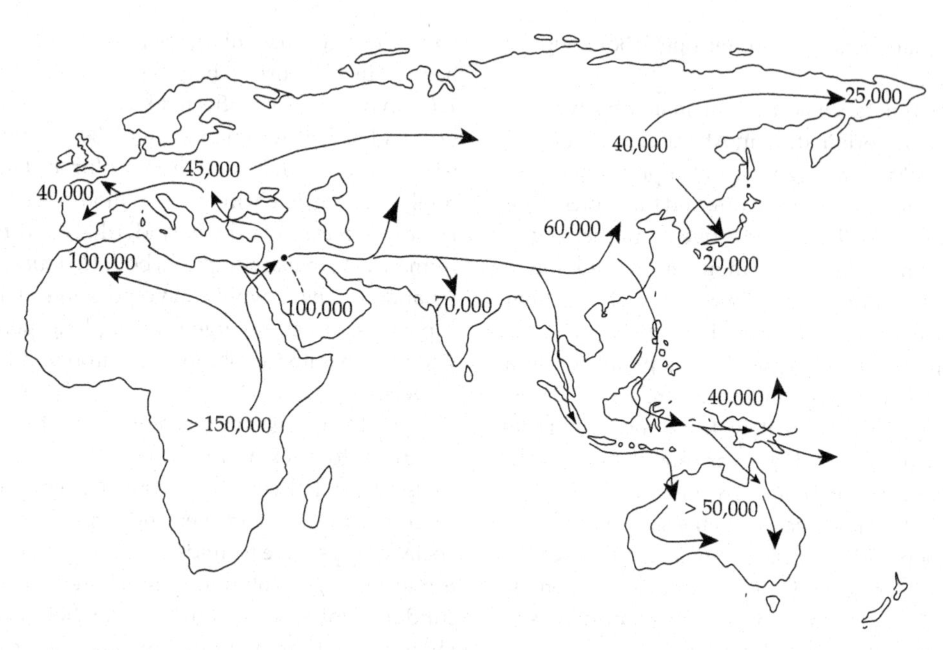

Figure 1.15 Modern humans originated in Africa more than 195,000 years ago, and by 100,000 years ago they had begun to colonize the rest of the world. Likely routes and dates of the earliest migrations are represented on the map.
Source: Based upon information in Stringer and Andrews (2005).

The evolution of human culture is beyond the scope of this book, but it is necessary to address the question of the likely cultural capacities of the earliest humans, in Africa more than 195,000 years ago. Could they speak? Did they possess the same degree of abstract reasoning, and aesthetic sensitivities as ourselves? Clearly, any insights concerning these questions may also affect judgements about the sexual lives, mating systems, and patterns of mate choice in early *H. sapiens*.

A great flowering of late Palaeolithic artistic activity is best documented for the period beginning 40,000 years ago, among those humans whose ancestors had emigrated from Africa and entered Europe. To them we owe a legacy of skillfully painted cave art, finely made tools, and even sculpture, including the enigmatic 'Venus' figurines, which will be discussed in Chapter 7. The question of whether these cultural attainments were the result of gradual, or sudden, change in Europe has occasioned much debate (Klein 2000; McBrearty and Brooks 2000; d'Errico et al. 2005; Mellars 2006). However, recent discoveries of shell beads made by *H. sapiens* in Africa at between 82,000–75,000 years ago make it clear that humans were creating symbolically significant objects, and were culturally advanced well before our species had entered Europe. At the Blombos cave, in South Africa, perforated beads, made from the shells of small marine snails, have been excavated from strata which are 75,000 years old (d'Errico et al. 2005). The presence of traces of red ochre on the shells may indicate that the wearers of the beads also adorned themselves with this pigment. Strikingly similar beads made using a species belonging to the same genus of snail (*Nassarius*) as at Blombos have also been excavated from the Grotte des Pigeons in Morocco. These beads are older (82,000 years) than at Blombos, but also bear traces of red ochre. Together with finds at other African and southeast Asian sites, they indicate that personal ornaments were made, and possibly were distributed over large distances 'at least 40 millennia before the appearance of similar cultural manifestations in Europe' (Bouzouggar et al. 2007).

The position taken here is that anatomically modern *H. sapiens*, originating in Africa at some period before 195,000 years ago, was also physiologically modern, as regards the structure and functions of the brain, as well as the reproductive and other organ systems. The origins of language are unknown, but Cavalli-Sforza et al. (1988) have shown that the degree of genetic relatedness between various modern human populations is paralleled to a remarkable degree by linguistic similarities. Thus, languages derive ultimately from common roots and from the original, African populations of *H. sapiens*. Cultural evolution would not have progressed at the same pace, or in the same way in the various populations which remained in Africa, or those which emigrated, and were subject to genetic bottlenecks and diverse environmental pressures in other parts of the world (Figure 1.15).

Culture did not begin with the origin of *H. sapiens*, and earlier species such as *H. heidelbergensis* and *H. erectus* would likely have possessed homologues of human culture, albeit at less complex levels. The gradual enlargement of the brain which has occurred during hominid evolution (Figure 1.10) may have been affected by both natural selection and sexual selection (Miller 2000). There is no reason to deny that some form of 'proto-language' might have existed in the common ancestor (e.g. *H. heidelbergensis*) of the Neanderthals and modern humans. On that assumption, both *H. neanderthalensis* and *H. sapiens* would have possessed language capacities, which diverged during long periods of widely separated evolution in Europe and Africa, respectively.

The intellectual skills used by the first *H. sapiens*, in Africa, to track and hunt prey, to forage, make tools, defend themselves against predators, and to communicate, survive, and reproduce within social groups were formidable. Although our modern technology is obviously much more advanced, fundamentally, as regards sexual behaviour we share many traits with our remote ancestors. The origins of some of these sexual traits, as determinants of the mating system, patterns of copulatory behaviour and mate choice are explored in the next eight chapters.

CHAPTER 2

Making Holes in the Dark

Sexual selection has played a major role in the evolution of the reproductive organs of animals. A full appreciation of this aspect of sexual selection is relatively new, however, and was unknown to Darwin at the time he formulated his theories concerning the pre-copulatory aspects of selection. Here, and in the following two chapters, I consider how sexual selection has affected the evolution of reproductive anatomy and physiology in mammals. Can meaningful correlations be established between measurements of genitalic traits and the types of mating systems displayed by various mammals? If so, how do measurements of these same traits in *Homo sapiens* fit within the broader, comparative analyses? Can this approach help us to strengthen interpretations of the likely origins of human mating systems based upon the limited fossil evidence (e.g. concerning sexual dimorphism in body size in extinct hominids), outlined in the previous chapter?

Consideration of these questions begins here with an examination of relationships between testes weight, body weight, and mating systems, as this also provides a useful entry point to the history of studies on sexual selection and genitalic evolution in mammals, including human beings.

Primates, like other mammals, exhibit large interspecific differences in their testes sizes, in relation to adult body weight. The first anthropologist to document the extent of these differences was Adolph Schultz who, in 1938, contributed a paper to the Anatomical Record entitled 'The relative weight of the testes in primates'. This presented data on testes weights and body weights for 82 individuals, representing 18 species of monkeys, gibbons, orangutans, chimpanzees, and human males. The bulk of this information had been assembled by Schultz during the 1937 Johns Hopkins Asiatic Primate Expedition (fifty-five specimens collected in the wild), and from specimens in the anatomical collections of Johns Hopkins University. The only relevant published account available to him at that time was by Hrdlička (1925), which included measurements of seven adult male howler monkeys. Schultz's data on the human male comprised just three cadavers of African-American men, obtained via the Anatomy Department at Johns Hopkins.

No one who has read Schultz's original account carefully could be under the illusion that human beings (as represented by his tiny sample of humanity) have large testes in relation to body size. He noted, for example, that a bonnet macaque weighing 8 kg had larger testes than a man weighing 69 kg. Some examples selected from Schultz's paper are shown (in ascending order of body weight) in Table 2.1 These differences were not due to any decrease in the bulk of sperm-producing tissue (seminiferous tubules) in those species having larger testes. On the contrary, Schultz was able to demonstrate that 'primates with relatively large testes (chimpanzee, baboon and

Table 2.1 Adolph Schultz's (1938) measurements of testes weights in primates

Species	Body weight (kg)	Testes weight (g)	Testes weight/ body weight	Number examined
Bonnet macaque	8.4	57.6	0.686	1
Rhesus macaque	10.43	76.0	0.729	2
Mandrill	31.98	88.9	0.278	1
Chimpanzee	44.34	118.8	0.269	3
Man	63.54	50.2	0.079	3
Orangutan	74.64	35.3	0.048	2

macaque) have a considerably greater proportion of sex-cell producing glandular tissue and a smaller proportion of connective tissue than have the primates (langur, gibbon and man) with comparatively small testes.' He calculated, for example, that a macaque has seventeen times more sperm-producing tissue in its testes than a langur of similar size.

Schultz discounted age differences or seasonal effects upon testes size as explanations for the remarkable differences he had discovered. At that time it was thought (incorrectly) that none of the Old World primates were seasonal breeders. Why chimpanzees or macaques should have larger testes and produce more sperm than langurs or men was a mystery. Schultz had provided valuable data, but he remained in the dark as to their functional and evolutionary significance. The darkness was to continue for several decades.

An apocryphal story relates how Robert Louis Stephenson, as a child in Edinburgh, was watching a lamplighter illuminating the gas lamps in the street outside his house. When asked by his mother what he was doing, he replied 'I am watching a man making holes in the dark.' Where reproductive biology and sexual selection are concerned, one of the most important lamplighters has surely been Geoffrey Parker, whose paper on 'Sperm competition and its evolutionary consequences in the insects' (Parker 1970) changed forever the way biologists view this field. When a female mates with two or more males, the capacity exists for competition to occur between the gametes of rival males for access to her ova. Following this line of reasoning, Parker was able to demonstrate that Darwin's concept of sexual selection also operates at the copulatory (and post-copulatory) levels via *sperm competition*, and that such competition encourages the evolution of mechanisms to optimize sperm numbers in the ejaculate and to enhance fertility.

As knowledge of primate mating systems and sexual behaviour increased, largely as a result of field studies conducted in the 1960s and 1970s, so the conditions were created to apply sperm-competition theory to the questions raised by Schultz's observations on testes sizes and sperm production in monkeys, apes, and human beings.

However, Parker's light did not penetrate all areas of biology immediately. Perhaps, because his work involved insects, and most notably the dung fly (*Scathophaga stercoraria*), its influence on vertebrate biologists was slow to take hold. Yet, like all truly great insights, the discovery of sperm competition has come to be applied throughout the animal kingdom, from dung flies to right whales, and from chimpanzees to humans. When Roger Short began to examine relative testes sizes in the great apes and man, working at the Medical Research Council's Reproductive Biology Unit in Edinburgh in the 1970s, he was initially unaware of Parker's work. Independently, Short reasoned that differences in testes sizes between the various apes and man might reflect variations in their mating systems and sperm production in relation to patterns of sexual behaviour. He was especially struck by the modest size of the testes in the gorilla, by comparison with those of its much smaller relative, the chimpanzee. Gorillas usually live in polygynous one-male units; females mate primarily with a single partner and copulations are relatively infrequent under natural conditions (Schaller 1963; Harcourt et al. 1980). A silverback gorilla, weighing more than 160 kg, had testes weighing about 30 g, by comparison with a male chimpanzee which at 45 kg had much larger testes (weighing approximately 120 g).

Short, unlike Schultz, had access to information about the sexual behaviour of free-ranging chimpanzees, derived principally from the field work of Jane Goodall and her students, at the Gombe Stream Reserve in Tanzania (Goodall 1965, 1968; McGinnis 1979; Tutin 1979, 1980). Chimpanzees live in multi-male/multi-female communities and females often mate with multiple partners during the follicular phase of the menstrual cycle, when the sexual skin is swollen and ovulation is most likely to occur. In her 1986 review volume on work at Gombe, Goodall records that 'I watched one party as it arrived at a new food source: the attractive female climbed into the tree along with eight bristling males, each of whom copulated with her in quick succession in a period of five minutes.'

These observations of the chimpanzee mating system have since been confirmed by fieldwork at other sites, in Tanzania (Mahale Mountains: Nishida 1990), Ivory Coast (Taï Forest: Boesch and Boesch 2000), and Uganda (Budongo Forest: Reynolds 2005). Sometimes male and female chimpanzees pair up to form temporary "consortships" (Tutin 1979) and the most dominant male in a community may monopolize access to particular partners

(Goodall refers to this mate-guarding tactic as 'possessiveness'). We still do not know how successful these alternative tactics are in terms of numbers of offspring sired by males across their lifespan in the community. It seems likely, however, that multiple-partner matings and selection for success in sperm competition have been powerful forces in shaping the evolution of chimpanzee reproductive biology. The increase in sperm-producing tissues in the chimpanzee testis, first measured by Adolph Schultz, makes perfect sense in this context.

Following Short's (1979) work on the great apes, Harcourt, Harvey, Larson, and Short (1981) went on to examine the question of testis weight, body weight, and breeding systems in primates, publishing a paper under this title in *Nature* in September 1981. Their principal findings are shown in Figure 2.1. This includes data on 33 species of monkeys and apes, as well as human beings. The findings embodied within this graph remain highly influential in discussions about sexual selection, sperm competition, and primate mating systems, so it is worthwhile to examine them in some detail. The double logarithmic plot of testes weights vs. body weights in Figure 2.1 shows that an allometric relationship exists between the two variables. In other words, testes size scales in

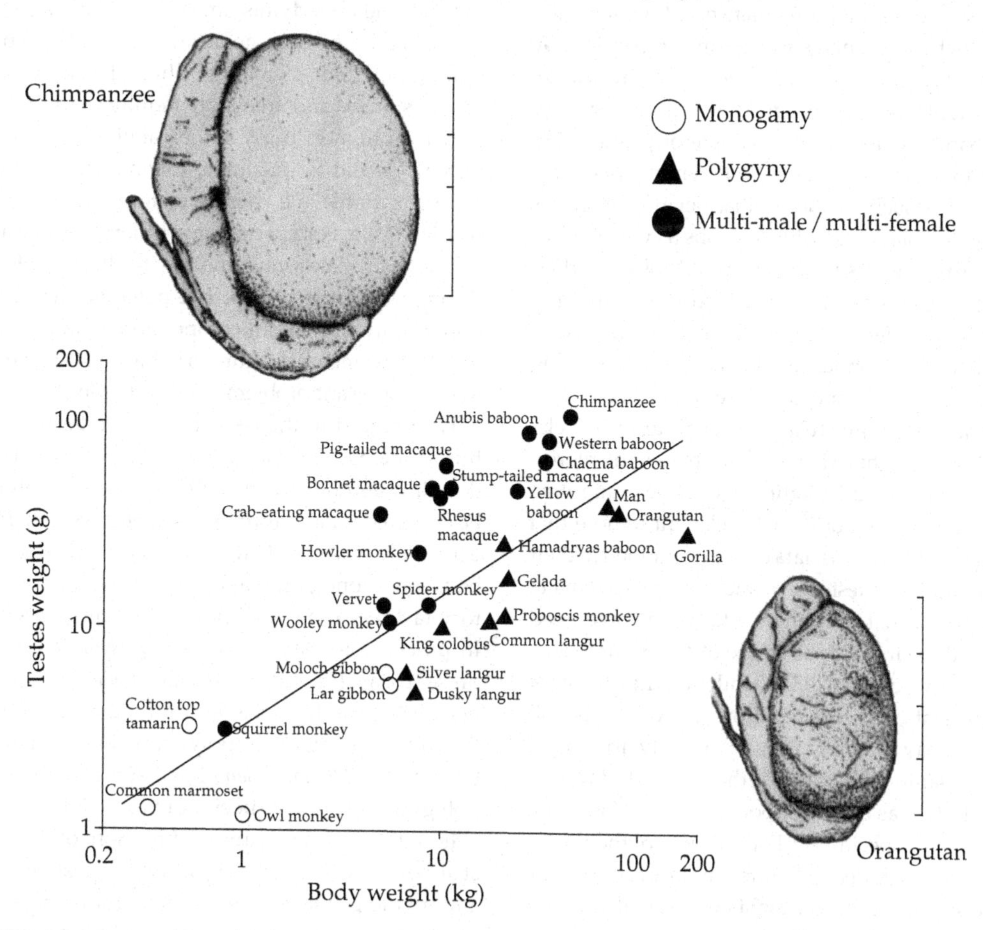

Figure 2.1 Relative testes weights and mating systems in anthropoid primates. A double logarithmic plot of combined testes weight versus body weight for anthropoids having monogamous, polygynous, or multi-male/multi-female mating systems.

Source: Based on Harcourt et al. (1981); modified from Short (1985).

a predictable way with body size but there is also considerable inter-specific variability in the relationship, as indicated by the distances of various species from the regression line which passes through the 33 data points. Adult males of those species which are situated above the regression line have larger testes than expected in relation to their body weights. As well as the chimpanzee, examples include various baboon and macaque species, howler monkeys and vervets, all of which have multi-male/multi-female mating systems. Sexual selection has favoured the evolution of larger testes in these species, because females commonly mate with multiple partners during the peri-ovulatory phase of the cycle. Sperm from rival males compete for access to ova within the female's reproductive tract. Males that are capable of maintaining high sperm counts in the ejaculate may therefore gain a reproductive advantage via sperm competition. By contrast, species whose data for body weight and testes weight plot below the regression line have smaller testes than expected for this sample of primates. In general these species include polygynous (one-male unit) species such as the gorilla, gelada, proboscis monkey, and black and white colobus as well as pair-living forms, such as the lesser apes (Moloch and lar gibbons), the owl monkey, and the common marmoset of South America. Sperm competition pressures are much less pronounced in these cases than in macaques, baboons, or chimpanzees.

Human males have relatively small testes in relation to body weight; 'Man' falls just below the regression line in Figure 2.1, slightly above the orangutan, but well above the gorilla. It is relevant to note that Harcourt et al. included data on only four men in their studies of relative testes size and mating systems in anthropoid primates. Human testes weight averaged 40.5 g in these individuals, three of whom had been measured by Schultz (1938), whilst the fourth set of measurements was obtained from a paper by Benoit (1922). It may seem unusual that only four men had been sampled. However, the focus of Harcourt et al.'s study was on anthropoids in general and not human beings specifically. For fourteen of the thirty-three species included in Figure 2.1, only one, two, or three sets of adult testes weights were available, less than the information on *Homo sapiens*.

The huge contribution made by Harcourt et al.'s work as well as some of its limitations were appreciated at the time. Thus, in the same issue of *Nature*, Martin and May (1981) contributed a 'News and Views' overview in which they pointed out that seasonal changes in testes size, which certainly occur in some monkey species, had not been taken into account. The relatively large size of the testes in the cottontop tamarin (a putatively monogamous species) by comparison with another New World species, the squirrel monkey (which lives in multi-male/multi-female groups), was surprising. Martin and May also drew attention to the relatively small data set employed, which did not include any of the Malagasy lemurs or other prosimian primate species. Other factors besides testes size and sperm production might also be important in determining male reproductive success in competitive situations. Sperm must pass through the epididymis after leaving the testis, and gametes are stored in the distal region (cauda) of the epididymis prior to mating. Thus, the dynamics of sperm storage and transport required study. Males also differ tremendously in their ability to gain access to females and to maximize mating activity during the most fertile (i.e. peri-ovulatory) period of the ovarian cycle. For sperm competition to occur at all, at least two males must mate with the same female during this fertile period. The potential 'window of opportunity' for sperm competition is quite narrow in primates, including human beings. We shall return to this important problem in later chapters.

With regard to the issue of relative testes size in human beings, Martin and May (1981) commented that 'it does not accord with the range of values to be expected for a multimale breeding system.' However, there was insufficient evidence to discriminate between monogamy and polygyny as contributory factors in the evolution of human testes sizes. In practice, both types of mating system occur in present day (or recent) human societies; thus 74 per cent of the 185 societies examined by Ford and Beach (1951) engage in polygynous marriages. For this reason, *Homo sapiens* has been classified as a polygynous primate species in Figure 2.1.

In the years since the publication of Harcourt et al.'s study of testes sizes and mating systems in the anthropoid primates, comparative studies have been conducted on many vertebrate groups. Among the mammals, sexual selection via sperm competition correlates with the evolution of larger testes sizes among the prosimian primates (Dixson 1987*a*, 1995*a*), bats (Hosken 1997; Wilkinson and McCracken 2003), whales and dolphins (Brownell and Ralls 1986; Connor, Read, and Wrangham 2000), eutherian mammals in general (Kenagy and Trombulak 1986), as well as in marsupials and monotremes (Rose, Nevison, and Dixson 1997; Taggart et al. 1998). The larger size of the testes in primates which have multi-male/multi-female mating systems is not significantly influenced by the occurrence of seasonal breeding (Harcourt, Purvis, and Liles 1995). As an example, a restricted mating season and heightened sexual activity in seasonally breeding macaques (e.g. the rhesus monkey) has not led to the evolution of larger relative testes sizes than in non-seasonal breeders with pronounced sperm competition (e.g. the chimpanzee).

This additional information can help us to reassess the accuracy and comparative significance of measurements of human testes size and its relevance to discussions of human evolution. Any such analysis is only as reliable as the data upon which it is based, however. It is necessary firstly to assemble a data set on testes weights and body weights for a large series of mammals and secondly to gain some clearer perception of individual and ethnic differences in human testes size. Figure 2.2 shows

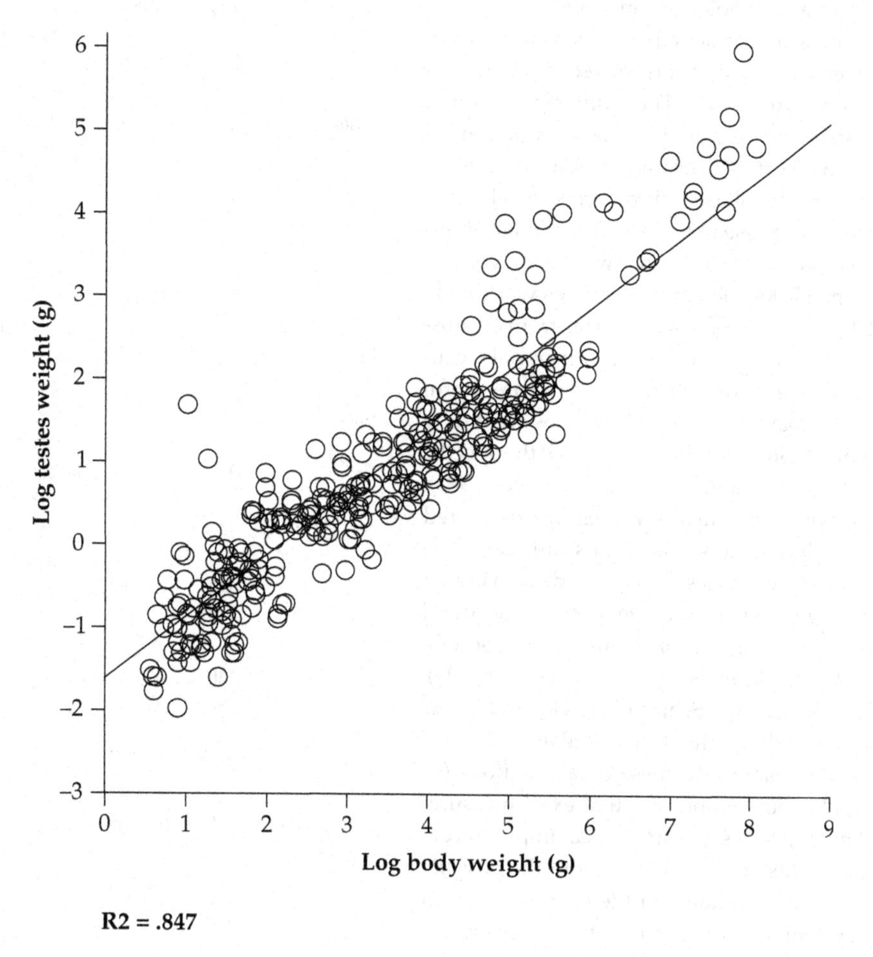

Figure 2.2 A double logarithmic plot of combined testes weight versus body weight for 339 species, representing a wide spectrum of mammals.
Sources: Brownell and Ralls (1986); Kenagy and Trombulak (1986); Harcourt, Purvis, and Liles (1995); Rose, Nevison, and Dixson (1997); Connor, Read, and Wrangham (2000); Wilkinson and McCracken (2003); Anderson, Nyholt, and Dixson (2004); Dixson, Nyholt, and Anderson (2004); Anderson and Dixson, (in press).

a double logarithmic plot of testes weights versus body weights for mammals. There are 339 species represented in this graph, which includes bats, rodents, primates, carnivores, artiodactyls, perissodactyls, proboscideans, and cetaceans as well as marsupials and monotremes. This is a respectable sample, but as more than 4,500 extant mammalian species have been named by taxonomists, even the information in Figure 2.2 is still far from complete.

Where do human beings lie within this enlarged analysis of mammalian testes sizes and body sizes? Data on testes weights of just four men (Harcourt et al. 1981, 1995) might be sufficient to explore this question if these measurements were representative of humanity as a whole. Unfortunately this is not the case. There are considerable differences in testes sizes between individuals and between human populations around the world. These differences cannot be accounted for solely on the basis of variations in body weight (Short 1984; Diamond 1986). To provide information on possible ethnic differences in human testes size, I have assembled from the published literature measurements of testes weight or volume for more than 7,000 men in 14 countries worldwide (Table 2.2). Even this information is far from exhaustive and must be viewed with appropriate caution. Measurements of testicular volume are useful because they may be converted to testes weight by using a correction factor (volume × 1.05; the specific gravity of tissue: Dahl, Gould, and Nadler 1993). However, when testicular volumes are measured by using orchidometers, the results are subject to considerable inaccuracies. This was demonstrated by Döernberger and Döernberger (1987) in a careful post-mortem comparison of 99 men, using sonography, water displacement (Archimedes principle), Prader's orchidometer, Schirren's circle, and linear measures (by sliding caliper) to calculate testes volume. Lest the reader find these details tedious (or obsessive!) I should emphasize that exact measurements of human testes size are of great importance if sound judgments are to be made about the possible effects of sexual selection and the origin of human mating systems. Doernberger and Doernberger found that orchidometers may over-estimate testes volume (by an average of 27 per cent using Prader's orchidometer and by 52 per cent using Schirren's circle) as compared to the volumes obtained

Table 2.2 Average combined human testes weights (grams) or volumes (millilitres) in fourteen countries, worldwide

Country	Sample size	Combined testes size	Comments	Sources
China (Hong Kong)	100	19.01 g		Chang et al. (1960)
	109	16.55 g	Age range 13–97 yrs	Short (1984)
Japan	?	30.0 ml	Prader orchidometer	Nakamura (1961)
	?	35.0 ml	Schirren orchidometer	Fujii et al. (1982)
South Korea	425	38.8 ml	Prader orchidometer	Kim and Lee (1982)
	1792	31.3 ml	Prader orchidometer	Ku et al. (2002)
India	325	35.4 g	Weights include epididymides	Jit and Sanjeev 1991.
Australia	222	48.2 ml	Sliding calipers; scrotal skinfold not measured	Simmons et al. (2004)
USA	132	36.3 g	Mainly Caucasian	Johnson, Petty, and Neaves (1984)
	3	50.2 g	African American	Schultz (1938)
UK	251	46.0 ml	Prader orchidometer	Jørgensen et al. (2001)
France	207	45.0 ml	Prader orchidometer	Jørgensen et al. (2001)
Finland	275	46.0 ml	Prader orchidometer	Jørgensen et al. (2001)
Denmark	349	47.0 ml	Prader orchidometer	Jørgensen et al. (2001)
	708	40.0 ml	Prader orchidometer	Andersen et al. (2000)
Sweden	140	42.0 g		Olesen (1948)
	54	41.5 ml	Linear measures	Lambert (1951)
Switzerland	1743	37.2 ml	Prader orchidometer	Zachmann et al. (1974)
Czechoslovakia	176	37.4 ml	Linear measures; scrotal skinfold not measured	Farkas (1971)
Nigeria	209	50.1 ml	Ages 19–23 yrs; linear measures; scrotal skinfold not measured	Ajmani, Jain, and Saxena (1985)

Note: g = weight in grams; ml = volume in millilitres.

by using Archimedes principle. Caliper measures are more accurate, and a simple formula allows the calculation of testes volume using length and width measurements. However, measurements of living subjects must also take account of scrotal thickness if accurate linear data are to be obtained. Failure to do so can result in overestimation of testes volume, by approximately 25 per cent according to my own calculations. Thus in Table 2.2 I have provided notes on methodology and other problems in relation to the individual reports listed there.

Cross-culturally there is a consistent difference in mean testis size; the right testis is on average 5.5 per cent larger than the left testis in men from eight countries (Table 2.3). Other authors have commented that the right testis tends to be larger but do not provide exact data. Statistically the effect is robust ($P< .01$ for data in Table 2.3, after conversion of volumetric data to weights). Why the effect occurs is unknown and nor does it necessarily apply to the great apes. Unfortunately relatively few chimpanzees, gorillas, or orangutans have been measured to estimate individual testes sizes, but there are no consistent effects for the available samples.

Table 2.3 Sizes of the right (R) and left (L) testes in various human populations

Country	Right testis	Left testis	R–L	%	Source
China (Hong Kong)	9.65 g	9.36 g	0.336 g	3.4	Chang et al. (1980)
	8.57 g	7.98 g	0.59 g	6.9	Short (1984)
South Korea	15.9 ml	15.3 ml	0.6 ml	3.8	Ku et al. (2002)
India	18.20 g	17.21 g	0.99 g	5.4	Jit and Sanjeev (1991)
Australia	24.3 ml	23.9 ml	0.4 ml	1.6	Simmons et al. (2004)
USA	19.0 g	17.3 g	1.7 g	8.9	Johnson, Petty, and Neaves (1984)
Denmark	21.6 g	20.4 g	1.2 g	5.5	Olesen (1948)
Czechoslovakia	19.59 ml	17.78 ml	1.81 ml	9.2	Farkas (1971)
Nigeria	25.66 ml	24.41 ml	1.25 ml	4.9	Ajmani et al. (1985)

Notes: ml = mean volume in millilitres; g = mean weight in grams. For information on sample sizes and methods, see Table 2.2.

Turning to measurements of combined testes size, there is a strong trend towards the occurrence of smaller testes in men from Asia (China, Japan, Korea, and perhaps in India) as compared to measurements from European and African populations, or their ethnic derivatives (e.g. in the USA and Australia). These differences are quite striking with respect to the information presented in Table 2.2. Smallest by far are the testes of 209 Hong Kong Chinese, weighed in two separate studies. Chang et al. (1960) weighed the testes of 100 men and found the mean combined weight to be only 19.01 g. Short (1984) reported even lower average weights (16.5 g) for 109 Hong Kong males on the basis of an unpublished study (Chan and Short). However, as these subjects varied considerably in age, from 13 to 97 years, it is probable that the inclusion of some adolescent males would have produced a slightly lower average for testes weight. In neither of these studies of Chinese males are their body weights known, but it is highly unlikely that smaller stature alone could account for such small testes sizes by comparison with data for European or African populations. Chan and Short's unpublished study does provide data on the height of the Hong Kong males for whom testes weights were calculated. The average height of their subjects was 157.9 cm (SEM± 1.5 cm). An interesting comparison is provided by Ajmani, Jain, and Saxena's (1985) measurements of Nigerian men. Part of their sample comprised 110 men who were less than 165 cm in height. Combined testes volumes averaged 43.9 ml in this group, which is equivalent to 46.0 g for average weight of the testes. Even if we allow for an overestimate of testicular volume in this study (due to a failure to control for scrotal thickness: see Table 2.2), Nigerian men have testes more than twice as large as those of Hong Kong Chinese subjects of similar height.

Although differences in body size may make some small contribution to the ethnic variations in testes size listed in Table 2.2, it is much more likely that fundamental differences in testes size exist in different human populations. Next smallest are the testes of Japanese men (averaging 30 ml and 35 ml in studies by Nakamura (1961) and Fujii et al. (1982)). The same is true for studies of Korean men. For all the studies in Japan and Korea, measurements were made using orchidometers, and may thus represent overestimates. In India, a study of 325 men yielded an average combined testes weight of 35.4 g but this included the epididymis and hence is also an overestimate. By comparison, the weights or volumes of testes of men in UK, France, Scandinavia and other parts of Europe, as well as in the USA, Australia, and Nigeria are larger on average than for men in those Asian populations for which data exist. Largest of all are the testes of 209 Nigerian medical students (50.1 ml, Ajmani et al. 1985) and the three black African Americans measured originally by Schultz (50.2 g).

In Figure 2.3 these data on human testes size in various populations have been log transformed and plotted against body weight in the same manner as for the much larger sample of mammalian species represented in Figure 2.2. The same regression line, calculated for mammals as a whole, has also been plotted on Figure 2.3. The data for the various human populations (represented by circles) are aligned vertically, below the regression line (exact weights of the men in each population are unknown, so an average weight has been used). Smallest of all are the relative testes sizes of Hong Kong Chinese; the circle at the foot of the row, which does not overlap with the others, represents this population. At the opposite end of the row, and closest to the regression line, is the circle representing relative testes size in Nigerian men. Other populations fall between these two extremes. How do these more

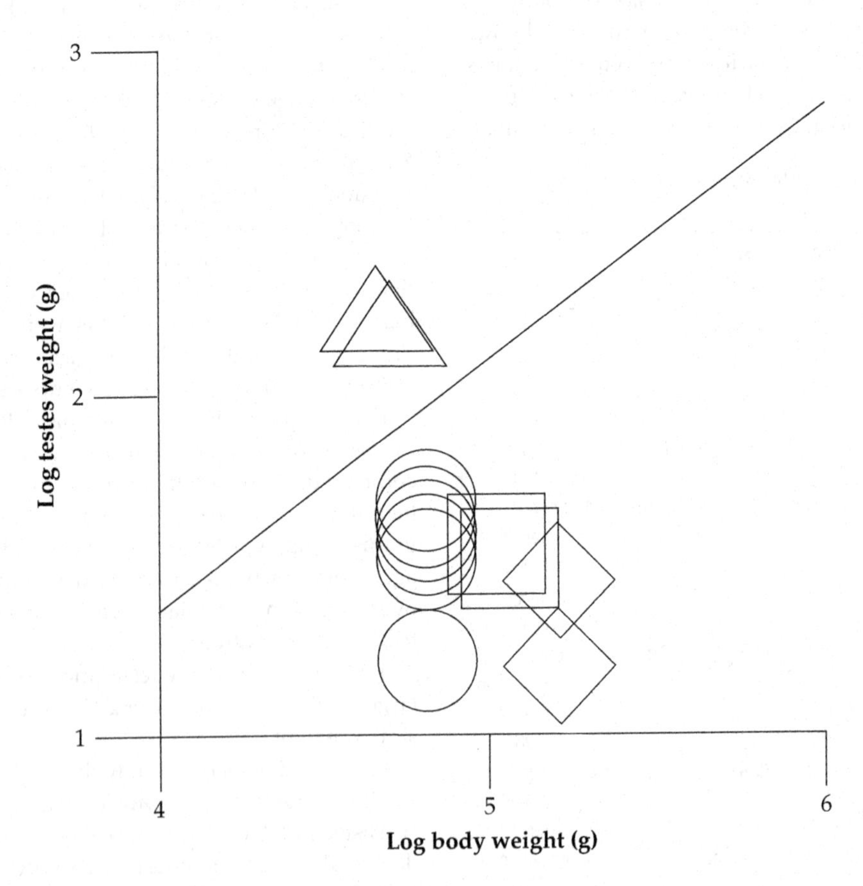

Figure 2.3 Relative testes sizes in human populations, as compared to the great apes. The regression line is taken from Figure 2.2. Data are from Tables 2.2 and 2.4. ○ = *Homo*; ◊= *Gorilla* (the western lowland gorilla is plotted separately and falls below the mountain gorilla on the graph); □ = *Pongo* (two species plotted); Δ = *Pan* (two species plotted).

extensive data on humans compare with those for the apes?

More up-to-date information on relative testes sizes in the apes has also been assembled (Table 2.4) and the great apes have been included in Figure 2.3, for comparison with *Homo sapiens*. Among the smaller, pair-forming (monogamous) gibbons, adequate data on testes size and body weight are only available for two species (*Hylobates lar* and *H. moloch*). They have small testes in relation to body weight, as originally reported by Harcourt et al. (1981). The term 'monogamous' does not necessarily denote a lifelong union and exclusive copulation with a single partner. Extra-pair copulations have been documented in *H. lar* (Palombit 1994; Reichard 1995). The genetic consequences of such behaviour in terms of offspring sired by males outside the family group are unknown, as no detailed DNA typing study of wild gibbon groups has yet been reported. The prediction is that extra-pair paternity rates are likely to be very low and that sperm competition has had little effect upon the evolution of reproductive physiology in gibbons. Males make a considerable investment in maintaining long-term relationships with individual females in these species and their reproductive success depends upon siring offspring and raising them within the family group. Their low relative testes size is commensurate with low sperm-competition pressure.

Table 2.4 Testes weights and body weights of the adult great apes

Species	Number of specimens	Mean combined testes weights (g)	Mean body weight (kg)	Sources
Pongo pygmaeus	Three (captive)	36.53	101.3	Dixson et al. (1982)
Pongo abelii	Nine (captive)	34.87	105.7	Dahl, Gould, and Nadler (1993)
Pan troglodytes	Thirteen (captive)	148.89	53.22	Schultz (1938); Dixson and Mundy (1994)
Pan paniscus	Two (captive)	167.7	47.95	Dixson and Anderson (2004) and unpublished data
Gorilla g. beringei	Three (wild)	28.96	164.66	Hall-Craggs (1962); Dahl (unpublished data)
G. g. gorilla	Nine (captive)	16.51	163.04	Dahl (unpublished data)

Even smaller, relative to their huge body sizes, are the testes of silverback male gorillas. In Table 2.4 and Figure 2.3 the mountain gorilla (*Gorilla g. beringei*) and the western lowland gorilla (*G.g. gorilla*) are shown separately. There are anatomical and ecological differences between these two forms, and some taxonomists accord the western lowland gorilla a specific rank. Mountain gorilla groups may sometimes include more than one silverback male, although there is usually a clearly dominant or leader male even in these circumstances (Schaller 1963). DNA typing studies of mountain gorilla groups have demonstrated that when two adult males are present, the dominant silverback sires 85 per cent of the offspring (Bradley et al. 2005). No comparable studies have been carried out on western lowland gorillas. Their groups normally contain a single silverback however, and there is less likelihood that sperm competition might occur. It is interesting that mountain gorillas also have relatively larger testes than western lowland gorillas. However, this finding may represent a sampling bias in the data. Only the testes of captive western lowland gorillas have been measured thus far and some captive males are infertile and exhibit testicular atrophy (Dixson, Moore, and Holt 1980; Enomoto et al. 2004). Individuals displaying extreme testicular pathology have been excluded from Table 2.4. Nonetheless, it is still possible that captive silverback western lowland gorillas are less fertile and have smaller testes than their counterparts in the wild. By contrast, the testes weights listed for mountain gorillas refer to free-ranging silverbacks. Perhaps their testes were larger due to better health and nutrition under natural conditions, or perhaps a true difference in relative testes size exists between the two taxa. More data, especially on free-ranging western lowland gorillas, will be required to resolve this question.

Gorillas are fundamentally polygynous in their mating behaviour and the differences between western lowland and mountain gorillas in relative testes sizes are minor and consistent with a mating system in which sperm competition plays a minimal role. By comparison, differences between gorillas and chimpanzees in relative testes sizes are dramatic, and it is now possible to confirm that the second chimpanzee species (the pygmy chimp or bonobo: *Pan paniscus*) also has extremely large testes in relation to body size (Table 2.4, Figure 2.3). Unlike chimpanzees, which may form consortships and engage in mate guarding (by alpha males: Goodall 1986) in addition to frequent multi-partner matings, the bonobo mating system is primarily a multi-partner arrangement in which sperm competition between rival males is likely to be intense. The fact that both *P. troglodytes* and *P. paniscus* are very similar in relative testis size indicates that sperm competition via multiple partner matings by females is the primary force which has driven selection for their exceptionally large testes. This in turn implies that male reproductive success resulting from consortships and mate guarding in *P. troglodytes* will likely be smaller, and ancillary to the major investment in sperm competition tactics. This question awaits resolution by detailed DNA typing studies to compare the success of the various sexual strategies employed by male chimpanzees across their lifespan. Where the bonobo is concerned, a DNA typing study of paternity in free-ranging animals by Gerloff et al. (1999) reported that dominant males, which are usually the sons of high-ranking females, have the greatest reproductive success. It must be said, however, that differences in social rank among male bonobos are not pronounced or linear. On the basis of agonistic interactions and displacement behaviour, Gerloff et al. placed three males in rank category 1, and another three males in the second ranking position. Although these authors state that 'two dominant males together attained the highest paternity success', the observed relationships between male rank and paternity were not pronounced. One of the rank 1 males sired no offspring, whereas all the rank category 2 males had fathered at least one infant. There were at least six potential sires in the group for the eight infants that were DNA typed. These results are more consistent with a multiple-partner, sperm-competition-driven mating system than with one governed by male dominance in relation to reproductive success. On that note, all eight males studied had high copulatory frequencies (ranging from 0.22–1.29 copulations observed per day) and none were completely excluded from mating with females in the community.

These observations of free-ranging bonobos receive support from recent studies of a captive group, housed at Twycross Zoo in the UK (Marvan et al. 2006). Two infants in the group were sired by two lower-ranking males, whereas the most dominant male in social contexts did not sire any offspring with the group's three adult females.

Turning to the orangutan, some refinements are necessary to the analyses of their relative testes sizes, compared to earlier accounts (Harcourt et al. 1981, 1995). Two species of orangutan are now recognized by some taxonomists (*Pongo pygmaeus* in Borneo, and *P. abelii* in Sumatra). Chromosomal and morphological differences between the two forms are significant, but some authorities remain doubtful about separating Bornean and Sumatran orangutans as distinct species (e.g. Courtenay, Groves, and Andrews 1988). Additional data on testes sizes are available for both types of orangutan, and these are included in Table 2.4. The number of specimens measured is still very small, however. I have followed Dahl, Gould, and Nadler (1993) by including only data for adult males weighing more than 85 kg. Males that are smaller than this may not be completely physically mature.

The social organization and mating system of orangutans is most unusual among the anthropoid primates, for they are comparatively nongregarious. Massive adult males, with prominent secondary sexual traits (cheek flanges, throat sac, and long hair) occupy very large, individual home ranges. Their world is centred within the forest canopy, for they only occasionally descend to ground level. Their large arboreal ranges overlap those of a number of adult females and their dependant offspring. Orangutans are the largest arboreal mammals in the world, and although they may congregate occasionally when favoured trees are in fruit, the sexes are normally separate and widely dispersed in the rainforest. Dominant *flanged* adult males are highly antagonistic to one another, and subordinate males exhibit socially mediated suppression of body growth and secondary sexual traits (Kingsley 1982; Maggioncalda et al. 1999, 2000). This alternative reproductive strategy may be maintained for years in the wild, until some opportunity arises for a *non-flanged* male to become dominant and transition to full physical development. Non-flanged adults attempt to mate coercively with females they encounter (Mackinnon 1974; Galdikas 1985*a*, 1985*b*). DNA typing studies have provided evidence that reproductive success is similar in non-flanged and flanged male Sumatran orangutans (Utami et al 2002; Utami Atmoko and van Hooff 2004), although only eleven infants were included in paternity analyses (Figure 2.4).

Given their widely dispersed social system, and slow rates of travel through the forest canopy, it is most unlikely that female orangutans encounter large numbers of adult males during the few days surrounding ovulation. There is some evidence that females prefer to approach and associate with a fully flanged male at this time (Schurmann 1982). Adult males are aggressive towards each other, and their impressive great calls and snag-crashing displays probably serve to maintain their spatial separation (Galdikas 1983; Mitani 1985). Non-flanged males, being more mobile and less conspicuous, can engage in additional copulations, however, so that some potential for sperm competition exists, with respect to the orangutan mating system.

In Figure 2.3 the relative testes sizes of both the Bornean and Sumatran orangutan fall just below the regression line for mammals in general, and are in general no smaller than those recorded for humans. Although the relative testes sizes of some men (e.g. in Nigeria) exceed those of orangutans, others do not, and are smaller in relation to body size (e.g. in Hong Kong, and in Japan). My interpretation of these more detailed comparisons is essentially the same as my view of Harcourt et al.'s (1981, 1995) original findings. Human testes sizes are unexceptional and consistent with an evolutionary history which involved pair formation or polygyny as the principal mating system. Sperm competition pressure would have been low under these circumstances. However, other authors including a number of evolutionary psychologists have interpreted differences between the relative testes sizes of men, the gorilla, and orangutan in a different way. They emphasize that human testes are larger than those of the gorilla and orangutan, and that sperm competition has shaped the evolution of human sexual

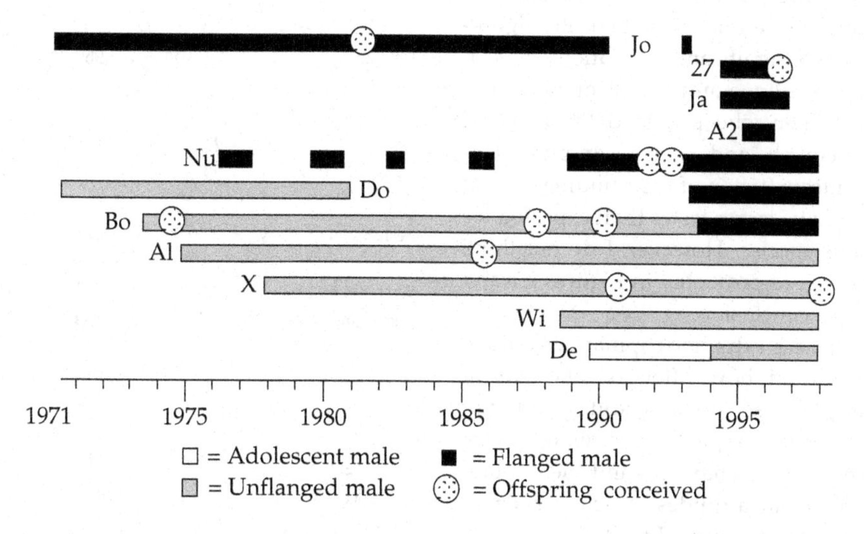

Figure 2.4 Paternity in free-ranging (flanged and non-flanged) male orangutans, as revealed by DNA typing studies of offspring and their potential sires. Both flanged and non-flanged males sired similar numbers of offspring. Two adult males, Do (Dobra) and Bo (Boris), developed from non-flanged to full expression of the cheek flanges and other secondary sexual traits during the course of the study.

Source: After Utami Atmoko and Van Hooff (2004).

behaviour (Smith 1984; Baker and Bellis 1995; Miller 2000; Buss 2003; Rolls 2005; Pound, Shackelford, and Goetz 2006). However, I believe this to be a case of wishful thinking; especially in the light of more detailed comparisons of human and great ape relative testes sizes (Figure 2.3). Yet, the issue will probably never be resolved by considering relative testes size in isolation. Relative testes size is, after all, only a crude index of sperm competition pressure. In the next chapter, comparative evidence on a much wider range of anatomical and physiological traits will be sifted in order to achieve a better perspective of the probable evolutionary basis of human mating systems and sexual behaviour.

An extreme view of the importance of human sperm competition is due to Baker and Bellis (1995), who claimed that sexual selection, via sperm competition, has played a crucial role in human evolution. As we shall see in the next two chapters, much of this work has been shown to be seriously flawed. Baker (1997) also reported that men rumoured to be more likely to engage in extrapair copulations had larger testes than those who were not rumoured to behave in this way. Yet in much larger and more carefully controlled studies, Simmons et al. (2004) found that men who reported engaging in extrapair copulations had, on average, smaller testes than men who did not (the difference was not statistically significant). There was a significant correlation between combined testes volume and sperm numbers per ejaculate in 50 men who provided the required testes measurements and semen samples. Simmons et al.'s interpretation of their findings is that human beings have larger testes than expected for a monogamous species. However, I do not think that the data (e.g. as presented in Figures 2.1 and 2.3) support this conclusion.

How frequent are extrapair copulations among human beings, and how often do they result in pregnancies? These are exceedingly difficult questions to answer, especially in view of the covert nature of such behaviour, and the cultural differences in sexual attitudes which pertain in different parts of the world. Events in modern-day New York or Paris probably bear little or no relation to patterns of extrapair copulation and paternity in remote ancestral populations of *Homo sapiens* or their African precursors. In their review of five studies conducted in UK, France, Australia, and the USA, Simmons et al. (2004) record that between 6.9–44 per cent of men and 5–51.7 per cent of women report having engaged in extrapair copulations. For people under 30 years of age the range is approximately 5–27 per cent. Of course, very few of these copulations may result in pregnancies and the birth of offspring. Rates of extrapair paternity in various human populations may range from as low as 0.03 to 11.8 per cent (see Table 2.5). The median value is 1.82 per cent, but very few studies have involved non-random samples of subjects or precise techniques to pinpoint paternity of offspring. The study by Sasse et al. (1994), conducted in Switzerland, involved 1,607 subjects and employed DNA fingerprinting techniques. Extrapair paternity was very low (0.7%) in this case. By contrast, Chagnon (1979), in a most interesting

Table 2.5 Rates of extrapair paternity in human populations

Population	% Extrapair paternity	*N*	Source
Michigan, USA	1.4	1417	Schacht and Gershowitz (1963)
Detroit, USA	0.21	265	Potthoff and Whittinghill (1965)
Oakford, California, USA	0.03	6960	Peritz and Rust (1972)
Hawaii	2.3	1748	Ashton (1980)
France	2.8	89	Le Roux et al. (1992)
Switzerland	0.7	1607	Sasse et al. (1994)
West Middlesex	5.9	2596	Edwards (1957)
Sykes family, UK	1.3	269	Sykes and Irven (2000)
UK	1.4	521	Brock and Shrimpton (1991)
Nuevo León, Mexico	11.8	396	Cerda-Flores et al. (1999)
South America, Yanomamo Indians	10.0	132	Chagnon (1979)

Source: From Simmons et al. (2004).

study of the Yanomamo Indians of South America, records extrapair paternity frequencies of 10 per cent for 132 cases, based upon blood group comparisons (Table 2.5).

Thus, rates of extrapair paternity are, in general, very low for the few human populations that have been sampled. The data are not sufficient to justify generalizations about the occurrence, or importance, of sperm competition in human reproduction. I shall return to this subject in Chapters 5 and 6, when the evolution of human copulatory behaviour is discussed in greater detail.

Ethnic differences in human testes sizes, discussed previously and summarized in Table 2.2, may result from a number of causes. Not all morphological differences between human populations are necessarily the result of natural, or sexual selection. As human beings dispersed across the globe, relatively small founder populations may have given rise to much larger numbers of modern day descendants, at least in some areas of the world. A limited founder gene pool may give rise to significant differences in morphological and physiological traits due to genetic drift. Genes may also influence a variety of *pleiotropic* effects. Mayr (1963) pointed out that 'a gene elaborates a gene product, which may be utilized in the differentiation of several organs (pleiotropy), and, conversely, any one character may be affected by many genes (polygeny).' In this regard it is interesting to consider the genetic link between gonadal size and fertility in males and females of the same species. Research on mice and sheep has demonstrated a connection between the occurrence of higher ovulation rates in females and larger testes sizes among males belonging to the same genetic strain. Short (1984) extended these findings on animals to propose that ethnic differences in human testes size might correlate with frequencies of dizygotic twinning among females belonging to the same populations. Smaller testes sizes of Hong Kong Chinese might thus be linked to lower frequencies of non-identical twin births among Chinese women, as compared to women in Africa or Europe. In Table 2.6, I have assembled data on human twinning rates (from Short 1984 and Diamond 1986) for comparison with the much larger body of data which is now available on testes sizes for human populations in nine countries. There is a positive correlation between dizygotic twinning rates and testes sizes (Spearman correlation coefficient = 0.8, $P < .02$). A word of caution, however, relates to the earlier discussion of the limits of error for some of these data, and especially those obtained by using orchidometers to assess testes volumes (Table 2.2). Unfortunately, if one attempts to correct for the various possible errors of measurement, using for example Döernberger and Döernberger's (1987) comparative studies of volumetric methods, then the correlation between dizygotic twinning rates and testes sizes shown in Table 2.2 loses statistical significance. Until the task of obtaining truly accurate data on testes weights for a large enough sample of human populations has been accomplished, some doubt remains as to whether the correlation with twinning rates is a genuine effect. The available data are suggestive of such a relationship. Might it be possible that, as human populations spread from Africa to Asia, and human physiques changed, twinning rates

Table 2.6 Human testes sizes and frequencies of dizygotic twinning: Data for nine countries

Country	Combined testes weight (g)*	Rank	Dizygotic twinning rates (per 1000 births)	Rank
Nigeria	52.62	1	40.0	1
UK	48.3	2	48.3	2
France	47.2	3	8.9	6
Sweden	43.6) 46.6 49.6)	4	8.6	3
Switzerland	39.06	5	8.1	4
Korea	32.9) 36.8 49.6)	6	5.1 5.8 7.9	8
India	35.4	7	6.8 7.3 8.1	5
Japan	31.5) 34.1 36.7)	8	2.3	9
China (Hong Kong)	19.0) 17.7 16.5)	9	6.8	7

*Testes weights from Table 2.2; volumes converted to weights by using a correction factor (× 1.05).

Source: Dizygotic twinning rates from Short (1984) and Diamond (1986).

underwent negative selection due to the smaller sizes of women and higher mortality rates caused by multiple births? This possibility is worthy of further study, and provides a feasible alternative to notions of decreased sperm competition and reductions in relative testes sizes in Asia.

Twinning occurs in some other primate species besides *Homo sapiens*. It is especially prevalent among the marmosets and tamarins (*Callitrichidae*) of South America. The cotton-topped tamarin, which Harcourt et al. (1981) reported as having surprisingly large testes (see Figure 2.1), normally gives birth to dizygotic twins. Larger than expected testes sizes in some tamarins and marmosets may be linked to the genetic predisposition to produce larger gonads and higher levels of reproductive hormones in both sexes of these monkeys.

Conclusions

1. Primates, like other mammals, exhibit large inter-specific variations in their testes sizes, in relation to body weight. The evolutionary significance of these variations only became apparent after Parker (1970) had developed the theory of sexual selection by sperm competition. Larger testes are required to contain the increased volume of seminiferous tissue necessary to produce higher sperm counts. Short (1979) and Harcourt et al. (1981, 1995) showed that relative testes sizes in primates correlate with their mating systems. The multi-male/multi-female mating systems of baboons, macaques, and chimpanzees are associated with large relative testes sizes. Females commonly mate with multiple partners during the peri-ovulatory period and sperm competition is pronounced in such species. By contrast, multi-partner matings are much less frequent in polygynous primates (e.g. the gorilla) or in pair-living (monogamous) forms such as gibbons. Relative testes sizes are significantly smaller in these cases.

2. Human relative testes sizes were shown to lie between these two extremes, being larger than those of the gorilla, but much smaller than the chimpanzee.

Some refinements to these conclusions emerge from the larger scale comparisons of mammalian relative testes sizes presented here. Thus, some Asiatic human populations have testes which are not larger, in relation to body weight, than those of orangutans or mountain gorillas. Yet, it remains the case that studies of testes size alone are unlikely to resolve debates about the likely importance of polygyny or monogamy in the evolution of human sexual behaviour.

3. Ethnic differences occur in human testes size. However, a consistent finding is that the right testis is, on average, 5 per cent larger than the left testis; this is confirmed here using comparative data from eight countries. Combined testes weights are smaller in men from Asiatic populations (China, Japan) than in their European and African counterparts. However, relatively few populations have been sampled and some data are subject to errors. This qualification applies particularly to volumetric measurements made using orchidometers as these often overestimate testes sizes.

4. Differences in testes sizes between human populations may be associated with genetic mechanisms which relate to gonadal function and fertility in both sexes (Short 1984). In mice and sheep, higher ovulation rates in females are associated with increased testes size in males of the same genetic strain. In human beings there is significant correlation between dizygotic twinning frequencies and testes sizes (data from nine countries). However, the caveat concerning possible errors of measurement of human testes sizes also applies to this finding.

5. Extrapair copulations and paternities occur in human beings but their current frequencies may tell us very little about selective pressures which shaped the evolution of human reproductive biology during the emergence of *Homo sapiens* from its African precursor, more than 195,000 years ago. It is noted, however, that frequencies of extrapair paternities are low in modern human populations (median value = 1.82 per cent for studies in six countries: Simmons et al. 2004).

6. A number of authors have overestimated the importance of relatively modest data on human testes size and frequencies of extrapair

copulation. Their conclusion that sperm competition played a significant role in the evolution of human reproduction is not justified on current evidence.

The next chapter examines the effects of sexual selection upon the evolution of a variety of reproductive traits, including sperm morphology, the sizes and functions of the accessory sexual organs, and phallic morphology. This broad approach, involving comparisons of many mammalian species with *Homo sapiens*, offers much deeper insights into the likely origins of human sexuality.

CHAPTER 3

Masculine Dimensions

'I've measured it from side to side
'Tis three feet long, and two feet wide.'
Wordsworth

The male genitalia evolved as an exquisitely complex, integrated system for the production, storage, and delivery of sperm (and accessory glandular secretions) to females during sexual activity. Studies of invertebrates have provided ample evidence for effects of sexual selection upon the evolution of male genitalic morphology and physiology (Eberhard 1985; 1996; Simmons 2001; Poiani 2006). The same evolutionary principles may be applied in comparative studies of the genitalia in the mammals, including human beings. The reproductive systems of male mammals are structurally diverse, with remarkable differences in their accessory reproductive glands and penile morphologies. At least some of this diversity may have been influenced by sexual selection, whether via sperm competition or cryptic female choice. The question of cryptic female choice will be deferred until the next chapter, where possible effects of sexual selection upon the evolution of the female genitalia are discussed. Here I shall explore the effects of sperm competition upon the evolution of the male reproductive system in mammals, in order to address a number of questions. Firstly, has sexual selection influenced the evolution of sperm morphology? Secondly, have there been measurable effects upon the evolution of the male ducts, such as the vasa deferentia, or upon the sizes and functions of the accessory reproductive glands which are highly developed in mammals? If comparative studies reveal significant effects of sexual selection, then how do the homologous structures of the human male compare with those of other mammals in relation to sperm competition pressures? Finally, given that the penis plays a crucial role in placing sperm within the female reproductive tract, has sexual selection influenced the evolution of penile complexity in mammals? If so, then how specialized, or complex, is the human penis by comparison with the intromittent organs of other primates, or mammals in general? Accurate answers to these questions should help to improve our understanding of the origins of human mating systems and sexual behaviour.

Evidence from mammalian sperm morphology

Sperm length

Mammalian sperm are relatively uniform in their basic structure (Figure 3.1), as each consists of a head (containing unique genetic material), a midpiece (containing mitochondria), and a flagellum, or tail, to provide motility. However, within this basic plan there are numerous inter-specific variations, such as differences in sperm length, in the shape and size of the sperm head (including the acrosome), and fine structural differences at the electron microscopic level in various taxa. Figure 3.2 provides some examples of these differences in sperm morphology among mammalian species; the human sperm is included along with those of various rodents, the hippopotamus, and some marsupials. In some murid rodents the acrosomal

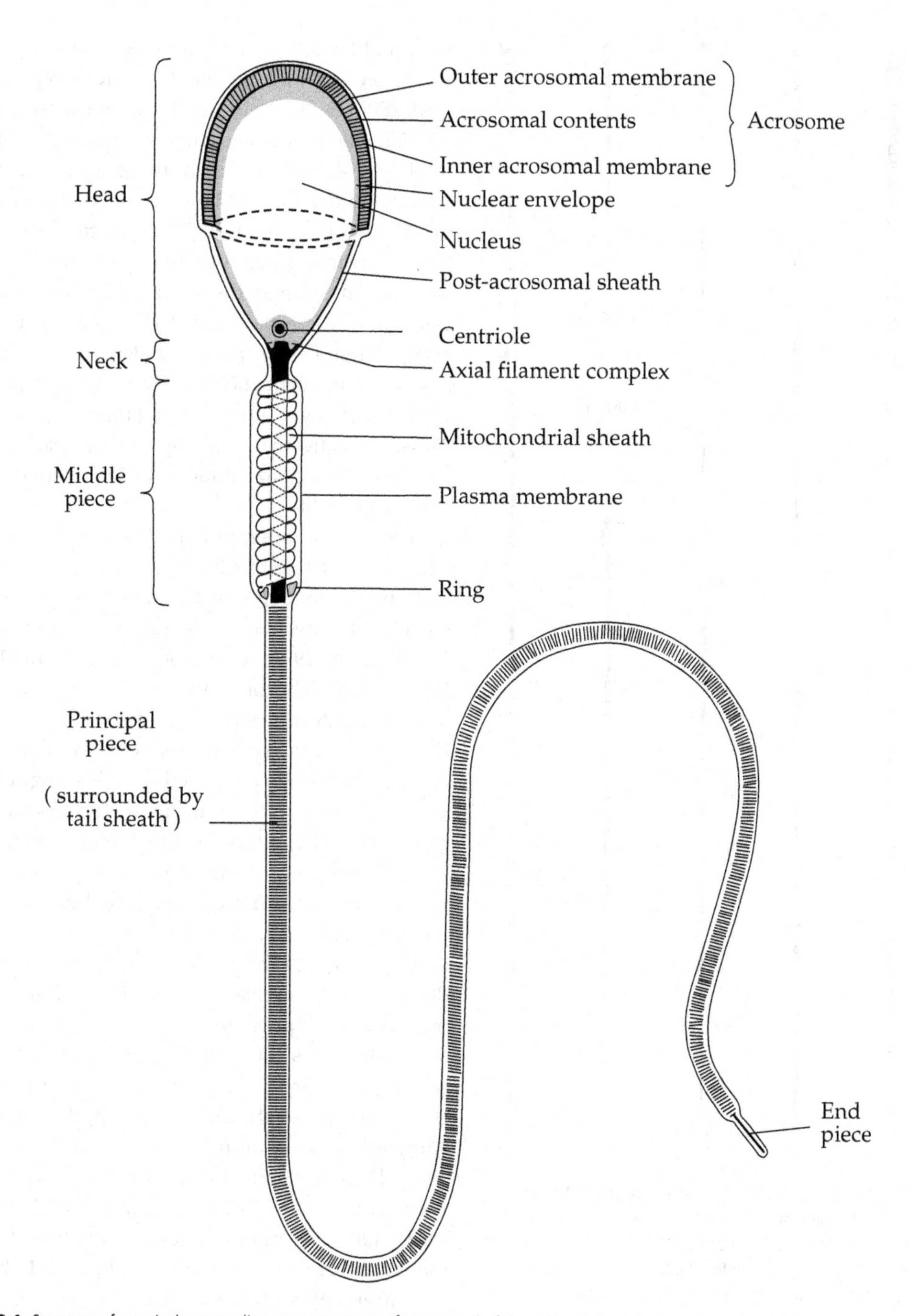

Figure 3.1 Structure of a typical mammalian spermatozoon, after removal of the cell membrane.
Source: After Fawcett (1975).

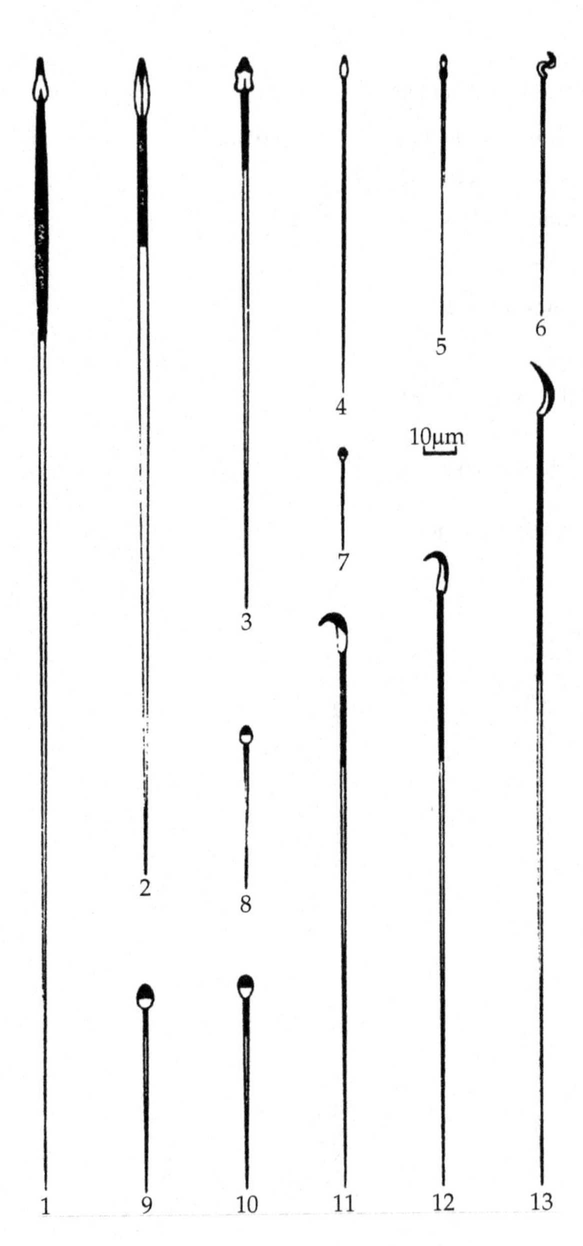

Figure 3.2 Morphological diversity of mammalian spermatozoa. The acrosome and midpiece of each sperm is shaded. 1. Honey possum (*Tarsipes rostratus*); 2. Marsupial rat (*Dasyuroides byrnei*); 3. Short-nosed brindled bandicoot (*Isoodon macrourus*); 4. Tammar wallaby (*Macropus eugenii*); 5. Brush-tailed possum (*Trichosurus vulpecula*); 6. Koala (*Phascolarctos cinereus*); 7. Hippopotamus (*Hippopotamus amphibius*); 8. Human (*Homo sapiens*); 9. Rabbit (*Oryctolagus cuniculus*); 10. Ram (*Ovis aries*); 11. Golden hamster (*Mesocricetus auratus*); 12. Laboratory rat (*Rattus norvegicus*); 13. Chinese hamster (*Cricetulus griseus*).

Source: After Setchell (1982).

region of the sperm head is hooked, and especially so in species with large relative testes sizes (Immler et al. 2007). Sexual selection may have favoured this specialization because such sperm form cohorts, hooked together and swimming more efficiently as a group than would be possible for each gamete on its own (Moore et al. 2002). In the majority of South American marsupials the sperm are joined together in pairs at the acrosome, and remain so until they reach the oviduct (Rodger and Bedford 1982). These *binary sperm* are also thought to possess more efficient motility, especially when moving through a viscous medium (Taggart et al. 1993). Thus, in some mammals sperm competition has resulted in specializations of sperm morphology to enhance motility.

The largest spermatozoon among mammals occurs in the diminutive honey possum (*Tarsipes rostratus*). Indeed, overall sperm length is not correlated with body size among mammals as a whole (Gage 1998; Anderson, Nyholt, and Dixson 2005). Might the observed differences in sperm length observed among mammals also be related in some way to sperm competition? Gomendio and Roldan (1991) proposed that this might be so, and that longer sperm occur in those species of primates and rodents, where females mate with multiple partners during the fertile period. Long sperm may swim more rapidly within the female reproductive tract, and hence they might secure some advantage in sperm competition. However, despite early support for this hypothesis, based on studies of primate sperm lengths and relative testes sizes (Dixson 1993), subsequent analyses of much larger data sets have failed to reveal any correlation between sperm length and sperm competition in mammals (Harcourt 1991; Gage 1998; Dixson 1998*a*; Gage and Freckleton 2003; Anderson et al. 2005). Examples of Anderson et al.'s (2005) comparative measurements of mammalian sperm are provided in Table 3.1. Linear measurements of the individual compartments of mammalian sperm do not correlate with mating systems or relative testes sizes. However, studies of the volume of the sperm midpiece have provided significant comparative and evolutionary insights, as described below.

Table 3.1 Sperm measurements in anthropoid primates

Species (*N*)	Total length (μm)	Midpiece length (μm)	Midpiece Volume (μm³)
Vervet (5)	91.5	12.1	11.5
Patas (6)	92.8	14.5	6.5
Stump-tail (6)	78.8	10.5	10.2
Mandrill (4)	45.7	7.1	8.0
Hamadryas baboon (8)	77.2	12.7	8.5
Gelada (8)	88.2	12.3	6.5
Siamang (7)	68.9	8.2	6.8
Orangutan (9)	60.5	8.9	3.9
Chimp (8)	60.4	6.3	7.8
Bonobo (12)	68.1	8.9	9.3
Gorilla (15)	52.3	13.2	6.9
Human (5)	56.9	5.9	3.8

Source: Data from Anderson et al. (2005).

Sexual selection and sperm midpiece volume

The sperm midpiece contains a tightly packed, helical array of mitochondria, which surround the central strut, or axoneme, of the sperm. The mitochondria differ in size and number between species. As examples, Figure 3.3 shows electron micrographs of the sperm midpieces of the monkey and the squirrel chimpanzee in longitudinal section. The squirrel chimpanzee sperm midpiece contains fifty-three mitochondrial 'gyres' (i.e. spiral folds around the central axoneme) by comparison with only twenty-two in the chimpanzee. However, in the chimpanzee the mitochondria are much larger than those of the squirrel monkey. These mitochondria are the sites of oxidative phosphorylation and production of energy. They are not the only potential source of energy for sperm motility, as glycolysis also plays an important role in this process (Miki et al. 2004). However, it follows that differences in mitochondrial loading in sperm of different species may be significant in functional and evolutionary terms. Might it be possible, for example, that larger volumes of mitochondria occur in the sperm midpieces of mammals which engage in sperm competition? More vigorous or more sustained motility of sperm with larger midpieces might be significant in competitive contexts. Unfortunately, attempts to implicate linear measurements of sperm midpieces

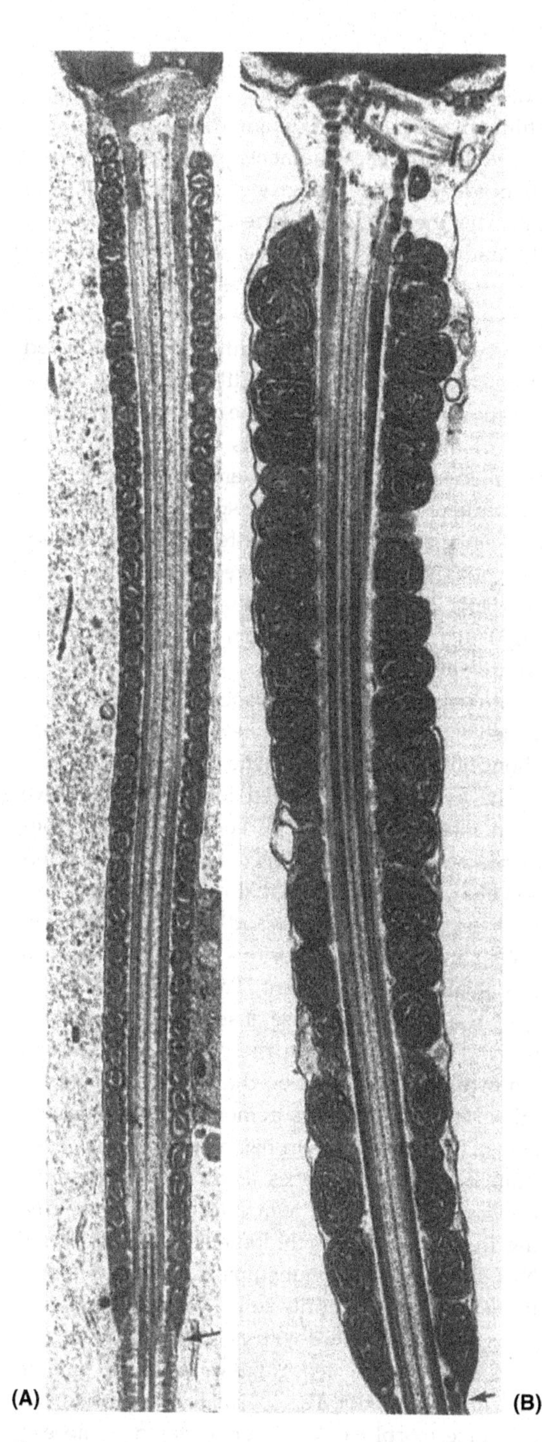

Figure 3.3 Electron micrographs of longitudinal sections through the sperm midpiece of (A) the squirrel monkey (*Saimiri sciureus*), and (B) the chimpanzee (*Pan troglodytes*). Note the pronounced differences in the numbers and sizes of the mitochondria.

Source: After Bedford (1974).

are of little value in addressing these questions (Malo et al. 2006). As indicated by the squirrel chimpanzee–monkey example discussed in Figure 3.3, volumetric measurements provide a more useful guide to the likely degree of mitochondrial loading. This method is not perfect, of course, because the midpiece consists of a dense sheath, enclosing the central axoneme as well as the mitochondria. However, if the midpiece is treated as a cylindrical structure, then its overall volume may be estimated from measurements of its width and length. Comparative measurements of the sperm of primates and of many other mammals confirm that larger midpieces occur in species where females mate with multiple partners, and where sperm competition is also associated with larger testes sizes (Figure 3.4). The positive correlation between sperm midpiece sizes and relative testes sizes is especially interesting. Thus, it appears that sexual selection, via sperm competition, has resulted not only in the ability to produce larger numbers of sperm, but also sperm which have larger midpieces and a greater mitochondrial loading for energetic purposes.

The findings summarized in Figure 3.4 derive from measurements of the sperm of 123 species of mammals (Anderson and Dixson 2002; Anderson et al. 2005) on substantial numbers of gametes (100 sperm for each male) and using a single, consistent methodology. Previous failure to establish this relationship (Gage and Freckleton 2003) may have resulted from the use of sperm measurements derived principally from the published literature. Unfortunately, this produced a dataset which contained inconsistent measurements of sperm dimensions. The same problem has hampered studies of human relative testes sizes, as was discussed in the previous chapter with regard to inaccuracies arising from the use of orchidometers (see Table 2.2). Nor do the findings on mammals presented in Figure 3.4 necessarily apply to all vertebrate groups. Thus, sperm midpiece volumes and relative testes sizes are not correlated in passerine birds (Immler and Birkhead 2007). The spermatozoa of birds are, of course, morphologically very different to those of mammals (Jamieson 2007).

Human sperm have relatively small midpieces, by comparison with those of many other primates, and in relation to testes size (Figure 3.4, upper part). Sperm midpiece volumes of human beings, the apes, and other anthropoids are included in Table 3.1. Human sperm midpieces are smaller than those of all the other hominoids, including the gorilla and orangutan. The human sperm midpiece contains fifteen mitochondria (Bedford 1974; Bedford and Hoskins 1990) and these are substantially smaller, in overall volume, than the mitochondria of the chimpanzee. Thus the data on human sperm morphometry indicate that sexual selection, via sperm competition, is unlikely to have played a significant role in human evolution. This finding should give pause for thought to those authors who maintain that human testes size reflects effects of sexual selection because it is larger than that of the gorilla, or slightly larger than that of the orangutan.

Recent research on sperm energetics using live gametes from chimpanzees and human beings has provided additional support for hypotheses concerning effects of sexual selection upon midpiece sizes and mitochondrial loading. Living sperm from four men and five chimpanzees were treated with JC-1, a fluorescent dye which stains metabolically active mitochondria red/orange. Intensity of staining provides a measure of mitochondrial membrane potential and has been shown to correlate with forward motility in human sperm (Marchetti et al. 2004). Comparisons of chimpanzee and human sperm show that fluorescence intensities in chimpanzees are markedly higher, due to bright red staining of mitochondria in the midpieces of their sperm (Anderson et al. 2007). Results are shown in Figure 3.5. These differences persist in sperm which have undergone (in vitro) *capacitation*, a process which normally occurs within the female reproductive tract and which prepares spermatozoa physiologically to fertilize ova. Whereas mitochondrial staining by JC-1 tended to decrease after capacitation in human sperm, this was not the case in chimpanzees (Figure 3.5). Capacitated sperm are capable of a distinctive kind of vigorous motility, often called *hyperactivated motility*, and which normally occurs in the oviduct prior to fertilization (Yanagimachi 1994). Whether the higher mitochondrial loading of chimpanzee sperm improves their motility in lower portions of the female tract, or their capacity for hyperactivated motility in the oviduct, requires further study. However, the JC-1

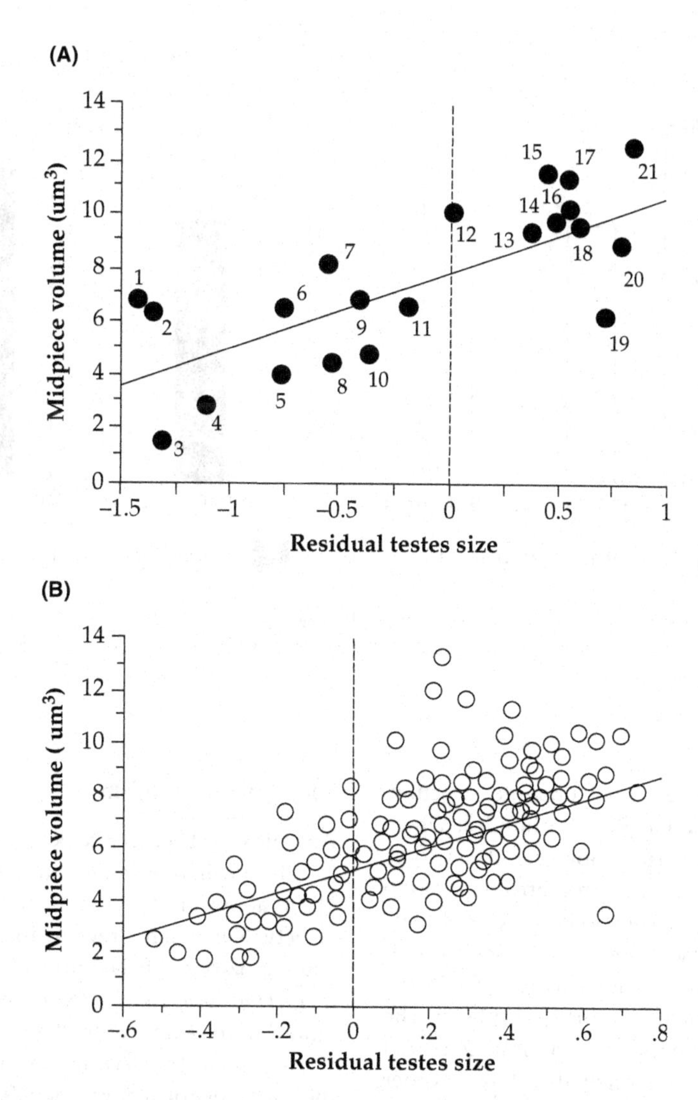

Figure 3.4 (A). Correlation between sperm midpiece volume (in μm³) and relative testes size in twenty-one primate genera ($y = 2.55x + 7.52$; $P < .0001$). 1. *Gorilla*; 2. *Erythrocebus*; 3. *Callithrix*; 4. *Hylobates*; 5. *Pongo*; 6. *Theropithecus*; 7. *Saimiri*; 8. *Cebus*; 9. *Galago*; 10. *Homo*; 11. *Cercopithecus*; 12. *Lophocebus*; 13. *Mandrillus*; 14. *Pygathrix*; 15. *Lemur*; 16. *Eulemur*; 17. *Papio*; 18. *Nycticebus*; 19. *Microcebus*; 20. *Pan*; 21. *Macaca* (B). Correlation between sperm midpiece volume and relative testes size in 123 species of mammals.
Source: After Anderson and Dixson (2002); After Anderson, Nyholt, and Dixson (2005).

fluorescence studies strengthen the conclusion that differences in midpiece volumes in human beings, and other mammals, do spring from functional differences in mitochondrial loading and sperm bioenergetics.

Two further points may be made as regards sperm midpiece volumes in mammals. Firstly, with rare exceptions, the mitochondria in the sperm midpiece do not participate in development of the fertilized egg. They are excluded, and it is only the mitochondria contained in the ovum, and deriving from the female parent, which contribute to embryonic development. This being the case, it might be argued that advances in mitochondrial efficiency

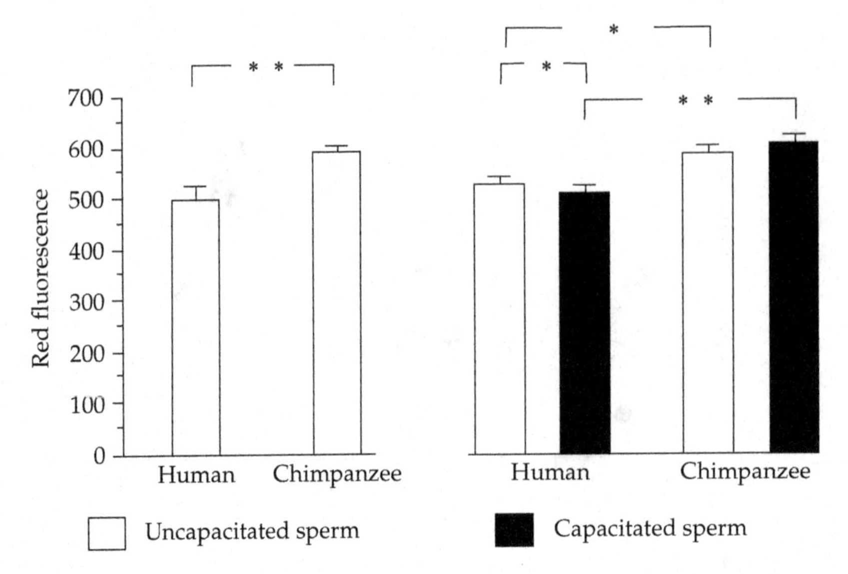

Figure 3.5 Left: Red fluorescence values for JC1-stained sperm samples from men and chimpanzees, as measured using flow cytometry. Right: Effects of in vitro induction of capacitation upon JC1 staining (red fluorescence) of human and chimpanzee sperm. $*P < .05$; $**P < .01$.
Source: After Anderson et al. (2007).

in the sperm midpiece cannot undergo selection, because it is only the female mitochondria which are transmitted to offspring. However, it has to be kept in mind that the genetic programs required to produce the morphological specializations of male gametes, including the volume of the sperm midpiece, certainly are transmitted to females in the haploid genome contained in the sperm head. The volume of the sperm midpiece can undergo sexual selection via sperm competition, therefore, in order to increase mitochondrial loading and improve sperm motility.

A second point relates to the variability of JC-1 fluorescence when human spermatozoa are stained with this dye. Staining intensity is much more variable among human gametes than is the case for chimpanzees (Anderson et al. 2007). This may relate to the greater morphological variability (pleiomorphism) of human sperm and the higher frequencies of abnormal sperm in human ejaculates. Because *sperm pleiomorphism* has been implicated in discussions of human sperm competition (Baker and Bellis 1988, 1995), I devote a section to this subject below, in order to place it in true perspective.

Sperm pleiomorphism

Antoni van Leeuwenhoek was the first person to observe human spermatozoa; he reported his observations in a letter to the Secretary of the Royal Society, Nehemiah Grew, in 1677. His drawings of spermatozoa, published in the *Philosophical Transactions of the Royal Society* (1679), show individual variations in the shape of the head, and other features (Farley 1982). These variations may reflect, in part, the limitations of Leeuwenhoek's microscope lenses, lenses which he ground himself and mounted in his pioneering instruments. Or, perhaps the individual differences which Leeuwenhoek saw (Figure 3.6) also related to genuine variations (pleiomorphisms) among these gametes. For human ejaculates commonly include a high percentage of abnormally shaped gametes, as is also the case among gorillas (Seuánez et al. 1977; Seuánez 1980) and cheetahs (O'Brien et al. 1985).

Baker and Bellis (1988, 1995) proposed that human sperm pleiomorphism results from selection for sperm morphs which fulfill different roles in relation to sperm competition. 'Kamikaze' sperm were posited to occupy strategic positions in the female

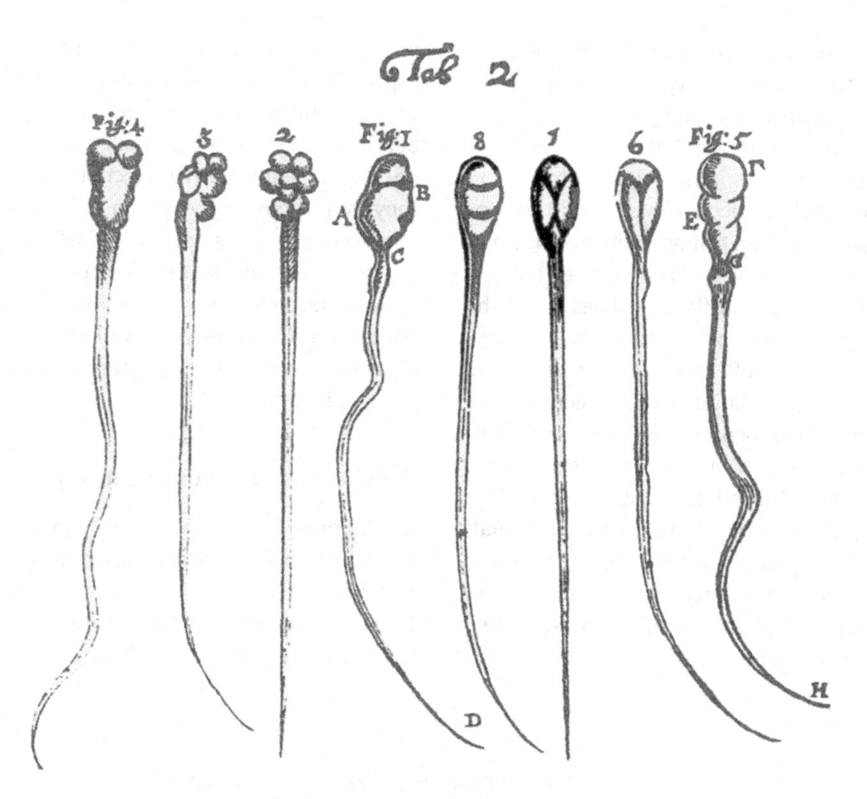

Figure 3.6 Leeuwenhoek's drawings of spermatozoa ('spermatic animalcules') as reported in the *Philosophical Transactions of the Royal Society* (1679).

Source: From Farley (1982).

reproductive tract and to block access to gametes of rival males. In mammals which produce copulatory plugs (e.g. rats, bats) it was proposed that the presence of kamikaze sperm functioned to form a meshwork and a focus for plug formation. By contrast, a smaller population of fertilizing or "egg-getter" sperm in human beings, and other mammals, was hypothesized to specialize in reaching the oviduct and effecting fertilization. In their book, *Human Sperm Competition* (1995), Baker and Bellis also proposed that oval-headed 'killer sperm' exist in the human ejaculate. When semen of two men were mixed, increased mortality and decreased motility were reported to occur. Baker and Bellis suggested that killer sperm released their acrosomal enzymes in order to attack the gametes of the second male in these mixed samples.

Although these startling claims achieved notoriety via the popular media, they have received little support scientifically. Moore, Martin, and Birkhead (1999) conducted careful quantitative studies of the effects of mixing ejaculates from different males and were unable to confirm Baker and Bellis's findings or to find any support for the existence of killer sperm in human beings. Much earlier, Harcourt (1989) had pointed out inconsistencies in Baker and Bellis's (1988) kamikaze sperm hypothesis, especially in relation to comparative data on mammalian gamete biology. Abnormal sperm, such as those which are trapped in the copulatory plugs of bats, may be so situated because they lack normal motility and not because they are specialized, kamikaze morphs. Harcourt also noted that, among the primates, the occurrence of pleiomorphic sperm, including abnormal sperm, is greatest among gorillas and human beings, and least in chimpanzees. This is surprising, because chimpanzees engage in frequent multiple matings and sperm competition. Thus, we should expect sperm pleiomorphism, and kamikaze morphs to be better developed in chimps than in gorillas or human beings. To make this point clearer, Figure 3.7 presents data on percentages of pleiomorphic sperm in the ejaculates of all the great ape species, as well as in human males. These data (from Seuánez 1980) make it clear that there is actually an *inverse* relationship between relative testes size, sperm competition pressures, and numbers of abnormal sperm in the ejaculates of men and the great apes. Indeed, it might be argued that it is the very absence of selection pressures relating to sperm competition which allows abnormal sperm to persist in the ejaculates of gorillas and human beings. The production of such abnormal sperm might relate to errors which occur during the meiotic cellular divisions of spermatogenesis, and especially to errors of chiasmata formation which underlie the 'crossing-over' of genes (Cohen 1967). Selection may fail to eliminate such errors if there are insufficient negative selection pressures. Hence they may persist in the gorilla and human male but not in chimpanzees or bonobos where sperm competition is of paramount importance. These observations lead to some additional considerations concerning the quality, as well as the numbers, of spermatozoa produced by human beings and other mammals.

Questions of sperm quality and quantity

In Chapter 2, it became clear that comparative measurements of relative testes sizes, viewed in isolation, are unlikely to resolve debates about the possible importance of sperm competition during human evolution. Although human testes are not

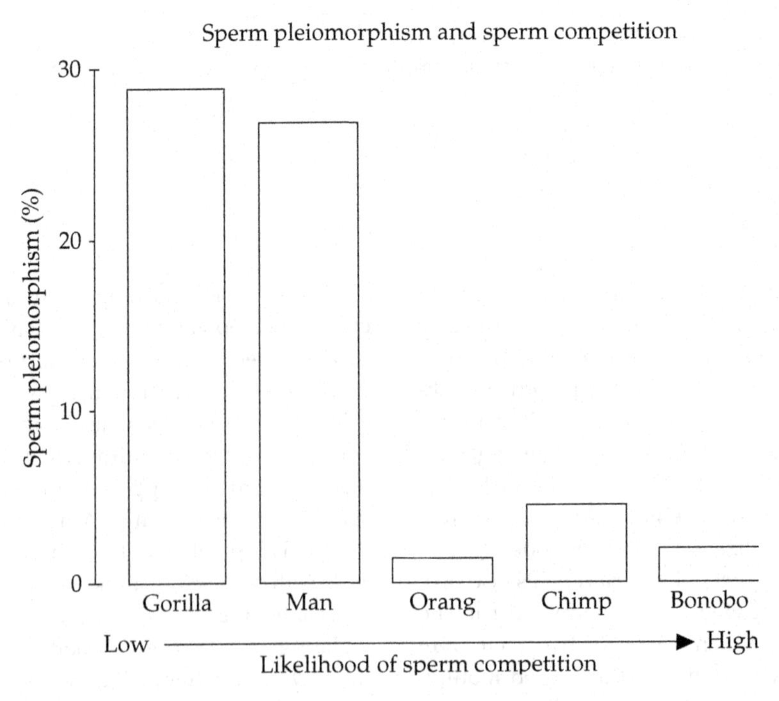

Figure 3.7 Percentages of morphologically abnormal (pleiomorphic) sperm in the ejaculates of man and the great apes. Abnormal sperm occur most frequently in man and in the gorilla. By contrast, species which are engaged in sperm competition (e.g. the chimpanzee and bonobo) have the lowest percentages of sperm abnormalities.

Source: Based upon data from Seuánez (1980).

large, in relation to adult body weight, they are larger than those of gorillas and orangutans (at least in men originating from Europe and Africa: see Table 2.2). It can always be argued that sperm competition may have played a significant role in the evolution of human reproduction, although it has been more important in chimpanzees or bonobos. However, given that comparisons of relative testes size provide only a relatively crude index of differences in sperm production, it is relevant to ask how *efficient* the human testes are at producing sperm, by comparison with the testes of other mammals? How many sperm are produced each day by each gram of seminiferous tissue (parenchyma) in the testes? After leaving the testis, spermatozoa must pass through the epididymis and be stored in its tail (cauda). How long are epididymal transit times and how many sperm are stored in the cauda epididymis in various mammals? How many sperm are contained in human ejaculates, and what effects do repeated copulations have upon sperm counts? If sperm competition has played a significant role in human evolution then we might expect to encounter the same kinds of specializations that occur in other mammals which engage in sperm competition.

Møller made valuable contributions to this field in several of his papers which deal with comparative studies of mammalian ejaculate quality and relative testes size (Møller 1988; 1989; 1991). He showed that a variety of measures of ejaculate quality tend to be correlated and proposed that they 'have been improved simultaneously, apparently by common mechanisms' (Møller 1991). Species with larger testes, in relation to body weight, produce larger numbers of spermatozoa in the ejaculate, and maintain larger reserves of sperm in the epididymis. They also produce a higher proportion of morphologically normal and motile sperm than mammals which have small testes in relation to body size. Interestingly, the number of spermatozoa produced per gram of parenchyma in mammalian testes does not correlate with relative testes sizes, and nor do epididymal transit times. Although Møller was able to assemble only a limited data set (nine species of mammals including *Homo sapiens*) his findings were statistically significant. Additional information has been published in the intervening years, so that now we are in an even stronger position to assess how sperm production and ejaculate quality in human beings compare with those of other mammals.

Table 3.2 provides data on rates of human sperm production, storage, and sperm numbers in the ejaculate. The total time required to produce a human spermatozoan is 74–76 days. Spermatogenesis takes longer in *H. sapiens* than in almost every other mammal for which measurements have been made. Sharpe (1994), in his detailed review of the regulation of mammalian spermatogenesis, lists durations in several species, as follows: 33.8 days (guinea pig), 35 days (mouse), 42 days (longtailed macaque), 44 days (rhesus and stump-tailed macaques), 48 days (rabbit and blue fox), 54 days (bull and coyote), and 57 days (olive baboon). The ultimate (i.e. evolutionary) mechanisms which may underly these species differences in the time required for the mitotic and

Table 3.2 Human spermatogenesis: Sperm production, epididymal transit time, storage, and numbers per ejaculate

Measurement	Duration/Number	Source
1. Duration of spermatogenesis	74–76 days	Sharpe (1994)
2. Daily sperm production/g	4.4 million	Sharpe (1994)
3. Total daily sperm production	20–270 million; mean 130 million	Johnson, Petty, and Neaves (1984); Sharpe (1994)
4. Sertoli cells/g	33–49 million	Johnson et al. (1984)
5. Spermatids/Sertoli	3.9±0.5	Sharpe (1994)
6. Epididymal transit	1–12 days (mean 5.5)	Orgebin-Crist and Olson (1984)
7. Epididymal sperm reserve	440 million	Møller (1989); Amann and Howards (1980)
8. Sperm per ejaculate	236.1±124.1 million	Pound et al. (2002)

meiotic cellular divisions and transition of spermatids to completed spermatozoa remain unclear. What is clear is that human males take longer to complete spermatogenesis than most mammals (Hochereau de Riviers et al. 1990; Sharpe 1994).

Møller (1989, 1991) pointed out that rates of sperm production per gram of parenchyma (seminiferous tissue) are not correlated with relative testes sizes in mammals. It is not the case that a species with large testes in relation to body weight, and a significant degree of sperm competition within its mating system, necessarily produces more gametes per gram of testis. There appear to be physiological constraints limiting the speed of sperm production within a given species; hence sexual selection has led to the evolution of larger testes, containing a greater volume of sperm-producing tissue. Nonetheless, it is intriguing that mammals vary considerably in their daily sperm production (DSP) per gram of parenchyma. Human males have the lowest recorded DSP among mammals (4.4 million/g). This compares to 13 million/g in the bull, 23 million/g in the rhesus monkey, and 25 million/g in the rabbit (Figure 3.8). Thus, if sperm competition has played a significant role in human evolution it is surprising that DSP rates are not higher in human males, or at least commensurate with those of other mammals.

Why, at the proximate level, are rates of sperm production per gram of testicular tissue so low in human males? One reason may relate to the functions of the Sertoli cells in the testis. Sertoli cells make up a substantial volume (11–40%) of the seminiferous epithelium in mammals (Russell 1998). First described in 1865 by Enrico Sertoli, when he was just 23 years old, these cells have been implicated ever since his discovery in the processes that control sperm production. These are very large and complex cells, each of which makes contact with the various germ cell types as they undergo their divisions and transition from spermatogonia, to spermatocytes, spermatids, and completed spermatozoa (Figure 3.9). Sertoli cells contain receptors for a variety of hormones and they orchestrate and co-ordinate the development of cohorts of spermatozoa (Bardin et al. 1994). It is not surprising, therefore, that there is a marked positive correlation between the numbers

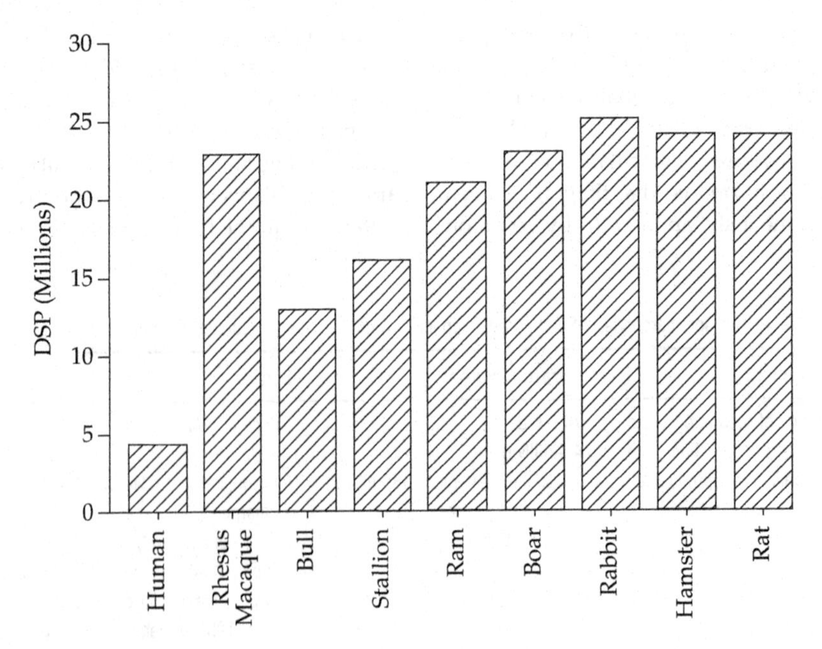

Figure 3.8 Daily sperm production (in millions per gram of testicular parenchyma) in human beings as compared to other mammals.
Source: Based upon data from Sharpe (1994).

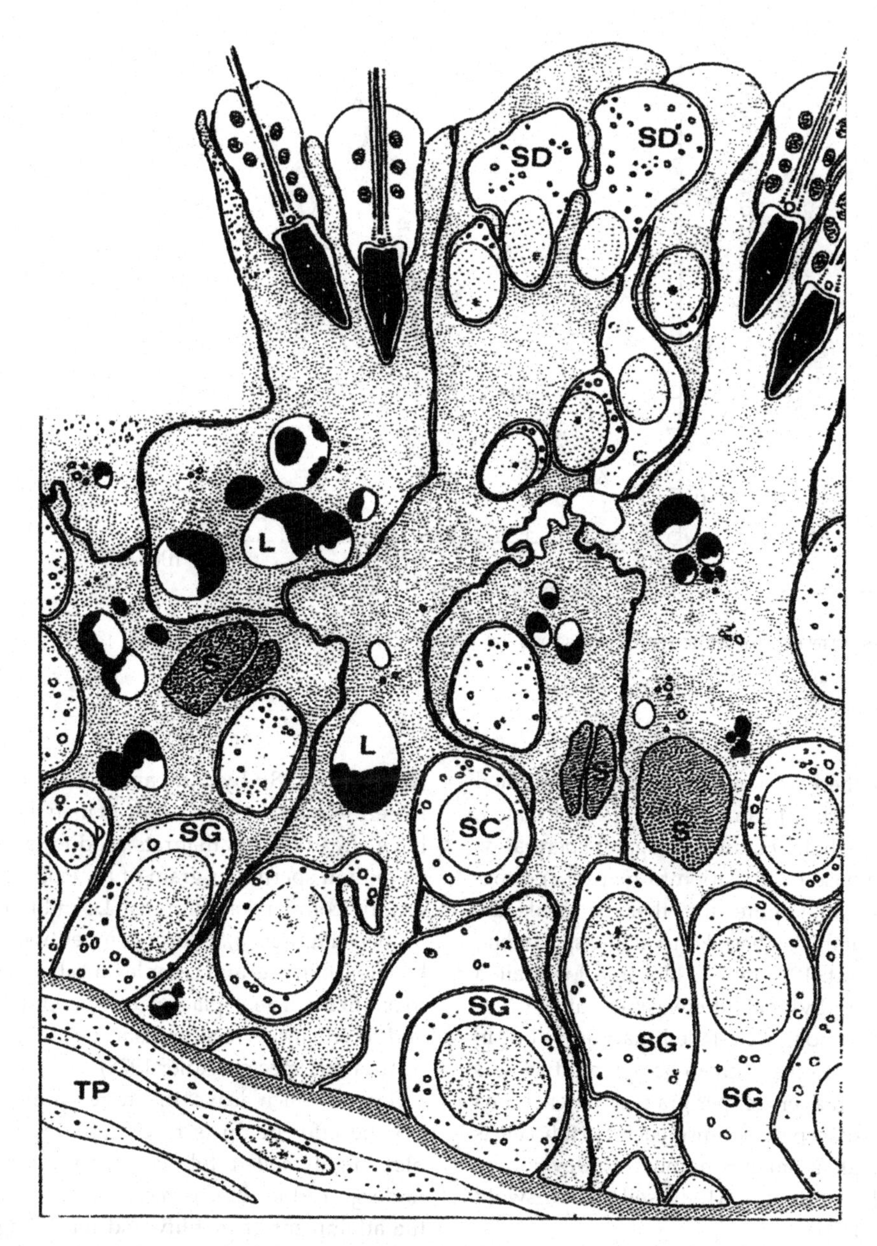

Figure 3.9 The human seminiferous epithelium, to show relationships between the developing germ cells and Sertoli cells. SG = spermatogonia; SC = primary spermatocytes; SD = spermatids; S = Sertoli cells, containing lipids (L). Specialized inter-Sertoli cell junctions are indicated by heavy lines.

Source: After de Kretser and Kerr (1994).

of Sertoli cells in the testis and rates of daily sperm production. Figure 3.10 shows the relationship between Sertoli cell numbers and daily sperm production rates in the testes of men (aged 18–71 years) studied in the USA by Johnson et al. (1984). These authors calculated that each human testis contains approximately 500,000 Sertoli cells, and that their numbers decline significantly with age. However, each human Sertoli cell is associated with the development of smaller numbers of germ

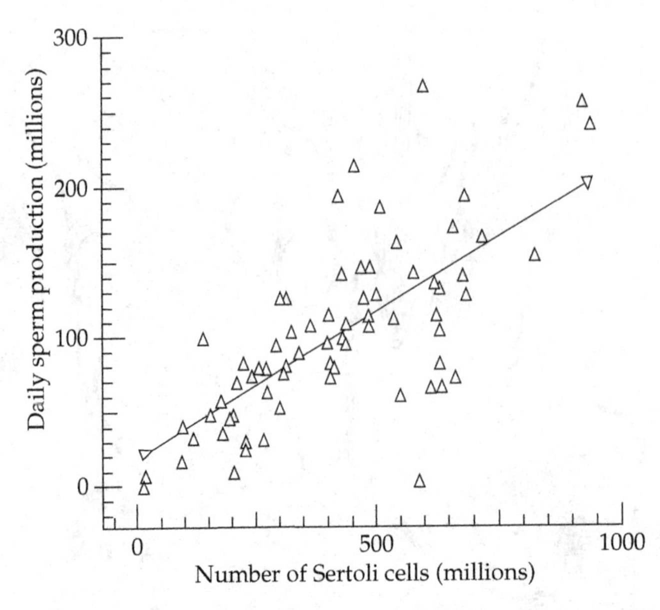

Figure 3.10 Positive relationship between Sertoli cell numbers and daily sperm production by the human testis ($y = 18.8 + 0.2 \times$ rho = + 0.7).
Source: Redrawn from Johnson et al. (1984).

cells than is the case for other mammals studied so far. Thus, each human Sertoli cell is associated, on average, with 3.9 mature (elongate) spermatids in the final phase of transition to mature spermatozoa. Comparable figures for other mammals are, for the orangutan: 5.7 (based on a single specimen), stallion: 11.5, rat: 10.3, rabbit: 12.2, and long-tailed macaque: 8.2. Individual differences certainly occur (for example, among long-tailed macaques) but in general it appears that human beings produce less spermatozoa on average, per gram of parenchyma in their testes, in part because each Sertoli cell is associated with a smaller number of developing gametes (Johnson et al. 1984; Russell and Griswold 1993; Sharpe 1994).

Given the comparatively low rates of sperm production achieved per gram of human testis, it is not surprising that total sperm production by human testes is much less than that measured for most other mammals (Møller 1989; Sharpe 1994). Body size is an important consideration, however, as testes size, and hence the bulk of sperm-producing tissue, is affected by body size. Male rats produce less spermatozoa each day than human males, (86 million vs. 130 million on average), but rat testes weigh only 3.8 g by comparison with the approximately 40 g testes of a European man. However, we have already noted that rats produce a remarkable 24 million sperm per gram of testicular tissue each day, by comparison with just 4.4 million per gram in men (Figure 3.8). Sperm competition is likely to be intense in rats and complex mechanisms exist to maximize efficient placement of copulatory plugs and to enhance sperm transport within the female tract in this species (Adler 1978).

Sperm which leave the testes are transported via the efferent ducts to the epididymis, a complex, highly coiled, tubular organ in which sperm undergo further biochemical changes required for the attainment of motility and fertilizing capacity. Epididymal transit times vary from 1 to 12 days in the human male, and approximately 440 million sperm are stored in the tail (cauda) region (Table 3.2). Human sperm reserves are small by comparison with those of other mammals (Møller 1989). Consider, for example, the epididymal sperm reserves of the rat (700 million), rabbit (2,200 million), rhesus macaque (13,000 million), or sheep (165,000 million). It is not surprising, therefore, that human sperm counts quickly begin to decrease as a result of repeated ejaculations. Freund (1962, 1963) showed that men's sperm counts decrease by 55 per cent when their frequencies of ejaculation increase from an average of 3.5 to 8.6 times per week. I shall return to this subject in Chapter 5, which deals with the evolution of human copulatory patterns.

In Table 3.2, a figure of 236 million is cited as the average number of spermatozoa in a human ejaculate. However, human sperm counts are exceedingly variable and it may be helpful to consider at least some of the factors involved. The vast majority of sperm samples used for clinical studies are obtained by masturbation. Men are asked to abstain from sexual activity prior to giving a semen sample, as sperm counts are intended to represent the maximum possible in a 'sexually rested' state. It is interesting that sperm numbers and sperm motilities are significantly greater in ejaculates collected by condoms as a result of copulation (Zavos 1985, 1988). Even those samples collected by masturbation contain larger numbers of sperm if the time taken to produce the sample is extended, and sexual arousal is prolonged (Pound et al. 2002). There are also considerable geographical variations in human sperm counts (Fisch and Goluboff 1996), and (controversially) declines in sperm counts worldwide have been reported, possibly as a result of environmental toxins (Carlsen et al. 1992). A host of factors including developmental, nutritional, and climatic variables may influence sperm counts in human populations round the world. Jørgensen et al. (2001), for example, recorded higher sperm counts in winter, than in summer, for men in Denmark, Finland, France, and the UK. Table 3.3 provides data on human sperm counts (per ml. of semen) in twelve countries worldwide. The various studies are arranged in chronological order, given concerns about possible declines in human sperm counts during the last 50 years or so.

Table 3.3 Geographic and temporal variations in average human sperm counts, per millilitre of the ejaculate

Location and year	No. of men	Sperm: millions per ml
USA: New York (1938)	200	120.6
USA: New York (1945)	100	134
USA: New York (1950)	100	100.7
USA: New York (1951)	1,000	107
USA: Washington State (1963)	100	110
Germany (1971)	100	74.4
USA: Iowa (1974)	386	48
USA: New York (1975)	1,300	79
Brazil (1979)	185	67.6
USA: Texas (1982)	4,435	66
France (1983)	809	102.9
Libya (1983)	1,500	65
Australia (1984)	119	83.9
Greece (1984)	114	72
Hong Kong (1985)	1,239	83
Thailand (1986)	307	52.9
Nigeria (1986)	100	54.7
Tanzania (1987)	120	66.9
United Kingdom (1989)	104	91.3
France (1989)	1,222	77.7

Source: From Fisch and Goluboff (1996); data from Carlsen et al. (1992).

Some useful general conclusions may be drawn with reference to Table 3.3. Just as human beings do not have large testes in relation to body weight, so sperm numbers in the human ejaculate are modest by comparison with most mammals. The sperm count in men from Hong Kong averages 83 million/ml, by comparison with 54.7 million/ml in Nigeria, and 66.9 million/ml in Tanzania. It will be recalled that Hong Kong Chinese have the smallest, and Nigerian males the largest testes sizes recorded so far (see Table 2.2). Higher sperm counts have been recorded in certain parts of the USA (e.g. New York: 100.7–120.6 million/ml). The same is true of some European studies, especially for samples collected during the winter months (e.g. Finland: 132 million/ml; Scotland: 119 million/ml; Jørgensen et al. 2001). However, setting aside the many and as yet unresolved variables affecting these sperm counts, none of them approach those recorded in species such as the chimpanzee, bonobo, rhesus macaque, or many other mammals.

Even though men do not produce large numbers of spermatozoa in their ejaculates, it is possible that they might 'allocate' sperm in differing numbers, depending upon the possible risks of sperm competition. Pound, Shackelford, and Goetz (2006) have reviewed evidence relating to this question, and I shall return to it in Chapter 5. For the moment, it is important to note that sperm counts may vary considerably among individual men, as well as between subjects in various studies. As an example,

Pound et al. (2002) found that among twenty-five regular semen donors (aged 22–44 years) concentrations of spermatozoa ranged from 12 to 156 million/ml (mean = 70.6 million/ml). Ejaculate volumes averaged 3.5 ml in these subjects, but again a considerable range was possible (0.9–9.0 ml) and total sperm counts in the ejaculate varied accordingly (from 26.4 to 834.4 million; mean = 236.1 million). All donors reported that they had maintained at least 3 days of sexual abstinence prior to giving their semen samples. Interestingly, there was a weak positive correlation between the time taken to produce a sample, sperm numbers per millilitre, and motile sperm concentrations. These effects applied to samples produced in less than 30 min (Figure 3.11). Pound et al. (2002) attribute their findings to heightened effects of sexual arousal and more prolonged penile stimulation in these subjects. Such behaviour may increase the strength and duration of smooth muscle contractions in the cauda epididymis and vas deferens, thus resulting in the transport of greater numbers of spermatozoa prior to ejaculation. This possibility raises the question of whether sexual selection has resulted in any measurable effect upon the muscular systems which transport sperm. If so, how does the control of sperm transport in human males compare with homologous mechanisms in other mammals and in relation to sperm competition pressures? These questions are explored in the next section.

Rapid sperm transport: the vas deferens and sperm competition

Given that sexual selection has influenced relative testes sizes and hence sperm production as well as the morphology of individual gametes (sperm midpiece volume) in mammals, we may further enquire as to whether other components of the male reproductive tract have been likewise affected by sperm competition. Rapid transport of many millions of spermatozoa during copulation is a complex process, and especially so for those species where the numbers of gametes ejaculated and frequencies of ejaculation may be greater due to competition between males for access to a given set of ova. Beyond the cauda epididymis, where sperm are pooled and stored, lays the vas deferens, the most muscular tubular duct in the human body (Figure 3.12). During sexual activity, sperm are transported rapidly through the vas deferens, by peristaltic contractions of its muscular walls. There are usually three muscle layers in the wall of the vas deferens; these can be seen in a transverse section of

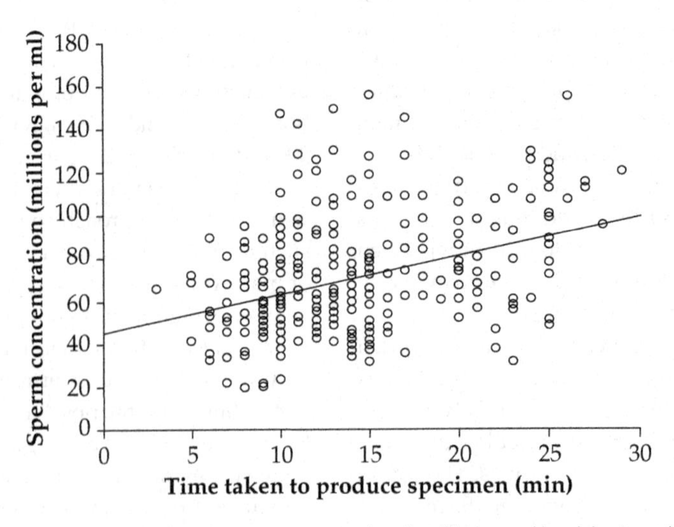

Figure 3.11 Scatter plot showing the relationship between sperm concentrations (in millions per ml.) and the time taken by semen donors to produce a semen sample (by masturbation) within a 30 min period. No. of specimens = 272.

Source: After Pound et al. (2002).

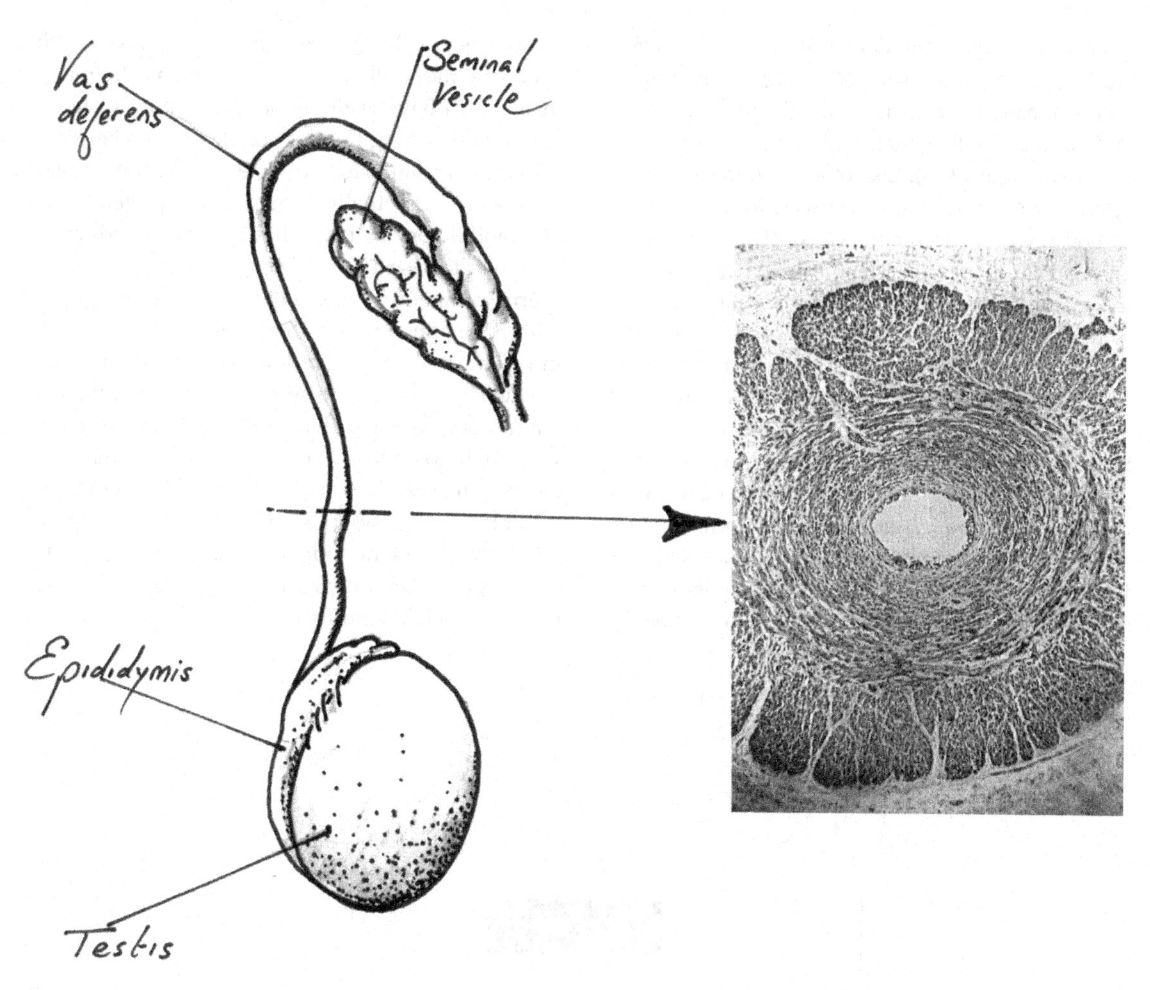

Figure 3.12 Diagram to show the anatomical relationships of the vas deferens. On the right is a transverse section of the vas deferens of a gorilla to demonstrate the powerful longitudinal and circular muscle layers surrounding the central lumen.

the duct (Figure 3.12). An outer layer of longitudinal muscles and an inner longitudinal muscle layer are separated by a central layer of circular muscles. However, thicknesses of these muscle layers may vary along the length of the vas deferens. In the distal region of the vas deferens, as, for example, in the stump-tail monkey, the inner longitudinal muscle layer may be absent. The central lumen of the vas deferens is narrow and is lined with ciliated, secretory epithelium. Some mammals have an expansion of the duct at its distal extremity, which is called the ampulla. The ampulla may be very large in certain species, such as the jackass, zebra and other equids, and absent or virtually so in others. The ampulla is of less concern to us here, as it is relatively small in the human male, but it is interesting to note that its functions still remain largely unknown. The richly folded secretory epithelium and expanded volume of the ampulla indicates that it may be an important glandular structure in some species (e.g. in the common shrew: Suzuki and Racey 1984).

Careful measurements of the length of the vasa deferentia and thicknesses of the various muscle layers in seventy genera of mammals, including *Homo*, have revealed some significant correlations between morphology, mating systems, and sperm competition

(Anderson, Nyholt, and Dixson 2004). The vas deferens is shorter, in relation to body weight, in mammals where females typically mate with multiple partners (MP systems) rather than with single partners (SP systems) (Figure 3.13). Like other mammals with SP systems, the human vas is relatively long in relation to body size. Muscular thickness of the vas deferens also varies in a consistent fashion depending on the mating system. The ratio of muscular thickness of the vas (all three muscle layers combined) to the width of the central lumen is greatest in mammals with MP mating systems. In such cases it averages 9.9, as compared to 6.8 in mammals which have SP mating systems (Figure 3.13). The human male, with a muscle wall/lumen width ratio of 6.3, thus displays an anatomical arrangement typical of mammals in which sperm competition pressure is low, as is the case in monogamous or polygynous mating systems. This finding is confirmed if we examine the thicknesses of the individual muscle layers which comprise the wall of the vas deferens. In mammals which have multiple-partner mating systems, both layers of longitudinal muscles are thicker than expected in relation to body size. The central circular muscle layer is actually thinner than expected, however, and it appears that it has undergone reduction as a result of sexual selection in mammals where sperm competition pressures are greatest. Species with SP mating systems have thinner longitudinal muscles than MP species, but their circular muscles are quite prominent and significantly thicker than in mammals where sperm competition is more marked. Examination of a transverse section of the human vas deferens reveals that it conforms to the pattern typical of SP mating systems among mammals (Figure 3.14). Thus, the longitudinal muscle layers are quite thin, and the central circular muscle layer is more pronounced in the human vas.

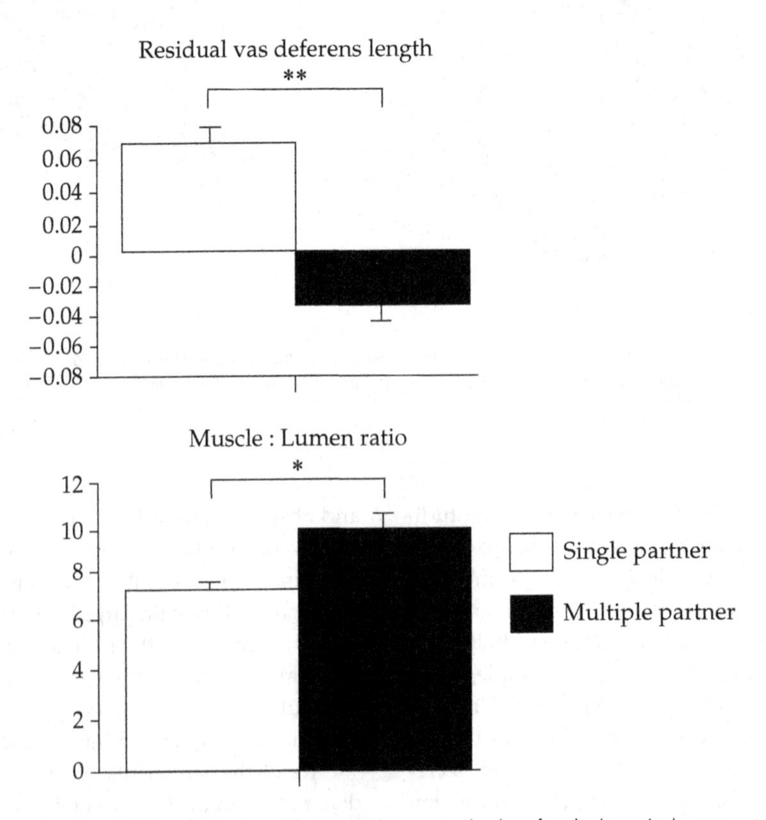

Figure 3.13 Vas deferens length and thickness (muscle wall/lumen ratio) in mammals where females have single-partner vs. multiple partner mating systems. $^{*}P < .05$; $^{**}P < .01$.

Source: Redrawn from Anderson et al. (2004).

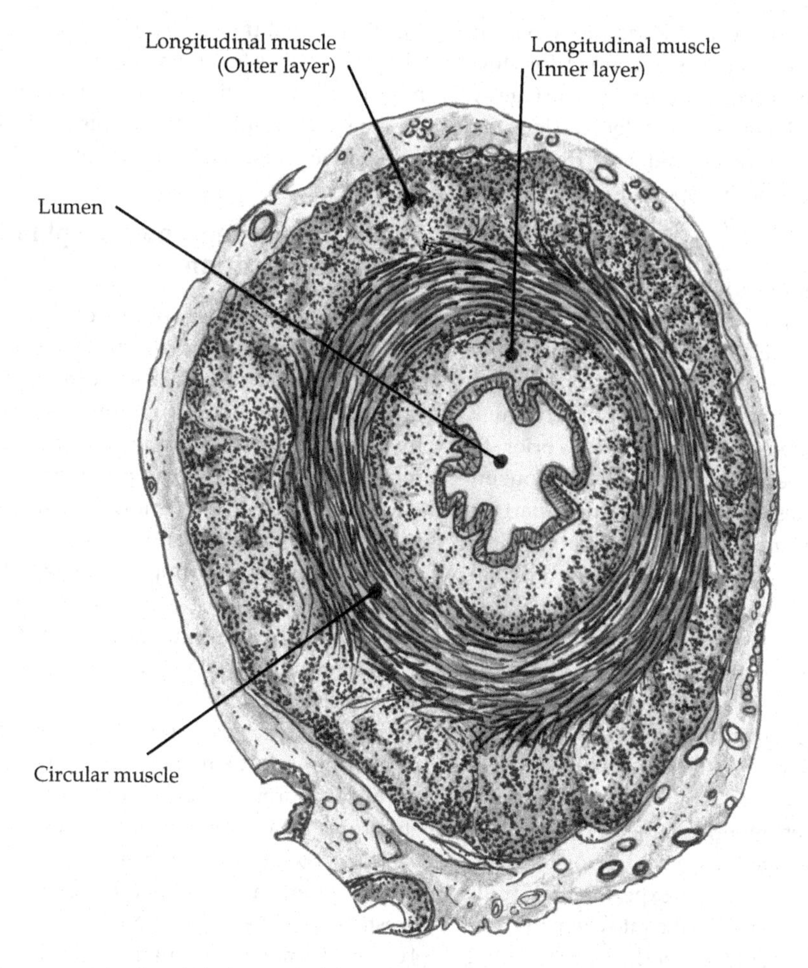

Figure 3.14 Transverse section through the human vas deferens, to show the thicknesses of its two longitudinal muscle layers, and the intermediate (circular) muscles. The thickness of the circular muscles (relative to the longitudinal muscles) is typical of those mammals which have monogamous or polygynous mating systems.

Source: Author's drawing, after Bloom and Fawcett (1962).

We do not know exactly how the muscle layers of the vas deferens operate during sperm transport. It may be useful to compare the functions of the vas deferens with those which characterize other tubular organs which exhibit peristaltic movements. One example is the human oesophagus, which contains a single, outer layer of longitudinal muscles, and an inner circular muscle layer. The outer layer of longitudinal muscles typically contracts sequentially in blocks, concentrating the activity of the inner circular muscles as they in turn contract to force a bolus of food down to the stomach. Thickening of the longitudinal muscles in the vas deferens of mammals where sperm competition pressures are greatest may increase the efficiency of the circular muscles in some way and thus make it possible to reduce their bulk. The actions of the inner longitudinal muscle layer of the vas are not fully understood. In this regard, it may be relevant that the vas deferens is capable of transporting sperm in both directions. In the buck rabbit, for example, experiments with a radio-opaque dye (ethiodiol) have shown that material is transported proximally, and returned to the cauda epididymis, from the vas after sexual activity (Prins and Zaneveld 1980). In this way spermatozoa might be recouped for storage in the cauda epididymis and undue wastage of gametes prevented. Likewise, a shorter vas deferens may be significant in terms of reducing the overall distance sperm must be transported during copulation. Even when possible biases in body size are controlled for, it is clear that the vas is significantly shorter in those mammals where females commonly engage in multi-partner matings, and its walls are 45 per cent thicker in relation to width of the central lumen.

In all mammals, many millions of sperm must traverse the vas deferens very rapidly prior to ejaculation. In a species like the rhesus monkey, with large relative testes sizes and multiple-partner matings by females, more than 1000 million sperm are transferred to the female during the ejaculatory mount. Interestingly, the male rhesus makes a series of mounts and intromissions prior to ejaculation. This specialized pattern of copulation is more commonly found in primates where sperm competition occurs (see Chapter 5, for a discussion of the evolution of copulatory patterns). It is not known whether spermatozoa are moved along the vas in cohorts during each mount of the series, so that numbers in the vas are maximized prior to ejaculation. This possibility requires experimental study, perhaps in a mammal such as the rat where the necessary experiments could be performed to count numbers of spermatozoa in the vas deferens. It may be that the vas deferens (and not just the cauda epididymis) is a temporary repository for gametes. Thus prolonged, or multiple, intromissions, may serve to fill the vas with gametes prior to the final ejaculatory phase. In the shrew (*Sorex araneus*) the tendency for the vas to store sperm has resulted in a more permanent arrangement, with the distal portion of the duct being distended to contain large reserves of sperm in the same way as the cauda epididymis (Suzuki and Racey 1984).

The possible effects of sexual selection upon the mammalian vas deferens will thus require deeper investigation in future. For the present, it is clear that the human vas deferens is typical of mammals which have monogamous or polygynous mating systems and where sperm competition pressures are low. This conclusion runs counter to Smith's (1984) hypothesis that the human vas deferens might display adaptations for sperm competition. At the time his paper was published, no comparative morphometric studies had been attempted, and the results summarized here were not available to him.

The accessory reproductive glands and sperm competition

Only 5 per cent of the volume of human semen is made up of fluids contributed by the testes and epididymis. Spermatozoa which leave the epididymis are transported rapidly through the vas deferens, and then mixed with the secretions of various accessory reproductive glands which provide the bulk of the ejaculate. Adult male mammals possess an impressive variety of these glands, including the seminal vesicles, prostate, and ampullary and bulbourethral (Cowper's) glands (Eckstein and Zuckerman 1956; Price and Williams-Ashman 1961; Hamilton 1990). Most relevant to the present enquiry are the seminal vesicles and prostate gland (Figure 3.15). The human seminal vesicles are tubular glands, each measuring approximately 15 cm when straightened, but normally configured as compact coiled structures measuring about 5 × 2.5 cm. The seminal vesicles produce approximately 60 per cent of the seminal fluid in the human ejaculate. The human prostate is a chestnut-shaped gland, approximately 4 cm in diameter, although it may be enlarged in older men. Prostatic secretions constitute approximately 30 per cent of human semen. The anatomical relationships of the vas deferens, seminal vesicles, and prostate gland are shown in Figure 3.15, which also shows the much smaller bulbourethral glands (approximately 10 mm in diameter in the human male).

A prostate gland is present in all mammals. It is represented by rudimentary ducts in the duck-billed platypus (a monotreme) and by much more complex, compartmentalized structures in the marsupials and eutherians. Seminal vesicles are present in the majority of eutherians, but they are absent in marsupials, in which the Cowper's glands may be greatly enlarged. Some eutherians retain only atrophic seminal vesicles, and the glands are absent in the majority of carnivores and in the cetaceans (Price and Williams-Ashman 1961).

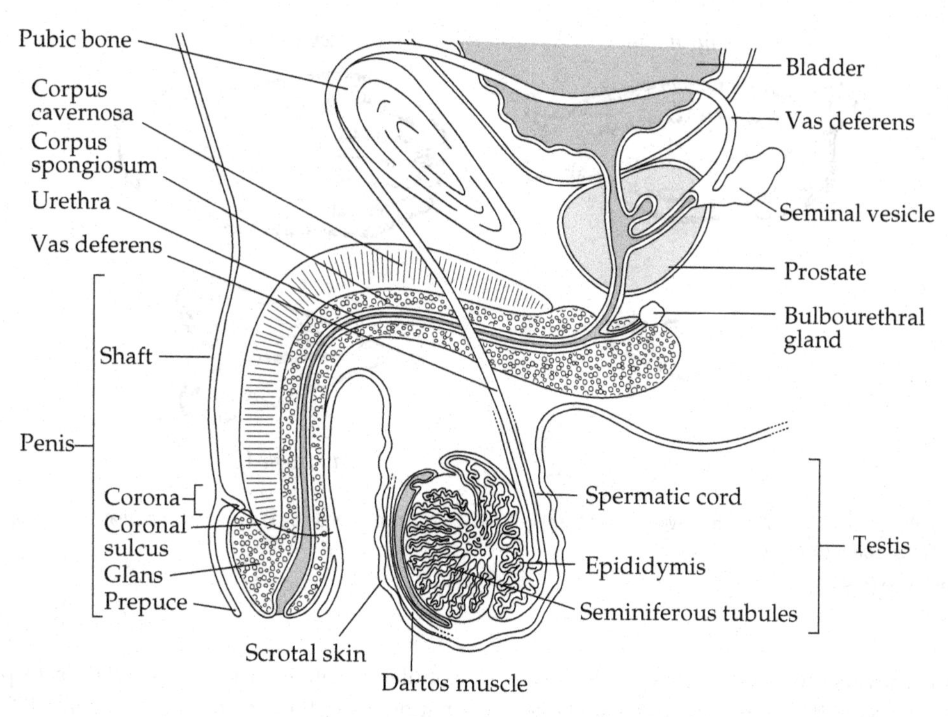

Figure 3.15 Diagram of the structure of the adult male reproductive tract in *H. sapiens*.
Source: After Piñón (2002).

The huge morphological diversity displayed by the seminal vesicles of mammals is intriguing, in view of the possible functions of seminal vesicular secretions in reproduction and sperm competition. As well as producing the bulk of the fluid in which human sperm are transported at ejaculation, the seminal vesicles also provide an alkaline secretion (pH 7.2–7.8) which may assist spermatozoa to survive in the more hostile acidic environment of the human vagina (Masters and Johnston 1966; Fox, Meldrum, and Watson 1973). Seminal vesicular secretions also contain an impressive variety of chemical constituents including sugars (notably fructose), prostaglandins (19-hydroxylated prostaglandins in human males), and proteins (Mann 1964; Mann and Lutwak-Mann 1981). Proteins are especially important for the coagulation of semen which occurs after ejaculation in many mammals. In primates, two proteins (semenogelin 1 and semenogelin 2) are involved in this process of coagulation, which is catalysed by an enzyme (vesiculase) produced by the cranial lobe of the prostate gland. Gertrude van Wagenen first described the effects of prostatic secretions upon seminal coagulation in the rhesus monkey, in a paper published in 1936. Coagulation produces a soft, whitish gelatinous material, but in some primates a more solid, rubbery copulatory plug is formed (e.g. in the chimpanzee, Figure 3.16). The functions of mammalian copulatory plugs have been much debated. Although it is tempting to speculate that copulatory plugs might act as physical (or chemical) barriers to the sperm of rival males, where mammals are concerned there is little experimental support for this view (Hartung and Dewsbury 1978). It seems more likely that copulatory plugs or coagulated semen may facilitate sperm retention and survival within the female reproductive tract (e.g. in the rat: Blandau 1945; in the rhesus macaque: Settlage and Hendrickx 1974). Copulatory plugs do not occur in human beings. For the moment, it is sufficient to note that in some primates and other mammals, one function of the seminal vesicles may be to assist coagulation of semen and to enhance male fertility.

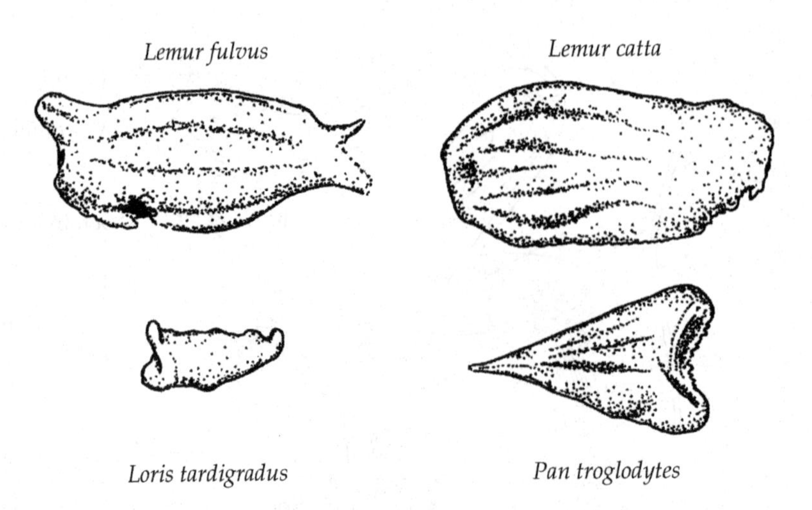

Figure 3.16 Primate copulatory plugs. The cervical end of the plug is shown on the right hand side in each case. Note that *Lemur fulvus* is now classified as *Eulemur fulvus*.
Source: After Dixson (1998*a*).

Given the diverse functions of the seminal vesicles, it is significant that these structures are largest in those primate species which have multi-partner mating systems and an increased likelihood of sperm competition (Dixson 1998*a*, 1998*b*). This is the case, for example, in the chimpanzee and bonobo, in the Old World macaques, mandrills and most baboon species, as well as in the spider monkeys and woolly spider monkeys (muriquis) of South America. By contrast, those primates which have primarily monogamous or polygynous mating systems also have relatively small seminal vesicles in relation to body size (Figure 3.17). Examples include the smaller apes (gibbons), the gorilla and gelada, as well as various New World monkeys which live in small family groups (e.g. the marmosets, owl monkeys, and titi monkeys). Reduction of the seminal vesicles may have occurred in some of these cases, because these structures are physiologically costly to maintain and are less advantageous in those mating systems where sperm competition pressures are low. The seminal vesicles are of medium size in *Homo sapiens*, not as large as in those primates which have multi-male/multi-female mating systems, but somewhat larger than in the typically monogamous species.

These findings are reinforced by the results of work to examine seminal coagulation and copulatory plug formation in relation to primate mating systems (Dixson and Anderson 2002). Using a four-point scale (from 1 = no coagulation, up to 4 = copulatory plug formation), ratings of seminal coagulation were obtained for twenty-six primate genera, including the apes and human beings. Coagulation ratings were greatest (mean = 3.64) for those genera in which females commonly mate with multiple partners, and lowest (mean 2.09) in monogamous and polygynous forms. This result was highly significant ($P < .001$) and the coagulation rating for *Homo sapiens* (2.0) was the same as those for the gorilla, and various gibbons and marmosets, as summarized in Figure 3.17. None of the primate species for which adequate information exists exhibits complete absence of coagulation. It appears that some degree of biochemical interaction between seminal vesicular and prostatic secretions is common to all primates. Based upon this common heritage, a more marked degree of seminal coagulation or copulatory plug formation may have evolved due to some reproductive advantage (e.g. in sperm competition). A coagulum, such as occurs in many macaques, baboons, and mandrills, may promote sperm survival by retaining gametes in an alkaline environment, rather than exposing them directly to the hostile, acidic conditions in the vagina. Under conditions

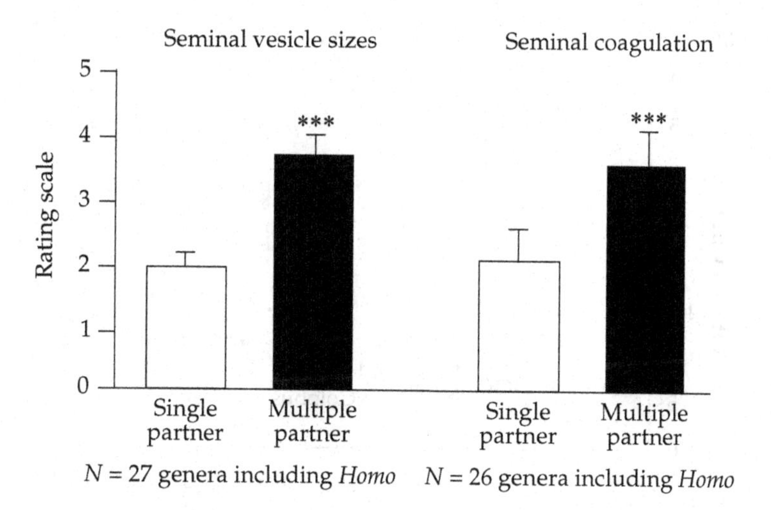

Figure 3.17 Seminal vesicle sizes and seminal coagulation ratings for primates in which females mate primarily with single partners, or with multiple partners, during the fertile period. $^{***}P < .001$.

Source: Based on data from Dixson (1998*b*), and Dixson and Anderson (2002).

of intense sperm competition, further specializations might have occurred to enhance coagulation and plug formation, thus maintaining a sperm-rich fraction of the ejaculate in close contact with the os cervix and facilitating sperm transport into the uterus. Such specializations are absent in *Homo sapiens* because sexual selection did not favour their evolution in the ancestral forms which gave rise to human beings.

Subsequent to the reports described above, molecular studies have shown that a relationship exists between sexual selection and evolution of the semenogelin genes in primates (Jensen-Seaman and Li 2003; Dorus et al. 2004). Dorus and colleagues compared rates of evolution of the semenogelin 2 gene in twelve primate species (including the great apes and human beings). They found that rates of evolution were accelerated in those species where sperm competition was most pronounced, as reflected by larger relative testes sizes, higher semen coagulation ratings, and more frequent multi-partner matings by females (Figure 3.18). The rate of semenogelin 2 gene evolution in *Homo sapiens* is comparable to that measured in other primates which are monogamous or polygynous, and notably less than that of macaques and chimpanzees, which have multi-male/multi-female mating systems.

Less information is available on the sizes and functions of mammalian prostate glands in relation to sperm competition, but recent work on rodents (Ramm, Parker, and Stockley 2005) and on a variety of other mammals (Anderson and Dixson, in press) confirms that prostate weights as well as weights of seminal vesicles are correlated with relative testes sizes and multi-partner mating systems. Both seminal vesicular weight and prostate weight in human beings are comparatively small in relation to body weight. By comparison with other mammals, the accessory reproductive glands are of modest size and do not provide evidence for significant effects of sexual selection during human evolution.

Penile morphology, sexual selection, and human evolution

It is an unfortunate circumstance that so many authors have applied hyperbole to descriptions of human penile morphology. Smith (1984) regarded the human penis as 'extraordinary relative to the other hominoids'. Baker and Bellis (like Smith) stressed that 'it is nearly twice as long and over twice as wide as that of the chimpanzee'. Jolly (1999) states that a 'peculiarity of humans is that the penis is twice the size for body weight as that of any other primate.'

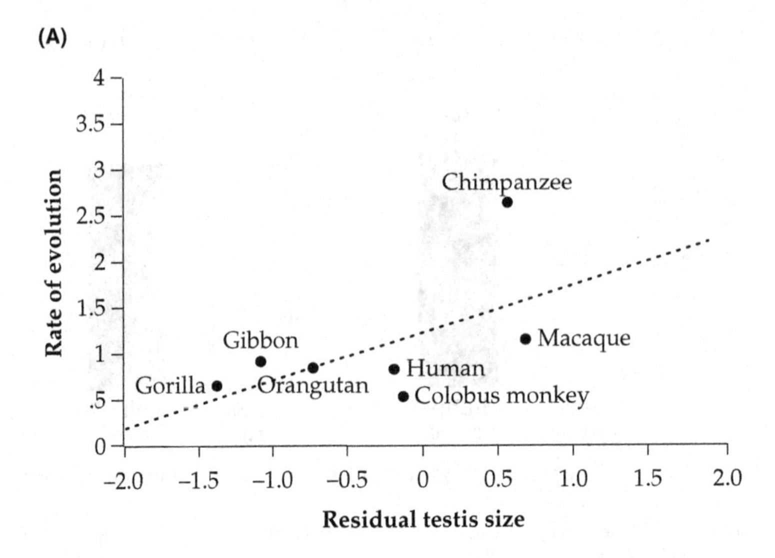

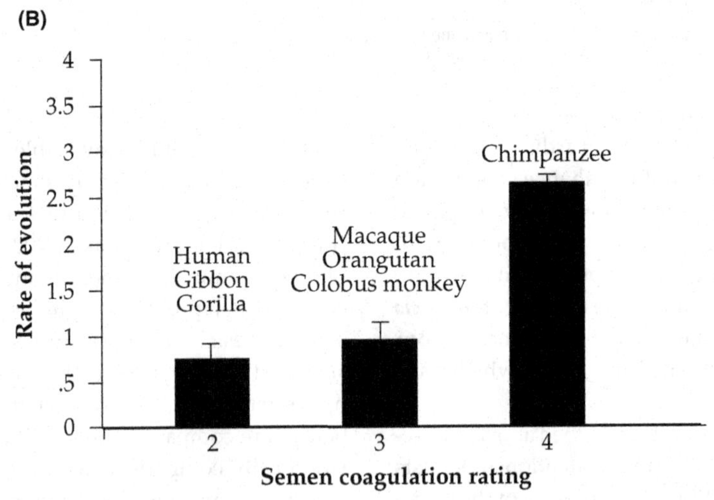

Figure 3.18 Correlations between the rate of evolution of the semenogelin 2 gene, and A. relative testes size, and B. semen coagulation ratings in primates, including *H. sapiens*.
Source: Redrawn from Dorus et al. (2004).

Miller (2000), in a book dealing with effects of sexual selection on evolution of the human brain, states that 'adult male humans have the longest, thickest and most flexible penises of any living primate'.

None of these statements, and many others of a similar nature which pervade the published literature, is accurate, with the exception of the observation that the thickness of the human penis exceeds that of the apes. It is regrettable that exaggerated accounts of human penile size and shape have been used to bolster the proposition that sexual selection has moulded the evolution of the human genitalia and the psychological mechanisms that influence copulatory behaviour. Four theories have been advanced to account for this:

1. The length of the human penis has been selected to deliver sperm as close as possible to the female's os cervix during copulation, in order to gain an advantage in sperm competition (Smith 1984).

2. The human penis acts as a 'piston' during copulation and its size and distal morphology

serve to displace and remove semen deposited during previous copulations, thus producing an advantage in sperm competition (Baker and Bellis 1995; Gallup et al. 2003; Gallup and Burch, 2004; see also Rolls 2005).

3. Penile size and shape evolved to impart pleasurable stimulation to the partner during copulation, and to induce female orgasm. Female orgasm influences sperm retention and transport in relation to sperm competition (Fisher 1982, 1992; Small 1993; Baker and Bellis 1993*a*; 1993*b*; 1995; Miller 2000).

4. The human penis is displayed more prominently than those of other primates and this might be partly a result of sexual selection to enhance a visual signal of attractiveness or status (Short 1980; Diamond 1997).

In order to evaluate these hypotheses, I will firstly present data on penile measurements in human beings and in other primates. Penile morphology is then considered in relation to primate mating systems and the possible effects of copulatory and post-copulatory sexual selection. Human penile size and shape will then be discussed in comparative perspective.

Table 3.4 provides data on penile dimensions in several human populations, together with reliable measurements of penile lengths in the great apes. On average, the human penis is 15–16 cm long and 10–12 cm in circumference, during tumescence. There is considerable individual variability in penile size, however, and these measurements are not correlated with men's height or body weight. There may be ethnic differences in penile dimensions but these remain uncertain, due to a lack of reliable comparative data. Potts and Short (1999) refer to a survey (conducted by Japanese prostitutes) reporting a slightly shorter average length of erection in Japanese males (13.75 cm) as compared to westerners. Rushton and Bogaert (1987) cite Nobile (1982) who used the Kinsey data to examine possible differences in penile size between American blacks and Caucasians. Measurements from Kinsey's surveys were published in a supplementary volume by Gebhard and Johnson (1979) which includes the lengths and circumferences of penes for some thousands of subjects. These data were self-ratings made to the nearest half inch (1.25 cm). Rushton and Bogaert emphasize that the length and circumference of the flaccid and erect penis is significantly larger in black males than in Caucasians. However, none of the differences they cite exceeds 0.5 in. and thus one wonders whether the methods used by Kinsey et al. might have produced

Table 3.4 Penile dimensions in man and the great apes

Species	Sample size	Measurement (cm)	Source
Homo sapiens			
USA	2,310	Length* 10.0	Gebhard and Johnson (1979)
		Erect 15.0	
		Circumference* 10.0	Wagner and Green (1981)
		Erect 12.5	
USA	80	Length* 8.9	Wessells, Lue, and McAninch (1996)
		Erect 12.9	
		Circumference	
		Erect 12.3	
USA	52	Length† 12.2	Spyropoulos et al. (2002)
Czechoslovakia	177	Length* 7.2	Farkas (1971)
		Range 4.5–11.0	
		Circumference* 9.5	
		Range 7.7–12.0	
Nigeria	20	Length* 8.1	Ajmani, Jain, and Saxena (1985)
		Circumference* 8.8	
Pan troglodytes			
	6	Length† 14.0	Dahl (1988)
	11	Erect 14.4	Dixson and Mundy (1994)
		Range 10.0–18.0	
Pan paniscus			
	1	Length† 17.0	Dahl (1988)
Gorilla gorilla			
	1	Length* 5.0	Dixson (1998*a*)
	1	Erect 6.5	Short (1980)
Pongo spp.			
	4	Length† 8.75	Dahl (1988)
		Range 7.5–10.0	

*penis flaccid.
†penis flaccid and stretched (which approximates to the length when erect).

differences due to subjective bias. In the only study I have been able to locate which was conducted in Africa, Ajmani et al. (1985) recorded that Nigerian medical students had penes averaging 8.16 cm in length and 8.83 cm in circumference when flaccid. Rushton and Bogaert cite the equivalent Kinsey measurements for American Caucasians as follows: length: 3.86 in. (9.65 cm) and circumference: 3.16 in. (7.9 cm). As matters stand, therefore, it would not be justifiable to conclude that robust ethnic differences in penile size have been demonstrated in human populations. Baker and Bellis's (1995) inclusion of a smaller range of erect penile lengths in Mongoloids (10–14 cm) is likewise not supported by a critical reading of the source cited in their book (Rushton and Bogaert 1987).

All of the great apes have longer penes than originally reported by Smith (1984) in his highly influential review on human sperm competition. Smith, citing the limited information then available, reported penile lengths of 8 cm (chimpanzee), 4 cm (orangutan), and 3 cm (gorilla). As can be seen in Table 3.4 all of these length measurements are inaccurate, by more than 100 per cent in the case of the orangutan and gorilla, and 75 per cent in the case of the chimpanzee. Human penes are not longer than those of chimpanzees, therefore, although they are markedly thicker (and hence have a greater circumference) than those of the great apes. No accurate data on penile circumferences are available for apes, or for the vast majority of primate species. The erect penis was 10–18 cm long in eleven chimpanzees measured in Gabon (Dixson and Mundy 1994) and averaged 14.4 cm, close to the 14 cm reported by Dahl (1988) for measurements of the flaccid (stretched) penes of chimps at Yerkes Primate Center in the USA. Dahl (1988) also made careful measurements of orangutan penes, finding them to be much longer than generally reported, with the exception of a long-neglected account by Hill (1939). Thus orangutans have penes ranging from 7.5 to 10 cm in length (Table 3.4). In the gorilla, the penis is approximately 6.5 cm long when erect; this measurement was made from an illustration in Short (1980) which includes a centimetre scale. I have been able to record penile length in only one gorilla; it measured 5.0 cm, in the flaccid condition, in a mature western lowland specimen (Dixson 1998*a*).

The human penis is thus not 'twice as long' as that of the chimpanzee, nor is it exceptionally long in relation to those of other primate species, especially when their body sizes are taken into account. Figure 3.19 presents data on penile lengths and body weights for fourteen primate species, including

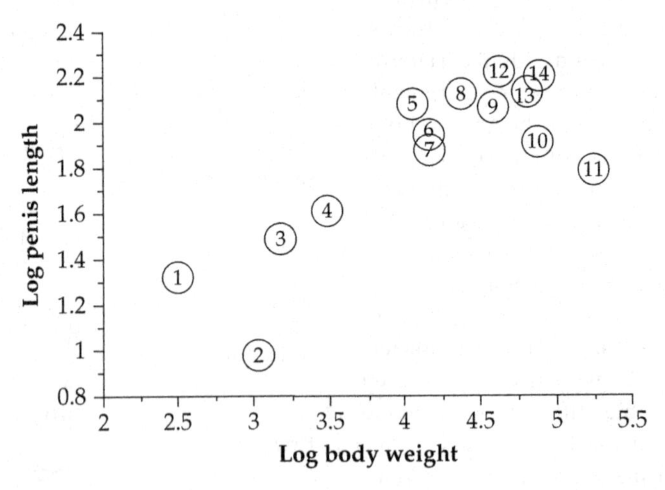

Figure 3.19 Relationship between body weight and length of the erect penis in fourteen primate species, including *H. sapiens*. 1. *Callithrix jacchus*; 2. *Aotus lemurinus*; 3. *Otolemur garnettii*; 4. *Lemur catta*; 5. *Lophocebus aterrimus*; 6. *Macaca mulatta*; 7. *M. arctoides*; 8. *Papio hamadryas*; 9. *Mandrillus sphinx*; 10. *Pongo pygmaeus*; 11. *Gorilla gorilla*; 12. *Pan paniscus*; 13. *P. troglodytes*; 14. *Homo sapiens*.

Figure 3.20 Large and morphologically complex penes occur in some of the non-human primates, so that human beings are not unique in this regard. Left: a woolly monkey (*Lagothrix*); Right: A spider monkey (*Ateles*).

Source: After Dixson (1998*a*) and Campbell (2007).

the ring-tailed lemur and greater galago, as well as seven species of monkeys, the great apes, and human males. The erect human penis is comparable in length to those of other primates, in relation to body size. Only its circumference is unusual when compared to the penes of other hominoids. However, even this comparison may not hold true for some primates, such as the spider monkeys (*Ateles*) which have large and thick penes. Primate penes are rarely depicted in the erect state in the anatomical literature, and the morphology of the erect organ often differs dramatically from its flaccid condition (examples are shown in Figure 3.20). Bowman (2008) suggests that the greater thickness of the human penis may be linked to changes which have occurred in the female pelvis and vagina during evolution: 'As the diameter of the bony pelvis increased over time to permit passage of an infant with a larger cranium, the size of the vaginal canal also became larger.' Coevolution meanwhile selected for thickening of the penis, to facilitate 'a satisfactory fit' during coitus. As we shall see, this mechanical explanation is more credible than hypotheses concerning the effects of sexual selection via sperm competition upon human penile morphology.

Is there any evidence that sexual selection has influenced the evolution of phallic morphologies in the non-human primates? The answer to this question is 'yes' and several traits are more highly developed in those species with mating systems which are conducive to the occurrence of sperm competition. Thus, primates with multi-male/multi-female, or dispersed mating systems (and larger testes sizes) have more complex penile morphologies than monogamous and polygynous forms (Dixson 1987*a*; 1987*b*; 1995*a*; 1998*a*). The differences include not only the tendency for the penes to be longer, but also morphologically more complex distally. In some cases elongation of the penile bone (baculum) also occurs.

Figure 3.21 shows representative examples of penile morphologies in non-human primates in which females have primarily multi-partner, or single partner, mating systems. Ratings of penile complexity (length, distal complexity, size of baculum, and penile spines) show some consistent differences between these mating systems, as can be seen in Figure 3.22. Using a 5-point scale, each trait was rated for species representing forty-eight primate genera. The data have been taken largely from my original report (Dixson 1987*a*) but with additions and alterations, because more information has become available during the intervening years. This revision also includes *H. sapiens*, so that it is now

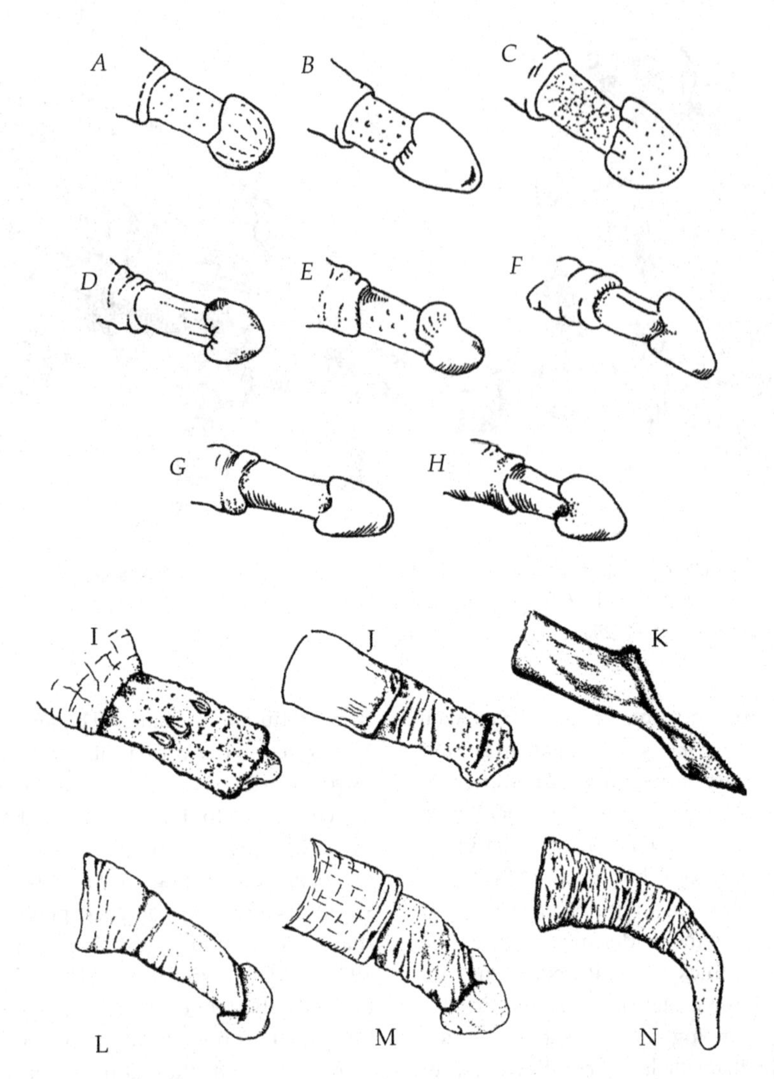

Figure 3.21 Examples of penile morphologies in primates which have primarily polygynous mating systems (A–H), or multi-male/multi-female mating systems (I–N). A. *Trachypithecus vetulus*; B. *Presbytis thomasi*; C. *Semnopithecus entellus*; D. *Cercopithecus mona*; E. *C. petaurista*; F. *C. albogularis*; G. *C. neglectus*; H. *Erythrocebus patas*; I. *Eulemur fulvus*; J. *Saimiri boliviensis*; K. *Macaca arctoides*; L. *M. fascicularis*; M. *Papio cynocephalus*; N. *Pan troglodytes*.

Source: From Dixson (1998*a*); A–H: After Hill (1958; 1966).

possible to compare human penile traits with those displayed by the non-human primates.

Primates in which the females typically show multi-partner matings (multi-male/multi-female and dispersed mating systems) have significantly longer and distally more complex penes than representatives of polygynous or monogamous genera (Figure 3.22). They also tend to have longer bacula (when present) and larger penile spines, but these traits are more variable. Larger penile spines, for example, are more common in prosimian primates than in the monkeys and apes. Nocturnal prosimians commonly exhibit non-gregarious (dispersed) mating systems, but their large penile spines may

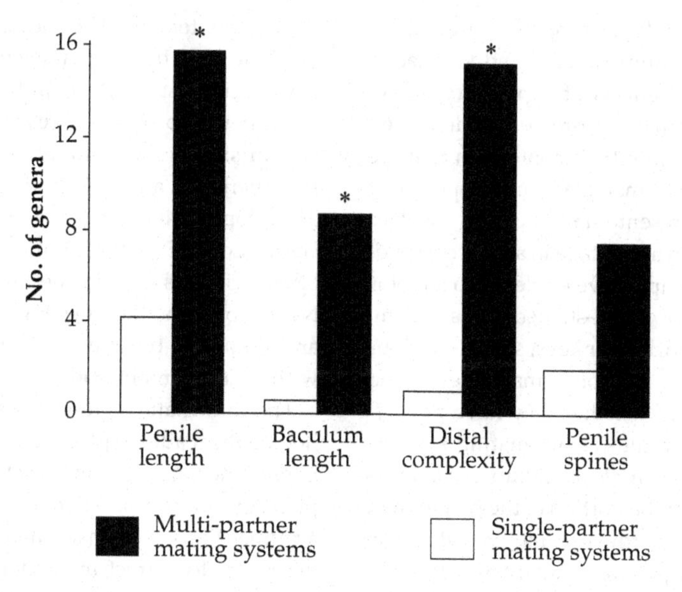

Figure 3.22 Penile morphologies are more complex in primate genera where females mate with multiple males (multi-male/multi-female and dispersed mating systems) than in those where single partners are the norm (monogamous or polygynous systems). Data are for forty-eight genera (including *Homo*), and show the numbers of genera which received the highest ratings (4 or 5 on a 5-point scale) for various traits (penile length, baculum length, distal complexity, and size of penile spines). $^*P < .05$. Further details are discussed in the text.

reflect a phylogenetic tendency among prosimians to exhibit this trait, and not just the effects of sexual selection (Harcourt and Gardiner 1994). *Homo sapiens* has been included in the comparative analysis as a polygynous primate species, on the basis that the majority of recent human cultures are at least partially polygynous (Ford and Beach 1951). The same rationale was applied to comparative analyses of relative testes sizes in human beings and non-human primates in Chapter 2. (Figure 2.1). However, I acknowledge that *H. sapiens* might equally be included with the monogamous primate genera for statistical purposes. Polygyny and monogamy involve predominantly single partner (SP) matings by females and lower relative testes sizes among males. Multi-male/multi-female and dispersed mating systems are characterized by occurrences of multi-partner (MP) matings by females, large relative testes sizes, and higher sperm competition pressures among males. Thus, Figure 3.22 shows the overall comparisons between SP vs. MP mating systems for their various penile traits.

Human penile morphology is not exceptional when compared to that of the prosimians, monkeys, and apes. The overall rating for all four traits analysed is 10 for *H. sapiens*. This is the same rating as scored by a number of putatively monogamous or polygynous primates (e.g. *Leontopithecus, Callimico, Erythrocebus, Theropithecus*) and less than the ratings given to twenty-seven of the forty-eight primate genera included in the study.

Human beings do achieve a maximum score for one trait (penile length). However, sixteen non-human primate genera also score equally high for this trait. Overall distal complexity of the penis (the shape and size of the penile glans or the equivalent if a glans is lacking) is rated as 3.0 for *Homo*; the same score as achieved by twenty-one other genera. Twenty genera score higher than *Homo* for complexity of the distal penis. In some of them, the glans is filiform or plunger-shaped, whilst the baculum may protrude distally, carrying the external opening of the urethra beyond the tip of the glans.

Gallup et al. (2003) tested Baker and Bellis's (1995) hypothesis concerning the 'plunger' action of the human glans penis in relation to sperm competition. They used models of human penes and vaginae to examine the putative effects of copulatory movements upon displacement of previously deposited (artificial) semen. Despite their contention that the large diameter of the human glans and its posterior margin (corona) represent adaptations to displace semen and provide an advantage in sperm competition, I can find no comparative evidence to support this view. A helmet or acorn-shaped glans is common amongst Old World monkeys, such as various colobines, macaques, baboons, mangabeys, and guenons, regardless of whether they have polygynous or multi-male/multi-female mating systems (see Figure 3.21 for examples). Gallup et al. stress that reduced length of the portion of the penile shaft covered by the prepuce (the *pars intrapreputialis*) in man is unusual, citing this as an adaptation to assist in removing semen from the vagina. However, this trait is shared by *H. sapiens* and the gorilla which, despite having a very small penis in relation to its body size, exhibits a distal morphology more similar to the human condition than is the case for other apes. In Figure 3.23, I have included an illustration of the external genitalia of a sub-adult male gorilla, from a rarely cited paper by Hill and Matthews (1949). The gorilla's genitalia are remarkably similar (in miniature) to the human condition.

Among the African apes, the chimpanzee and the bonobo have the most specialized and derived penile morphologies. A glans penis is lacking, and distally the penis is filiform and contains a very small baculum (6.9 mm in *P. troglodytes* and 8.5 mm in *P. paniscus*). The tip of the penis is distinctive in *P. paniscus*, terminating in a 'Y' shaped urethral aperture as compared to the simpler, slit-shaped opening of *P. troglodytes* (Izor, Walchuk, and Wilkins 1981). Measurements of penile and vaginal lengths in the chimpanzee indicate that the long and filiform penis probably co-evolved in association with the development of the large sexual skin swelling which characterizes this species. I shall return to this question in the next chapter, which deals with sexual selection by cryptic female choice.

The human penis lacks a baculum (except in rare cases) but the notion that its loss was associated with the evolution of morphological specializations for sperm competition (Smith 1984; Baker and Bellis 1995) is not tenable. The baculum is reduced in size in all the apes, by comparison with Old World monkey species, especially so in the case of the African apes. It was probably already undergoing reduction in size in the common ancestor of *Homo* and *Pan*, therefore, and was gradually lost, either in the australopithecines or in early members of the genus *Homo*. Loss of the baculum has occurred in some other primates (e.g. in the tarsiers, and in several New World monkey genera). The reasons for this are unknown, but have no established connection with sperm competition.

The information presented thus far has, I hope, helped the reader to place earlier accounts of human penile morphology and that of apes and other primates in comparative perspective. A detailed argument has been made here because I believe it necessary to correct misunderstandings about the role of sexual selection and sperm competition in relation to the evolution of the human genitalia. It is highly unlikely that penile size or shape in human beings has been influenced by sexual selection via sperm competition. Elongation of the penis, in order to place semen close to the os cervix (Smith 1984), has likely influenced evolution of the peculiar, filiform penes of the two extant *Pan* species. The reality is that in primates and other mammals the length of the erect penis and vaginal length tend to evolve in tandem. Whether or not sperm competition occurs, it is necessary for males to place ejaculates efficiently, so that sperm have the best opportunity to migrate through the cervix and gain access to the higher reaches of the female tract. Nor is there any credible evidence that human penes evolved as plungers to displace semen deposited by previous males. Gallup and Burch (2004) even go so far as to state that a second male's semen, removed in this way, might be transferred to another female by the male concerned, thus resulting in fertilization! Such Byzantine reproductive tactics are known to occur in certain invertebrates, as for example in the flour beetle (*Tribolium castaneum*) where males may remove sperm deposited in females by previous partners, and inadvertently transfer them during subsequent copulations (Haubruge et al. 1999). Significant numbers of offspring can result from this unusual circumstance. However, it would be simplistic in the extreme to create scenarios concerning

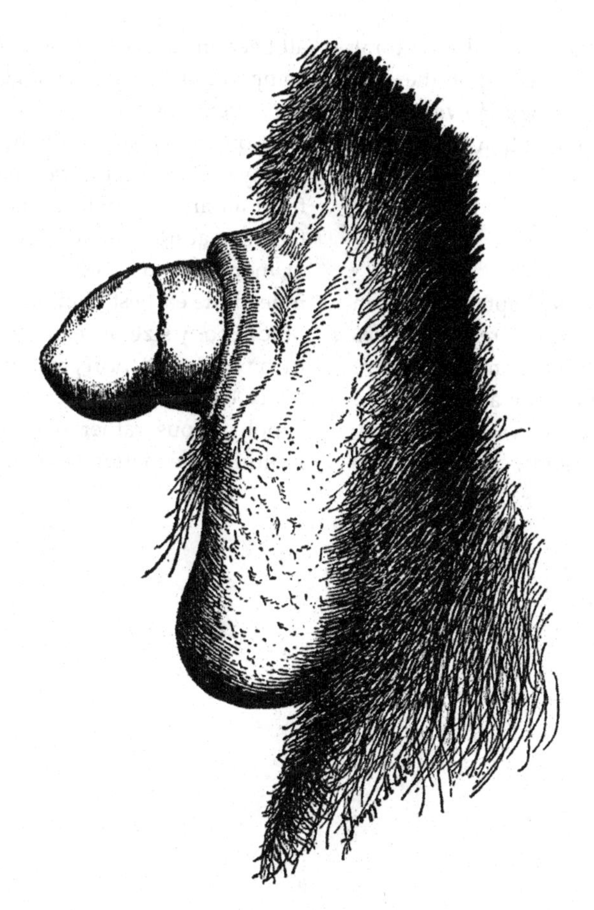

Figure 3.23 External genitalia of a sub-adult male gorilla (*Gorilla g. gorilla*).
Source: After Hill and Matthews (1949).

human copulatory behaviour and sperm competition based upon analogies with insects. As a further example, because damsel flies have evolved specialized penile structures to remove sperm deposited by previous males (Waage 1979), this does not mean that similar mechanisms necessarily exist in human males.

Of the four hypotheses listed above, to account for evolution of human penile morphology, the first two have been have been evaluated and may be discounted. The third hypothesis relating to effects of penile stimulation upon female orgasm and sperm transport mechanisms will be addressed in the next chapter, which considers the question of cryptic female choice and sexual selection during human evolution. The fourth hypothesis regarding the role of the penis as a visual signal (Short 1980; Diamond 1997) is very difficult to evaluate. It is certainly the case that the human penis is more prominently displayed than in many non-human primates, especially because humans are bipedal, and loss of the surrounding body hair renders the external genitalia and pubic hair more conspicuous. Cross-cultural studies have shown that altering the length of the non-erect penis in images of the human male does influence female ratings of attractiveness, although it is not necessarily the largest penis which is rated as most attractive (China: Dixson et al. 2007*a*; Cameroon: Dixson et al. 2007*b*). Štulhofer (2006) has also reported that many women rate penile length and girth as being a significant factor in judgements of their male partners. Thus it is possible that sexual selection as well as natural selection may have had some effect upon the evolution of human penile morphology. However, it is most unlikely that sperm competition has played any role in this respect.

Conclusion

The research reviewed in this chapter has taken us well beyond considerations of relative testes size in human beings and in other mammals. We have seen that sperm morphology as well as the structures and functions of the vasa deferentia, the accessory reproductive glands, and penis have all been influenced by sexual selection and sperm competition in mammals. However, in each case, the development of specializations for sperm competition are lacking in the human male. Thus, it is highly unlikely that *Homo sapiens* is *descended directly* from an ancestor which had a multi-male/multi-female mating system. Descent from a polygynous or monogamous precursor is much more likely. If indeed the earliest hominids were sexually dimorphic in body size, as many anthropologists suggest, then whichever form gave rise to *H. sapiens* is more likely to have exhibited such dimorphism within a polygynous rather than within a multi-male/multi-female mating system.

CHAPTER 4

Cryptic Female Choices

The last chapter focused on the effects of sexual selection, via sperm competition, upon the evolution of the male genitalia in mammals, with special reference to the question of human reproduction and evolution. However, in human beings as in other mammals, it is the female's reproductive tract which constitutes the arena for sperm competition. Spermatozoa must traverse anatomical and physiological sieves and barriers in the vagina, cervix, uterus, uterotubal junction, and oviduct before encountering an ovum, enclosed in its vestments (Figure 4.1). Once it is acknowledged that the female's anatomy and physiology may influence the fate of spermatozoa, it is logical to ask whether such factors might act differently upon the ejaculates of rival males. If, for example, one male possesses a more advantageous penile morphology, or copulatory pattern, or a biochemically more effective mixture of accessory glandular secretions, then the female's reproductive system might preferentially receive and transport his spermatozoa. It is this hidden female potential which William Eberhard refers to as *cryptic female choice*. Although Eberhard was not the first scientist to employ this term (see Thornhill 1983) he provided the major synthesis of this important concept (Eberhard 1985; 1996). Cryptic female choice may partially explain, for example, why sexual selection has favoured the evolution of complex phallic morphologies in animals in which females mate with multiple partners. The phallus may function as an *internal courtship device* under such conditions.

Much more is known about possible mechanisms of cryptic female choice in insects and some other invertebrates than is the case for mammals. As an example, in the beetle *Chelymorpha alternans* the male's intromittent organ bears a long flagellum. The flagellum is threaded into the female's spermathecal duct, and sperm pass along it to gain access to her storage organs. If a male's flagellum is shortened surgically, then the female ejects more of his gametes after mating has occurred. Males with longer flagella are more successful in fertilizing ova and achieve greater reproductive success (Eberhard 1996). The muscles of the female's sperm storage organ play a crucial role in selective ejection of spermatozoa (Rodriguez 1994; Simmons 2001). Mechanisms underlying cryptic female choice have been explored in several insect species (e.g. in the cowpea weevil: Wilson et al. 1997; damsel flies: Cordoba-Aguilar 1999; and the yellow dungfly: Ward, 2000). Although equivalent investigations of mammals are few in number and limited in scope, there are two issues which are worthy of consideration in relation to the origins of human sexuality. Firstly, has cryptic female choice played any role in moulding the evolution of phallic morphology in relation to stimulation of the female during copulation? Specifically, is orgasm in the human female significant in terms of sperm transport and fertility, and might penile morphology in the human male have evolved in relation to mechanisms of female orgasm? Secondly, does female genital morphology in mammals correlate in predictable ways with their mating systems and the likelihood of sperm competition between males? Specifically, is the length of the oviduct subject to sexual selection via cryptic female choice, and, if so, how does the size of the human oviduct compare with that of other mammals?

Phallic morphology and female orgasm

The correlations between primate mating systems and the morphological complexity of their penes discussed in Chapter 3 in relation to sperm competition are also consistent with Eberhard's hypothesis concerning cryptic female choice (Eberhard 1985; 1996). Thus, it is the case that, in those primates in which females have multiple partner mating systems, males have the longest and distally most complex penile morphologies (see Figure 3.22). Selection for more complex penes may also have occurred in response to pressures dictated by the female's anatomy and physiology, and not solely by sperm competition. Relationships between sperm competition and cryptic female choice are likely to be complex, however, and difficult to distinguish in practice (Eberhard 1996; Telford and Jennions 1998; Dixson and Anderson 2001; Birkhead and Pizzari 2002; Reeder 2003). In the chimpanzee and bonobo, for example, the male's elongated and filiform penis is well adapted to negotiate the depth of the sexual skin swelling and vagina of the female. Figure 4.2 shows vaginal lengths in female chimpanzees at minimum and maximum sexual skin swelling, compared to lengths of the erect penes of adult males. When a female's sexual skin is flat, most male chimpanzees might be capable of depositing a copulatory plug at the cervical os. However, the vast majority of matings occur when the female's sexual skin is at full swelling, and when ovulation is most likely to occur. Under these conditions, the operating depth of the vagina increases by up to 50 per cent

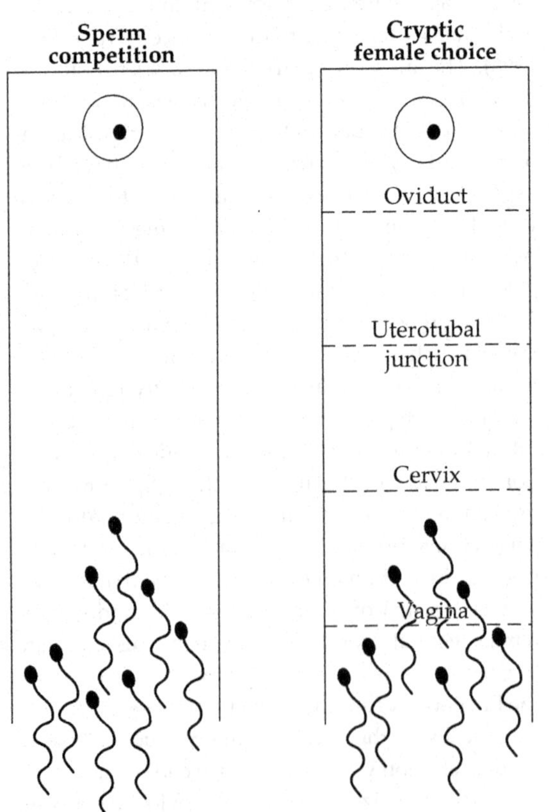

Figure 4.1 Schematic representation of sperm competition and cryptic female choice within the reproductive tract of a female mammal. As well as competition between the gametes of rival males for access to a given set of ova (on the left of the diagram), the female's anatomy and physiology may exert cryptic influences upon the fate of spermatozoa (as indicated on the right hand side of the diagram).

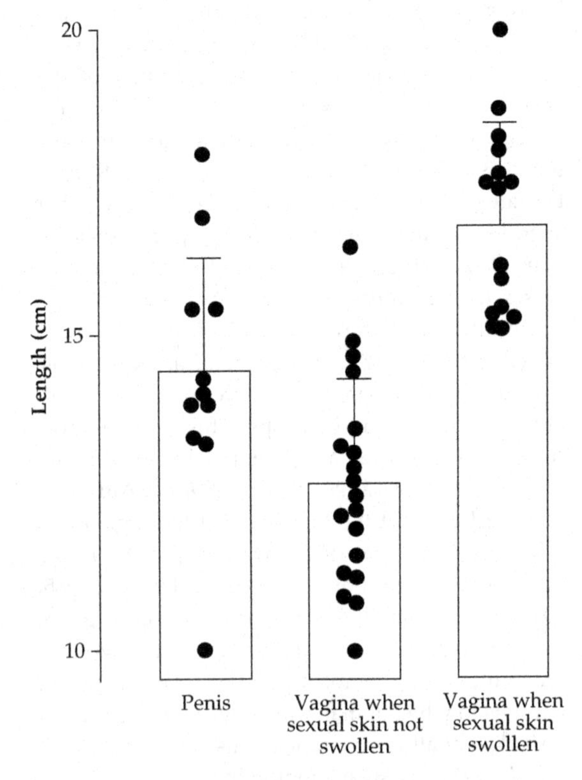

Figure 4.2 Measurements of the erect penis and vaginal length (when the female sexual skin is not swollen, and when it is at maximal swelling) in chimpanzees (*Pan troglodytes*). Data are for individual animals, with the overall means and standard deviations indicated by the histograms and bars.

Source: From Dixson (1998*a*); after Dixson and Mundy (1994).

in some females. Males having shorter penes are less able to place copulatory plugs against the female's cervix under these conditions (Dixson and Mundy 1994). It is possible, therefore, that the size of the female's swelling and the length of the male's penis have co-evolved in chimpanzees. The same might be true of some other Old World primates, such as mandrills, mangabeys, and red colobus monkeys, in which the female's swelling is exceptionally large at mid-cycle. Unfortunately, the necessary genital measurements of sufficient numbers of species which possess (or lack) sexual skin swellings are not yet available and so this hypothesis requires further testing.

The older anatomists were aware that some relationship might exist between female swelling and male phallic morphology. Hill (1970) commented that in baboons

> the great length of the penis appears to be an adaptation to permit penetration in the presence of the enormous catamenial swelling of the female.

In the mangabeys, Hill (1974) again noted that

> as observed by Pocock, the penis in *Cercocebus* is large and long in adaptation to the catamenial swelling of the female, thus ensuring penetration during copulation, which takes place usually during the period of turgidity in the female.

The human penis is long, but not exceptionally so in relation to body weight, as we saw in the last chapter (see Figure 3.19). Irrespective of whether sperm competition occurs frequently or not, it is logical to suggest that penile and vaginal lengths have co-evolved in mammals. Selection has presumably favoured efficient placement of the ejaculate close to the cervical os, in order that spermatozoa may migrate (or be actively transported) into the uterus. The human vagina is approximately 5–6 in. (12.5–15 cm) long (Dickinson 1949) and is capable of dilation to accommodate the penis during intercourse (Masters and Johnston 1966). Among the primates as a whole, there are only nine species for which both vaginal lengths and (erect) penile lengths are known, including *H. sapiens*. For this small sample there is a highly significant correlation between these two genital measurements (Figure 4.3, $r_s = 0.85$; $P < .01$). The sample includes species which are monogamous (marmoset), polygynous (gorilla), and others which have multi-male/multi-female mating systems (e.g. macaques). There is nothing exceptional about vaginal or penile length in human beings. Pawlowski (1999*a*) cites 'the considerable length of the penis in *Homo*' as evidence in support of the notion that human ancestors possessed a sexual skin swelling. However, this hypothesis is founded upon the false premise that

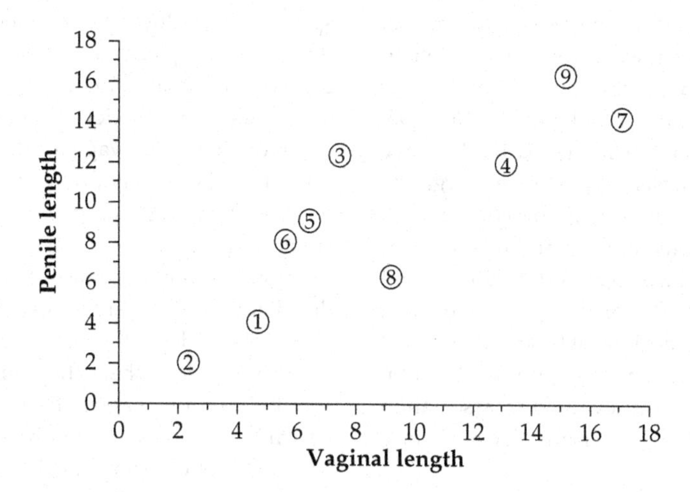

Figure 4.3 Relationship between vaginal length and length of the (erect) penis in nine species of primates, including *H. sapiens*. 1. *Lemur catta*; 2. *Callithrix jacchus*; 3. *Lophocebus aterrimus*; 4. *Mandrillus sphinx*; 5. *Macaca mulatta*; 6. *M. arctoides*; 7. *Pan troglodytes*; 8. *Gorilla g. gorilla*; and 9. *H. sapiens*.

human penile length is exceptional among the primates. This is not the case.

Does the thickness of the human penis have any significance as regards sexual selection during human evolution? If human penile morphology evolved to promote pleasurable stimulation of the female partner, there might be at least two avenues of selective advantage. Firstly, if enduring relationships between the sexes with long-term reproductive benefit in terms of offspring survival are facilitated by pleasurable sexual activity, then penile morphology might be adaptive in this context. However, it will be recalled that in the monogamous and polygynous non-human primates, males tend to have the least specialized penile morphologies, despite the occurrence of long-term sexual relationships in such species. There is some evidence that women rate the thickness and length of a partner's penis as significant factors in their sexual satisfaction (Štulhofer 2006). Human mate choice and long-term relationship decisions are immensely complicated, however. Cross-cultural studies indicate that qualities such as kindness, a good sense of humour, and ability to provide resources are valued more highly by women than physical attractiveness alone (Buss 2003). This is not to deny the importance of physical attractiveness in human evolution and I shall return to this subject in Chapter 7.

A second possible function of pleasurable penile stimulation might be to promote female orgasm, which in turn might play a role in relation to sperm transport and sperm competition (Fisher 1982; Small 1993; Baker and Bellis 1993*b*; 1995; Miller 2000). Baker and Bellis in particular have advanced the hypothesis that women's orgasms play an important role in sperm competition. However, the functions and evolution of orgasm in the human female have been much debated and remain controversial. The reader is referred to an excellent analysis and critique of this field by Lloyd (2005). Here I shall concentrate upon the question of whether female orgasm evolved in association with sperm competition and cryptic female choice in human beings.

Baker and Bellis (1993*b*) proposed that 'by altering the occurrence, sequence and timing of the different types of orgasm [nocturnal, masturbatory and copulatory orgasms], the female can influence both the probability of conception in monandrous situations and the outcome of sperm competition in polyandrous situations...much of this influence will be cryptic to the male partner(s).' These ideas arose from the notion that female orgasm might produce an 'upsuck' response, drawing sperm through the cervix and into the uterus (Fox, Wolff, and Baker 1970; Singer, 1973). Baker and Bellis conducted experiments with human subjects in order to estimate sperm numbers in 'flowbacks:' the fluid ejected from the vagina after copulation. Their goal was to define the effects of female orgasm upon sperm retention by women. The highest sperm retentions occurred when women experienced orgasm during a window of time extending from 1 min before ejaculation, until 45 min afterwards. Estimates of sperm retention were significantly greater for women who engaged in extra-pair copulations, as compared to estimates involving their usual (long-term) partners. This increase was attributed to a higher incidence of sperm retention orgasms during (or after) extra-pair copulations.

Baker and Bellis also attributed important functions to female orgasms which occurred during intervals between matings. However, these 'intercopulatory orgasms' were viewed as activating the upsuck mechanism to draw acidic (vaginal) material into the cervix and to negatively affect existing sperm or to reduce the likelihood of sperm retention at a subsequent mating. Even in the absence of copulation, female orgasms were posited to have a beneficial (antibiotic) effect, by drawing acidic secretions from the vagina into the cervical canal. Thus during pregnancy or in young (virgin) females masturbatory orgasms might assist in 'combating cervical infection'.

Before examining some of these findings in more detail, it will be helpful to consider the whole question of the inelegantly named 'upsuck hypothesis'. What evidence is there that female orgasms draw sperm from the vagina into the cervix, or that sperm transport is affected by orgasms in women or females of other mammalian species? This is an old theory and Dickinson provided some interesting comments on it in his (1949) volume on *Human Sex Anatomy*.

> Suction pump action in orgasm is accepted behavior in lay literature. It may some day be proven to be usual, but calls for confirmation. The mucus held within or produced for the occasion by the glands of the healthy cervix is asserted to be extruded during the peak of the climax in order to tanglefoot the sperms and then is said to be wholly dragged back through the external os.

However, Dickinson also noted a lack of connection between orgasmic responsiveness and fertility in women:

> Insuck is supposed to depend on orgasm. Orgasm is absent in the frigid. Yet in my study of a thousand office patients the frigid were not notably infertile, having the expected quota of living children, and somewhat less than the average incidence of sterility.

In 1950, Grafenberg published the results of a study to determine whether female orgasm might affect transfer of fluid from the vagina and through the cervix during copulation. Grafenberg asked women to wear a plastic cervical cap throughout a menstrual cycle; the cap was filled with a radiopaque oil. Orgasm occurred as a result of sexual intercourse in these subjects, but X-ray examinations failed to reveal any transfer of the radiopaque oil into the cervix or uterus. Masters and Johnson (1966) also used radiopaque fluid held in position using contraceptive diaphragms. However, they attempted to produce a fluid which more closely matched human semen in its physical properties. Six subjects participated in experiments to determine the effects of (manually induced) orgasms upon fluid movement. As in the Grafenberg studies, however, X-ray examinations revealed no effects. Masters and Johnson were able to take X-rays during orgasm, as well as 10 minutes afterwards.

All experiments have methodological limitations, especially so where clinical research and sexology are concerned. Thus, it might be objected that the use of a contraceptive cap could interfere with the natural orgasmic mechanism and transfer of fluid from the vagina into the cervix. Fox et al. (1970) criticized the results of earlier studies and examined effects of sexual intercourse upon intra-uterine pressure in a single female subject. They employed telemetry to measure a marked post-orgasmic fall in intrauterine pressure. However, no attempt was made to ascertain whether female orgasm, and associated changes in uterine pressure, had any effect upon fluid transport from the vagina into the cervix. Lloyd (2005) in her detailed critique of theories of human female orgasm, points out that Fox et al.'s data are therefore limited and do not address the major question of whether orgasm influences sperm transport. Given the considerable individual variability in orgasmic responsiveness which occurs among women, it is also regrettable that only one subject was studied by Fox et al., and that only two experiments were reported. Yet when Smith (1984) wrote his influential review on human sperm competition he cited Fox et al.'s studies as important evidence for the notion that female orgasm results in semen transport and facilitates fertilization. Subsequently, Baker and Bellis (1993*b*; 1995) continued to cite this study in support of the upsuck hypothesis. The negative findings of Grafenberg (1950) and Masters and Johnson (1966) were accorded much less weight.

Rapid transportation of spermatozoa to the oviducts has been recorded in various mammals and (non-orgasmic) contractions of the female reproductive tract are important in facilitating this process. Table 4.1 provides information on the time taken for sperm to reach the oviduct after copulation or (artificial insemination) in ten species, including *Homo sapiens*. Thus, in the absence of orgasm, sperm may reach the oviduct within 5 min (rabbit), 15 min (rat, mouse, pig), or 5–68 min (women).

Table 4.1 Time taken by spermatozoa to reach the oviduct after copulation or artificial insemination in mammals

Species	Time	Region of tube
Human	5–68 min	Oviduct
Mouse	15 min	Ampulla
Rat	15–20 min	Ampulla
Hamster	2–60 min	Ampulla
Rabbit	Several min	Ampulla
Guinea pig	15 min	Ampulla
Domestic dog	2 min to several hours	Oviduct
Pig	15 min	Ampulla
Domestic cattle	2–13 min	Ampulla
Sheep	6 min–5 h	Ampulla

Source: Data from Harper (1994); after Dixson (1998*a*).

Various physiological mechanisms have been proposed to account for sperm transport under these conditions, including coitus-induced release of oxytocin in women (Carmichael et al. 1987; Komisaruk, Beyer-Flores, and Whipple 2006) and the presence of prostaglandins in semen (which is unlikely to affect sperm transport: Mortimer 1983). Wildt et al. (1998) conducted experiments to determine the possible effects of oxytocin upon sperm transport through the reproductive tract of the human female. They positioned pressure recorders in the uterus and showed that rhythmic contractions occurred (3 times per minute) prior to any hormonal treatment. Administration of oxytocin resulted in an increase in the strength and frequency of uterine contractions; the highest pressures occurred lower in the uterus (near the cervix) and lower pressures were recorded close to openings of the oviducts. Under these conditions, artificially formulated semen containing radiopaque particles was rapidly transported from the vagina, through the cervix and into the lumen of the uterus. This occurred promptly after administration of oxytocin (intravenously or by nasal spray) and in the absence of sexual stimulation or orgasm. Lloyd (2005) has summarized these findings as follows:

> Wildt et al attribute rapid transport of sperm to the peristaltic contractions of the uterus—much like the contractions documented in dogs and cows—and to the muscular layers of the fallopian tubes. The relevant peristaltic contractions occur with a frequency of 2 to 5 per minute in healthy women at all times.

Release of oxytocin in women may occur as a result of vaginal dilation, cervical stimulation, and coital stimulation in the absence of orgasm. Thus, as Wildt et al. (1998) point out, there is no requirement to posit that female orgasm induces the genital contractions which influence sperm transport.

The absence of robust evidence for the occurrence of uterine 'upsuck' in relation to female orgasm casts considerable doubt upon the existence of 'sperm retention orgasms' in human females (Baker and Bellis 1993*b*, 1995). Many women require additional (manual) stimulation to achieve orgasm and do not necessarily achieve orgasm during (or after) intercourse in the absence of such stimulation (Kinsey et al. 1953; Fisher 1973; Hite 1977). Cross-cultural evidence also casts doubt on the proposition that female orgasm is a widespread or inevitable part of human sexual response (Mead 1967). Baker and Bellis (1993*b*; 1995) did not measure directly the effects of female orgasm on female fertility and reproductive success. Numbers of sperm lost in flowbacks were measured, but sperm numbers inseminated and retained after orgasm were only *estimated* mathematically. The accuracy of these procedures is highly questionable, however. Thus, Figure 1 in Baker and Bellis' (1993*b*) paper in *Animal Behaviour* includes examples where numbers of sperm in flowbacks actually exceed the numbers estimated at insemination! Lloyd (2005) critiques the selective use of data in these studies and concludes that 'there are such serious problems with the fundamental data set on flowbacks used by Baker and Bellis that it fails to meet basic scientific standards of evidence.' For example, of 11 couples who contributed data on flowbacks, a single couple accounted for 73 per cent of the 127 measurements analysed, whilst four couples supplied only one measurement each.

Given such serious criticisms and doubts concerning Baker and Bellis's reports, it is regrettable that their results have often been cited in the literature as if they represented the established facts of human physiology. This is not the case, and they do not support arguments concerning possible effects of cryptic female choice or sperm competition during human evolution.

Why does orgasm occur in human females and what is its evolutionary basis? Although answers to these questions are unlikely to advance our understanding of the origins of human mating systems, I believe that some discussion of these questions may be helpful to readers. Firstly, female orgasm is not confined to *Homo sapiens*. Putatively homologous responses been recorded in a number of non-human primates, including stump-tail and Japanese macaques, rhesus macaques, and chimpanzees (Dixson 1998*a*). Pre-human ancestors of *Homo sapiens*, such as the australopithecines, probably possessed a capacity to exhibit female orgasm, as do various extant monkey and ape species. The best documented example concerns the stump-tail macaque (*Macaca arctoides*), in which orgasmic uterine contractions have been recorded during female–female mounts (Goldfoot et al. 1980) as well as during copulation (Slob et al. 1986). Not every mount is accompanied by a female orgasm, however. De Waal (1989) estimates that female stump-tails show their distinctive *climax face* (which correlates with occurrence of uterine contractions) once in every six copulations. Vaginal spasms were noted in two female rhesus monkeys as a result of extended periods of stimulation (using an artificial penis) by an experimenter (Burton 1971). Likewise, a female chimpanzee exhibited rhythmic vaginal contractions, clitoral erection, limb spasms, and body tension in response to manual stimulation of its genitalia (Allen and Lemmon 1981). Masturbatory behaviour, accompanied by behavioural and physiological responses indicative of orgasm, has also been noted in Japanese macaques (Wolfe 1991) and chimpanzees (Goodall 1986).

In human beings, men and women experience similar sensations during orgasm; the process appears to be physiologically homologous in the two sexes. Male and female descriptions of orgasm are often indistinguishable, except for references to sexually dimorphic structures, such as the genitalia (Vance and Wagner 1976; Bancroft 1989). The main sensate focus for tactile stimulation resulting in orgasm is a homologous structure; the penis in the male and the clitoris in the female. Interestingly, Wallen and Lloyd (2008) have shown that variability in the length of the human clitoris greatly exceeds variation in penile or vaginal length. They conclude that this is due to co-evolution of penile and vaginal length, and a relative absence of selective pressure upon the evolution of clitoral size in humans.

Orgasm occurs during ejaculation in men, and is followed by a quiescent (refractory) period during which penile detumescence occurs and sexual activity usually ceases. In many non-human primates ejaculation is accompanied by behavioural responses (cessation of thrusting movements, body tension and tremor of the limbs, and changes in facial expression and vocalization) indicative of orgasm. A refractory period occurs after ejaculation in monkeys and apes, just as in males of many other mammals (Dixson 1998*a*). In women (as in females of other primate species) orgasm is not followed by a refractory period. Thus, some women are capable of repeated (multiple) orgasms, whereas such capacities have rarely been recorded in men.

Given the homology of the physiological and anatomical substrates for orgasm in the two sexes, and given also that these homologues exist in younger (prepubertal) individuals as well as in adults, Symons (1979) proposed that female orgasm might represent a non-selected homologue of a primarily masculine response. Thus orgasm in males is associated with important reproductive functions in adulthood (ejaculation, refractory period) and has been selected for on this basis. The female response might represent a homologue of the male process. By analogy, males possess homologues of female traits (e.g. nipples), which play a vital role in human reproduction (e.g. lactation) but have no function in males. Although Symons (1979) uses the male nipples as a case of non-adaptive homology, there are other examples. Thus human males and male anthropoids such as the macaques and guenons possess a vestigial and functionless homologue of the female uterus: the *utriculus prostaticus* or *uterus masculinus*. This tiny blind-ended pocket of tissue, adjoining the prostatic urethra, is all that remains in adult males of the Müllerian ducts, which give rise in females to the uterus and upper portion of the vagina during embryonic development (Gray 1977).

Symons (1979) cites Beach (1976*a*) to the effect that neural mechanisms mediating sexual patterns typical of one sex may also be capable of expression in the opposite sex, given favourable circumstances for their elicitation. Symons regards human female orgasm as a 'byproduct of female bisexual potential'. This view is not unrealistic or disrespectful to female sexuality, despite feminist criticisms of Symons' hypothesis on the basis that it is 'androcentric' (Wasser and Waterhouse 1983), in 'denial of the significance of female sexual pleasure' (Caulfield 1985) or that it smacks of 'a gentlemanly breeze from the nineteenth century' (Hrdy 1979). Everyone, of both sexes, is entitled to seek a loving, mutually respectful (and pleasurable) sexual relationship with their partner. In certain human cultures, women are reported to attain orgasm regularly during copulation and men are expected to learn the sexual skills required for its elicitation (e.g. among the people of Mangaia, in the Cook Islands: Marshall and Suggs 1971; Le Vay and Valente 2002). This is by no means the rule in human cultures, however. It should be kept in mind that Symons (1979) was examining the evolution of human sexuality, and was not seeking to define a culturally or gender-biased ideal of how people should behave.

The occurrence of orgasm in female non-human primates is better documented than it was in 1979, when Symons's book was published (Slob et al. 1986; Wolfe 1991; Dixson 1998*a*; Campbell 2007). The factors which influence the expression of female orgasm are still poorly understood, however, and further primatological research on this topic would be beneficial. For example, Troisi and Carosi (1998) reported that lower-ranking female Japanese macaques exhibit orgasms more frequently when copulating with a high-ranking male. The non-human primate data do not support the notion that female orgasm improves 'bonding' and pair formation between the sexes, however. Indeed, the best documented examples of female orgasmic responses occur in species with multi-male/multi-female mating systems (e.g. macaques and chimpanzees) rather than in monogamous forms such as gibbons (Dixson 1998*a*).

Forty years ago, Morris (1967) advanced the peculiar argument that orgasm exhausts women, and makes it more likely that they will rest after intercourse, rather than walking upright with the resulting loss of semen. There is no comparative depth to the argument, however, as monkeys and apes exhibit female orgasm in association with dorso-ventral copulatory postures and an absence of post-mating rest periods. Lloyd (2005) carefully analysed nineteen adaptive accounts of female orgasm, including Morris's (1967) ideas, and found that all of them are inadequate to explain its evolution in human beings. I agree with Lloyd that the most parsimonious explanation for the available evidence concerning the evolution of female orgasm in *H. sapiens* is the *byproduct hypothesis* (Symons 1979). The byproduct hypothesis has little to tell us about the origins and evolution of human mating systems, however. So, we must leave matters there in order to explore another avenue of enquiry: the anatomy and functions of the oviduct in mammals, and its relationship to cryptic female choice.

The mammalian oviduct, sperm competition, and cryptic female choice

It is of the greatest interest, given that hundreds of millions of spermatozoa are deposited in the vagina at ejaculation, that only a few hundreds or thousands of these gametes are recoverable from the oviduct. In Table 4.2, examples are provided of numbers of sperm deposited during copulation, and numbers recovered from the oviducts of various mammals, including *Homo sapiens*. These numbers do not represent the *only* spermatozoa to reach the oviducts. Rather they constitute 'snapshot' counts of a small, transient population of sperm which pass through the oviduct, and which are replenished by gametes moving upwards through the uterus and uterotubal junction, to enter the isthmus (Figure 4.4). In the isthmus, spermatozoa typically adhere for a time to the oviductal epithelium before migrating to the upper part of the oviduct (ampulla) where fertilization takes place (Harper 1994; Yanagimachi 1994). This transitory association between spermatozoa and the isthmic epithelium is of great importance for fertility. A temporary pool of gametes is created, in which spermatozoa may accumulate until ovulation occurs and ova are transported to the ampulla (Figure 4.4). During their sojourn in the isthmus, spermatozoa may complete the changes (capacitation) required to enable them to move rapidly through the oviduct (hyperactivated motility) and penetrate the egg and its vestments (Yanagimachi 1994).

Table 4.2 Numbers of spermatozoa ejaculated, sites of deposition, and numbers of sperm arriving in the oviduct in mammals

Species	No. of spermatozoa per ejaculate	Site of sperm deposition	Sperm nos. in ampulla
Human	280 million	Vagina	200
Mouse	50 million	Uterus	<100
Rat	58 million	Vagina	500
Rabbit	280 million	Vagina	250–500
Ferret	—	Uterus	18–1600
Guinea pig	80 million	Vagina and uterus	25–50
Domestic cattle	3000 million	Vagina	A few
Sheep	1000 million	Vagina	600–700
Pig	8000 million	Uterus	1000

Source: Data from Harper (1994); after Dixson (1998*a*).

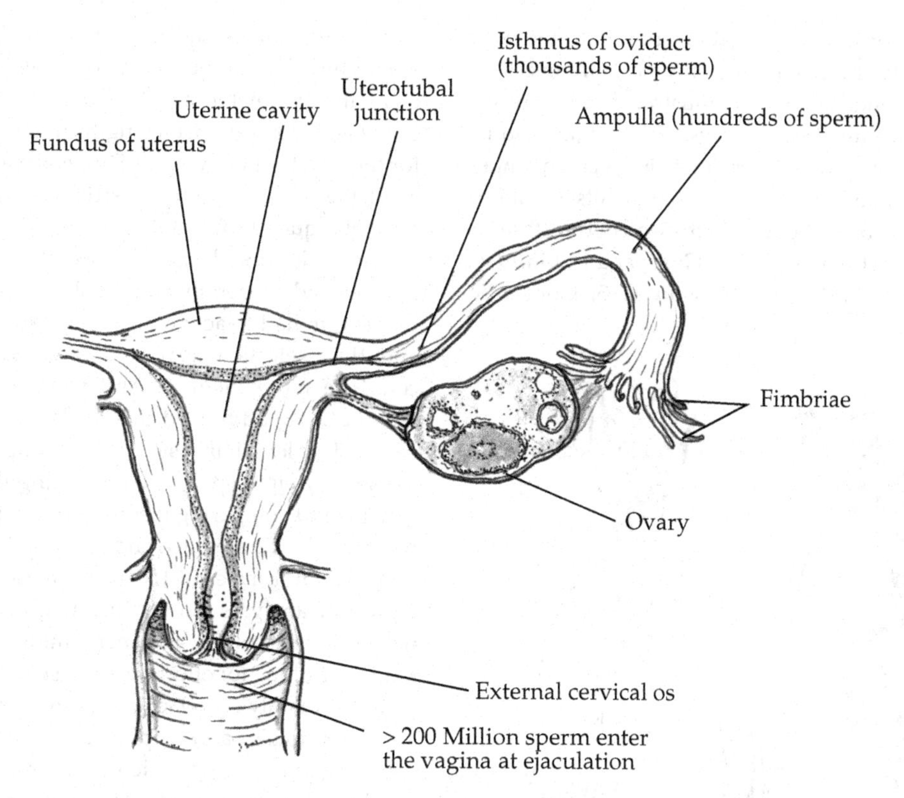

Figure 4.4 The human female reproductive tract and post-coital sperm numbers recorded at various levels within the tract.

For those mammals in which females mate with multiple partners during the fertile period, it is not known to what degree the spermatozoa of rival males are represented in the oviductal populations listed in Table 4.2. It would clearly be valuable to know if all the males that mate with one given female contribute gametes to this oviductal sperm population and, if so, in what proportions their gametes are represented. In fruit flies (*Drosophila*) elegant techniques have been developed to label and identify the sperm of rival males, and to verify a *last male advantage* in siring offspring (Civetta 1999). In mammals, sperm competition is thought to represent a 'raffle' in which males allocating the greatest numbers of sperm are likely to experience an advantage in sperm competition, irrespective of the order of mating (Parker 1998). Thus it has been established that single litters of offspring may have multiple sires, as, for example, in Belding's ground squirrel (Hanken and Sherman 1981), the common shrew (Stockley et al. 1993), the black bear (Schenck and Kovacs 1995), the Ethiopian Wolf (Sillero-Zubiri, Gottelli, and MacDonald 1996), and the cheetah (Gottelli et al. 2007).

Given the importance of sperm competition, might the oviductal environment filter the gametes of rival males in some way, so that some sperm secure an advantage over others in gaining access to ova? If so, are there structural differences between the oviducts of those mammals which commonly engage in sperm competition, and those which are monogamous or polygynous? In this regard, it is noteworthy that mammalian oviducts exhibit great differences in morphology, including their length (Hunter 1988). Some mammals have relatively long and convoluted oviducts, whereas in others the ducts are quite short. There are differences also in the morphology of the ciliated fimbria (which guide the ova into the entrance of the oviduct) as well as in internal features such as the structure of the uterotubal junction (Hunter 1988; Harper 1994). Some interspecific differences in oviductal morphology are illustrated in Figure 4.5, which includes *Homo sapiens*. The human oviduct is approximately 8–15 cm long (its average length is 11 cm) in its extra-uterine extent, whilst the intramural portion embedded in the wall of the uterus is about 1.5–2.0 cm long (Lisa, Gioia, and Rubin 1954; Pauerstein and Eddy 1979). In 1993, Gomendio and Roldan made the important discovery that sperm numbers in the ejaculates of eleven species of mammals (including *H. sapiens*) are positively correlated with oviductal lengths (after controlling for the effects of body size). The small sample sizes available to these authors restricted further work, but subsequently Gomendio et al. (1998) pointed out that 'it would make sense if polyandrous females had longer, more convoluted oviducts, so that sperm had to actively swim greater distances, and thus selection against less vigorous sperm could be more intense.' Anderson, Nyholt, and Dixson (2006) addressed this question by measuring oviduct length in a substantial sample of mammals (forty-eight species, representing thirty-three genera) and comparing residuals of oviduct length to residuals of testes size and sperm midpiece volumes in adult males of the same species. Multiple regression analyses showed that oviduct length correlates with testes size and sperm midpiece volume after controlling for effects of body size. The sample included mainly artiodactyls, carnivores, primates, and marsupials, and appropriate statistics were used to control for possible phylogenetic biases in the data set. Some results are shown in Figure 4.6 and data for *H. sapiens* have been included for comparative purposes. The human oviduct is relatively short, in relation to female body weight, and in relation to relative testes size and sperm midpiece volume in men.

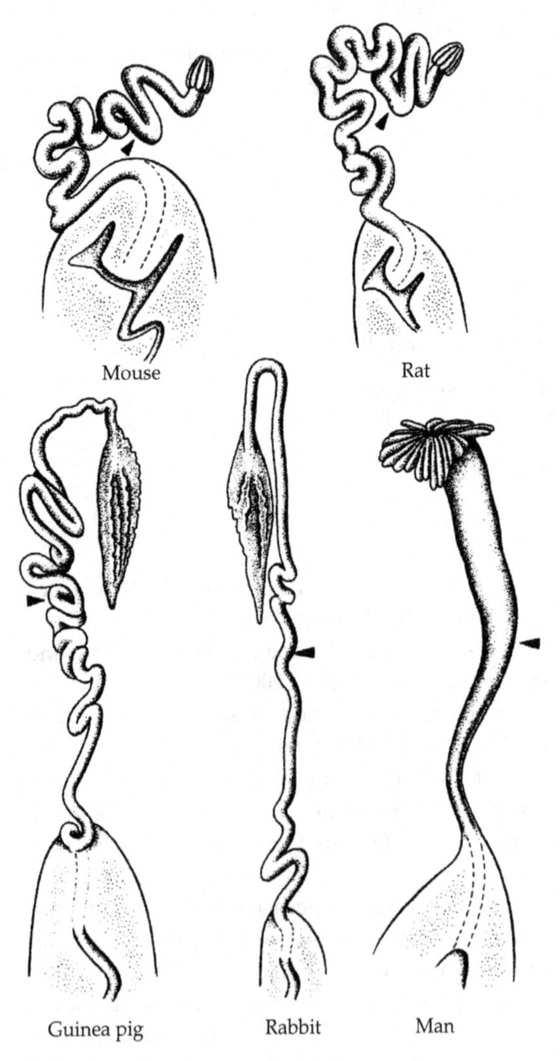

Figure 4.5 Examples of oviductal morphologies in mammals, to illustrate differences in the length and degree of coiling of the Fallopian tubes. The arrows indicate the approximate position of the junction between the isthmus and ampulla.

Source: From Harper (1982).

A longer and more convoluted oviduct represents an additional challenge to sperm as they ascend the duct in order to fertilize the ova, either in the ampulla or at the ampullary–isthmic junction. When the sperm of rival males are present, an elongated oviduct may aid the female in selecting gametes of males with the greatest reproductive potential. It is the female's anatomy and physiology which challenges sperm in these ways; hence elongation of the oviduct in mammals where sperm competition is most likely may represent an example of sexual selection by cryptic female choice. The relatively short oviduct in *H. sapiens* is indicative of an evolutionary history involving low selection pressures for genital specialization and accords with the low values for relative testes size and sperm midpiece volumes in human beings.

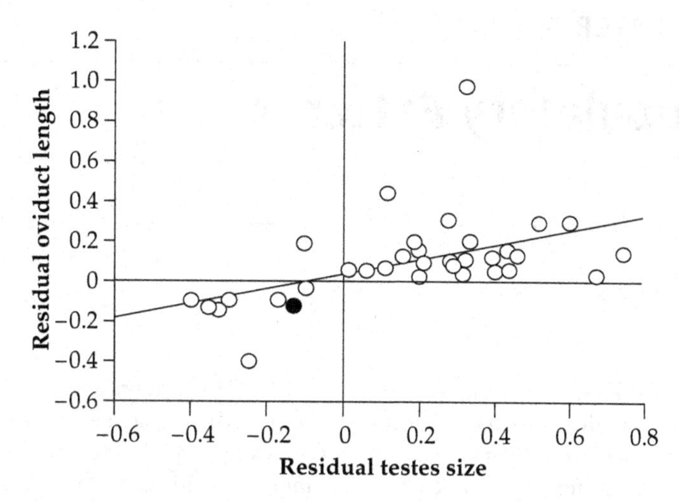

Figure 4.6 Correlation between oviductal length and relative testes size in mammals. Data for *Homo sapiens* (•) are included for comparative purposes.

Source: Adapted from Anderson, Dixson, and Dixson (2006).

Conclusions

Much less is known about mechanisms of cryptic female choice in mammals than is the case for sperm competition. However, the limited evidence presented in this chapter is consistent with the conclusions reached in Chapter 3 concerning sperm competition and the origins of human sexual behaviour. Thus, there is little credible evidence that orgasm in women or in females of non-human primate species plays any role in sperm transport or fertility. Likewise, the evolution of human penile morphology and copulatory patterns is not connected to any requirement to induce female orgasm, in order to increase the likelihood of fertilizing ova. Rather, it appears that female orgasm, which occurs in a number of anthropoids besides *Homo sapiens*, represents a functionless homologue of responses which accompany ejaculation in the male. Functionless homologues of female structures (such as the nipples and *uterus masculinus*) also occur in men and in males of other mammals.

Recent work on the evolution of oviductal length in mammals is reviewed here. The extramural portion of the oviduct is significantly longer in those mammals in which females mate with multiple partners during the fertile period. Interestingly, relative testes sizes, sperm midpiece volumes, and relative oviductal lengths are positively correlated in mammals. Given that the oviduct is a final and critical arena for sperm competition, it is possible that longer and more convoluted oviducts serve to 'test' gametes, and represent a mechanism of cryptic female choice. In this regard it is significant that the human oviduct is relatively short by comparison with the oviducts of many other mammals. There is no evidence that cryptic female choice has affected the evolution of human oviductal morphology.

CHAPTER 5

Copulatory Patterns

The desire for novelty and a capacity for inventiveness pervades human sexuality, just as it affects other areas of our lives. Such inventiveness existed in bygone ages, as captured, for example, by the erotic stone images of couples which adorn ancient Indian temples. These cultural expressions of human sexual diversity, however, can easily obscure those fundamental aspects of copulatory behaviour which humans share with other primates, as well as with mammals in general. In this chapter, I shall attempt to identify homologues of human copulatory patterns, and those which occur in non-human primates. The purpose of this exercise is to address three sets of questions. Firstly, what is the evolutionary basis of the various copulatory postures displayed by people in various parts of the world? Face-to-face (ventro-ventral) copulatory positions are found in most of these populations. How did they originate? Secondly, is prolonged copulation an ancient trait of *Homo sapiens* and why do only some other primate species engage in prolonged copulatory patterns? Thirdly, why are frequencies of mating so variable among the primates, and how do frequencies of copulation in human beings compare with those observed in lemurs, monkeys, and apes?

Copulatory postures

Anthropological studies indicate that face-to-face copulatory positions predominate among the majority of human societies for which reliable information exists. Usually, the woman lies on her back with the man on top of her; the ventro-ventral or so-called *missionary position* (Figure 5.1). In their study of *Sexual Behavior in the Human Male*, Kinsey, Pomeroy, and Martin (1948) reported that 70 per cent of North Americans used only the missionary position during sexual intercourse. Setting aside the sexual inhibitions and reluctance to explore different positions in some of these subjects, there appears to be a fundamental human preference for face-to-face copulatory behaviour. In China, it is the major pattern used, although 54 per cent of couples may sometimes try other positions (Liu et al. 1997). Ford and Beach (1951) found that face-to-face postures are widespread in many indigenous societies, and represent the most frequently recorded position among the 185 societies they examined worldwide. Variations of such positions involve the man squatting or kneeling before the woman, with her legs straddling his thighs. The man may then pull the woman up, so that the couple embraces, face-to-face. Ford and Beach noted that this behaviour is characteristic of Pacific Island peoples, such as the Trobrianders and the Balinese. It appears to be a common pattern among those peoples who colonized the South Pacific, beginning more than 40,000 years ago. Marquesan Islanders, for example, have a term for this position, *haku noho* (sitting style), as distinct from other positions used for intercourse (Suggs 1966).

Another type of copulatory posture, the *female superior position*, involves the woman sitting astride the man, as he reclines on his back (Figure 5.1). Kinsey, Pomeroy, and Martin (1948) record that this is the most common alternative to the missionary position among North American couples. It also occurs as a secondary posture in many other societies around the world (Ford and Beach 1951).

Finally, there are variants of the *dorso-ventral copulatory posture* (Figure 5.1). These are typical of the great majority of mammals, in which the male mounts the female from the rear. Copulatory positions involving

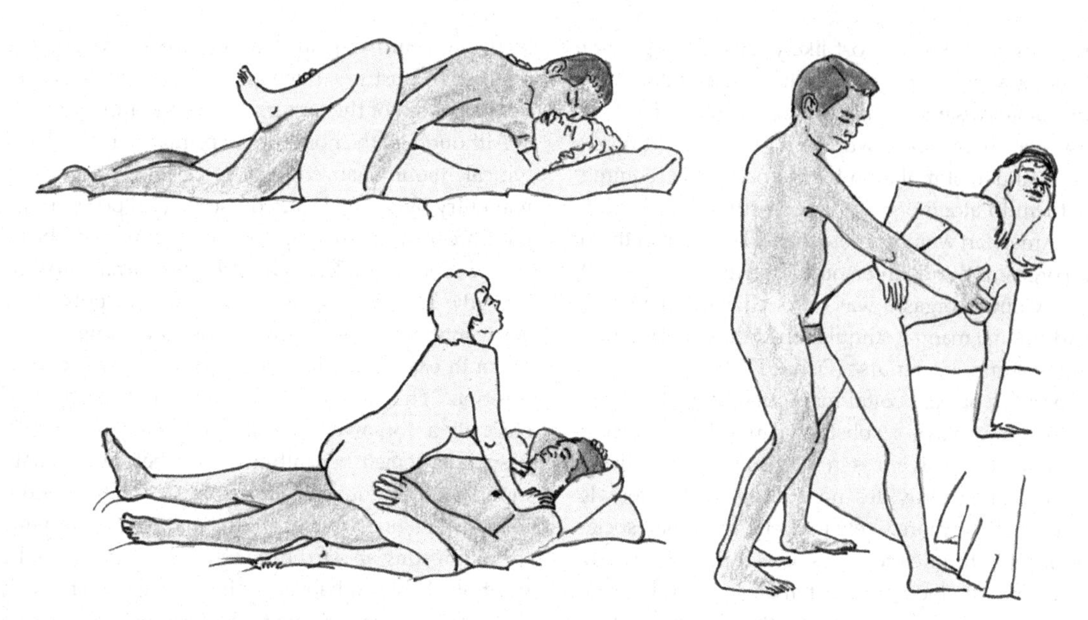

Figure 5.1 The principal human copulatory positions: Ventro-ventral ('missionary position'), female superior, and dorso-ventral.
Source: Author's drawings from photographs in LeVay and Valente (2002).

rear entry did not represent 'the usual practice' among the societies considered by Ford and Beach. However, they noted twelve examples in which intercourse occasionally takes place while the couple lies side by side, with 'the man entering the woman from the rear.' In eight societies (the Crow, Dobuans, Hopi, Kurtatchi, Kuroma, Lepcha, Marshall Islanders, and Wogeo) full dorso-ventral positions are sometimes used, 'with the man standing behind the woman as she bends over or rests on her hands and knees'. Suggs (1966) also records this position for the Marquesans, who refer to it as *patu hope*. The Marquesans also use a position in which the female lies on her back, and the man faces her while lying on his side. They refer to intercourse in this side position, as *haka ka'aka* or 'in the gecko-lizard manner'. How frequently dorso-ventral copulatory postures are used by people of the various western cultures is not known. However, such positions are frequently depicted in sex manuals and in the literature on sexual behaviour in North America and Europe.

Ford and Beach (1951) speculated that the evolution of ventro-ventral and female superior copulatory positions in human beings might be connected with the requirement to achieve maximum stimulation of the clitoris during intercourse, and thus to facilitate female orgasm. The evidence available to Ford and Beach indicated that, among the primates, the occurrence of female orgasm is unique to human beings. Moreover, it was thought that all the non-human primates copulate in dorso-ventral positions, and that postures involving rear entry by the male might not be conducive to clitoral stimulation and the occurrence of female orgasm. Ventro-ventral copulatory postures had been observed in orangutans and in immature chimpanzees, but only in captive animals. It seemed to Ford and Beach that the natural sexual behaviour of free-ranging apes might well be different, and more typical of the situation observed in other primates.

It is now known that neither female orgasm nor ventro-ventral and female superior copulatory postures are unique to *Homo sapiens*. In the last chapter, orgasm in female stump-tail macaques, Japanese macaques, chimpanzees, and other species was discussed, and compared to putatively homologous responses in the human female. A ventro-ventral copulatory position is not a prerequisite for the expression of such responses, as they occur in female monkeys and chimpanzees during dorso-ventral contacts with males. The capacity for female orgasm appears to be an ancient trait among anthropoid primates, and would most likely have been present in ancestral hominids, such as the australopithecines. This capacity is not equally expressed among females, however, so that some women report that additional (e.g. manual) stimulation by the partner is required for them to attain orgasm. In one study of 300 married American women, more than 90 per cent of them had orgasms during intercourse but, for 95 per cent of these women, orgasm was only achieved as a result of additional manual stimulation of the clitoris, either before or after intercourse (Fisher 1973).

Foreplay or post-coital manipulation of the clitoris therefore plays a substantial role in enhancing orgasmic responsiveness in women. The missionary position alone provides no guarantee that female orgasm will be more likely to occur. In human societies where attitudes towards sexual intercourse are repressive, the incidence of female orgasm is probably low. For example, among the Yolngu of Arnhem Land, in northern Australia, it is customary to arrange marriages between young women and much older men. Among the Sambia of Papua New Guinea, men frequently marry women from other villages, and these 'in marrying' women are often not trusted or treated affectionately by their husbands. In both these cultures, intercourse is said to be peremptory in nature and often lacking in orgasm for the female partners. We saw in the last chapter that female orgasm serves no established function in sperm transport or fertility in human beings. Rather, it is most likely to represent a non-adaptive homologue of orgasm during ejaculation in the male (Symons 1979; Lloyd 2005). Thus, when the sexual relationship lacks sufficient rapport, or where psychological factors such as sexual guilt interfere with such rapport, then women may be inhibited from displaying orgasms. This kind of inhibition also occurs in some men who, despite prolonged copulatory stimulation, are unable to achieve orgasm and to ejaculate intra-vaginally. Inhibition of orgasm in the female also has its (rarer) male counterpart, therefore (Masters and Johnson 1970).

If the ventro-ventral and female superior copulatory postures depicted in Figure 5.1 did not evolve to enhance the likelihood of female orgasm, what is the origin of such behaviour in human beings? To answer this question, it will be helpful to conduct a comparative survey of the copulatory postures used by the non-human primates, and to enquire whether distinctive postures have arisen in particular families or genera of the lemurs, monkeys, and apes.

Although all the non-human primates make use of typical mammalian copulatory postures, involving rear entry by the male and dorso-ventral positioning, there are some interesting variations around this basic theme. For example, among the nocturnal prosimians, the lorisines, including the African potto and angwantibo as well as the slender and slow lorises of south east Asia, all copulate in an 'upside-down' position. This unusual copulatory posture is still basically a dorso-ventral one, but the female usually suspends herself beneath a branch before the male mounts her (Figure 5.2). However, among the closely related galagines (bushbabies), mating occurs with the pair resting above the branch in the usual, upright position. The bushbabies are highly mobile animals; they use vertical clinging and leaping to traverse their home ranges and to avoid potential threats. The lorisines are very different, being anatomically specialized for slow-moving and cryptic behaviour. Their inverted mating postures may, therefore, represent part of their adaptations for gripping tightly to substrates and avoiding detection by predators. Despite the large geographical separation between the African and Asian lorisines, their copulatory postures are as distinctive, in evolutionary terms, as their locomotor and other anatomical specializations.

A similar diagnostic test of evolutionary relationships may be made by examining in detail the copulatory postures of certain New World monkeys. Among the spider monkeys and woolly monkeys, which belong to the sub-family Atelinae, a remarkable 'leg-lock' posture occurs during copulation. The male mounts the female from the rear, but then places his legs over the female's thighs (Figure 5.3).

Mounts are prolonged in these monkeys, lasting 6–35 min in spider monkeys, for up to 18 min in woolly monkeys, and more than 8 min in muriquis (woolly spider monkeys). The evolution of the leg-lock may be connected with prolonged patterns of copulation in these large-bodied monkeys. The posture allows the male to sit snugly behind the female, holding her in position. Both sexes also use their prehensile tails to anchor themselves to branches during extended copulatory bouts. The howler monkeys also belong to the Atelinae but they do not

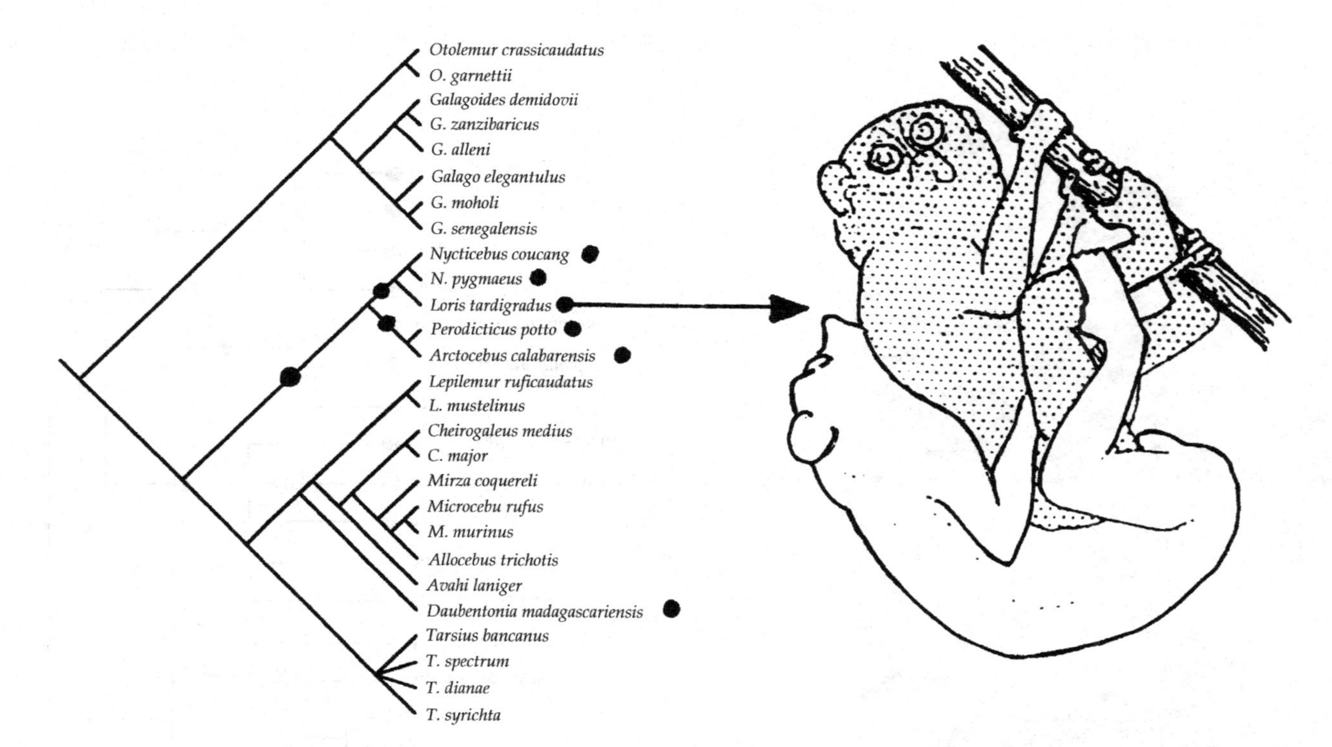

Figure 5.2 Relationships between phylogeny and copulatory postures in nocturnal prosimian primates. The inverted copulatory posture occurs in the lorisines of Africa and Asia and is likely to represent an ancient behavioural trait for this group. One of the Malagasy lemurs (the aye-aye, *Daubentonia*) may also copulate in an inverted ('upside down') position.

Source: The phylogeny is adapted from Kappeler (1995). The illustration of slender lorises mating is from Schulze and Meier (1995).

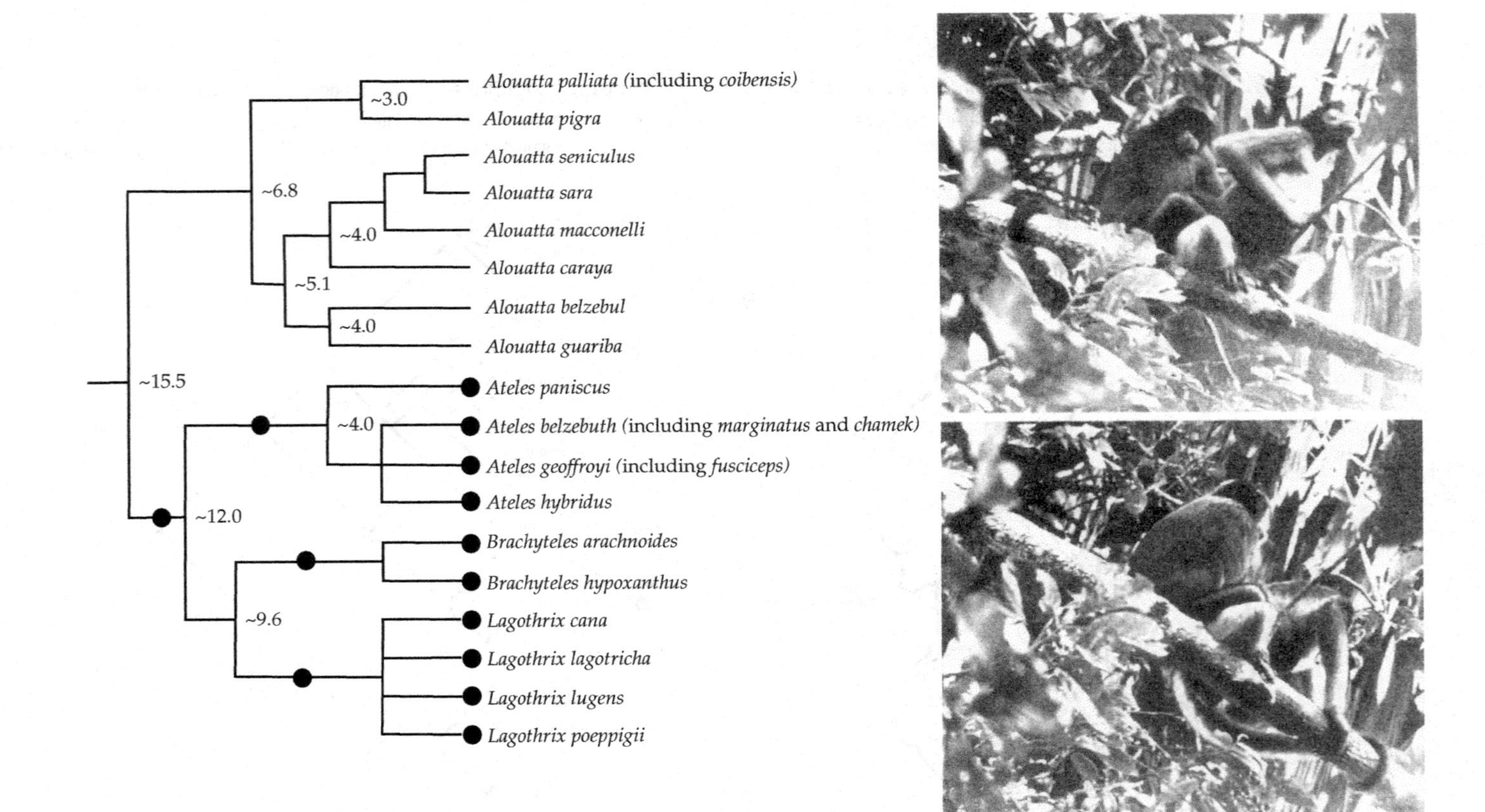

Figure 5.3 Phylogeny of the South American ateline monkeys, and occurrences of the specialized 'leg-lock' copulatory posture in members of three genera (*Ateles, Brachyteles*, and *Lagothrix*). *Source*: The phylogeny is from Di Fiore and Campbell (2007). The illustrations of copulation in *Ateles* are from Dixson (1998*a*), after Klein (1971).

copulate for prolonged periods, nor do they exhibit the leg-lock mating specialization (Figure 5.3). None of the other New World primates has been observed to use leg-lock mounting postures.

As regards the hominoids, it is significant that although the apes often copulate in the typical primate (i.e. dorso-ventral) manner, ventro-ventral or female superior copulatory postures have also been recorded in various species, both in the wild or in captivity. Few records are available for gibbons; however, Koyama (1971) observed ventro-ventral copulation in free-ranging siamangs (*Hylobates* (*Symphalangus*) *syndactylus*) at Fraser's hill, in Malaysia. Among the great apes, there have been numerous accounts of ventro-ventral or female superior mating postures. Rijksen (1978), for example, recorded wild Sumatran orangutans mating in a variety of such positions, in association with their versatile quadrumanous climbing and suspensory behaviour. The male orangutan sometimes reclines on his back, with his penis erect, as a sexual invitation to the female. The female may then initiate intromission, by sitting in the male's lap, facing towards him, and making pelvic thrusting movements (Figure 5.4). Schurmann (1982) has observed this kind of behaviour in wild orangutans. In captive Bornean orangutans, Nadler (1977; 1988)

Figure 5.4 The pre-copulatory penile display and female superior copulatory posture of the orangutan (*Pongo pygmaeus*).
Source: From Dixson (1998*a*), after Nadler (1988).

found that the male pattern of reclining and penile display occurs most often in situations where females have some control over mating interactions. Nadler fixed a barrier between the male and female cages, such that only the smaller female orangutan could pass underneath and gain access to her partner. During these *restricted access pair tests*, females were much more likely to initiate copulation at mid-menstrual cycle, and males reclined, displayed, and invited females to approach and mount them.

Turning to the African apes, the pigmy chimpanzee (bonobo) mates in both dorso-ventral and ventro-ventral positions. The common chimpanzee mates only in the dorso-ventral position, at least as far as is known from field observations of several populations in eastern and western Africa (Goodall 1986; Boesch and Boesch-Achermann 2000). Ventro-ventral copulations have been observed in both free-ranging and captive bonobos; the ventro-ventral pattern may be more typical of sub-adult or young adult animals (Figure 5.5). In studies conducted in the Lomako forest (in what was formerly Zaire), 26 per cent of the seventy copulations observed by Thompson-Handler et al. (1984) took place with the partners facing each other. The majority of the matings were of the usual, dorso-ventral type. Takayoshi Kano's (1992) more extensive field studies of bonobo behaviour, at Wamba, revealed that

> every copulation is invariably preceded by female presentation, which takes two forms. In one form the female stands quadrupedally;...... this is followed by dorso-ventral copulation. In the other form, the female lies on her back in front of the male, spreads her thighs, raises her buttocks, adjusts herself and shows her sexual organs; the male approaches and copulates ventro-ventrally.

Kano also found that ventro-ventral copulations were less frequent (29.1 per cent of observations) than dorso-ventral ones, and 'more frequent in male and female adolescents than in full adults'. Ventro-ventral mounts also occurred frequently between female bonobos, the partners embracing face to face and thrusting their genitalia together (the so-called *G–G rubbing display*: Figure 5.5).

Western lowland gorillas have occasionally been observed to mate face-to-face, although the majority of their copulations take place in the dorso-ventral position (Figure 5.6). Thus, a number of authors, including Schaller (1963), Hess (1973), and Nadler

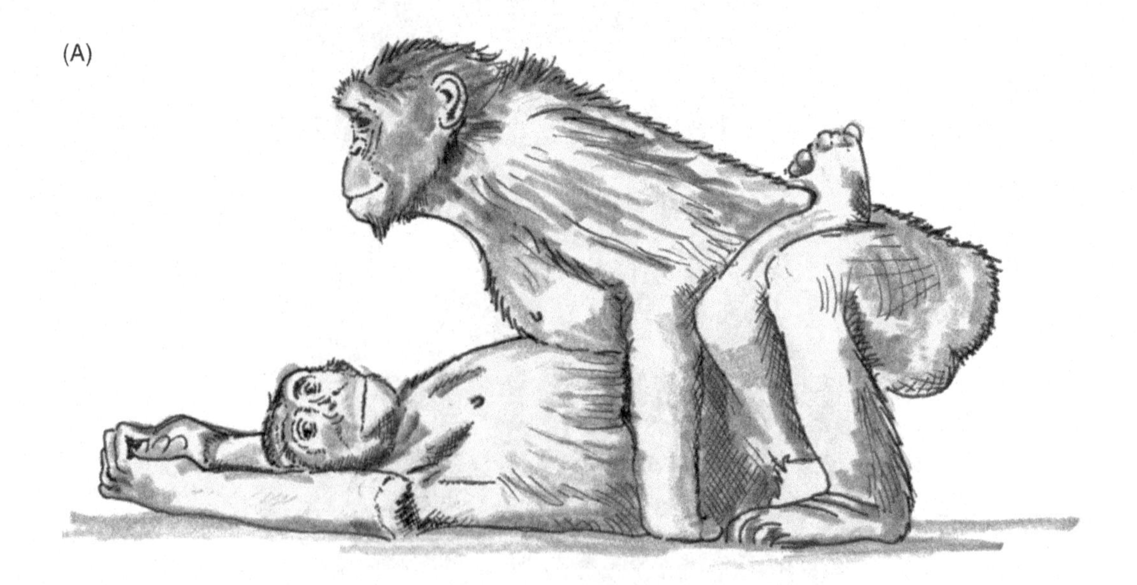

Figure 5.5 (A). Ventro-ventral copulatory posture of the bonobo (*Pan paniscus*). (B). Two female bonobos embrace, and engage in 'G-G rubbing'.

Source: Author's drawings, from photographs by De Waal (2003).

Figure 5.5 *continued*

(1980) provide descriptions and illustrations of captive western lowland gorillas mating in these ways. Although Schaller observed only two copulations in mountain gorillas, he supplemented his study by observing the behaviour of captive western lowland gorillas at the Columbus Zoo. These animals occasionally exhibited ventro-ventral patterns of copulation. More recently, it has been possible to observe western lowland gorillas in the Congo, in an area where groups emerge into open, swampy areas to feed. Copulation in a face-to-face, seated posture has recently been photographed there. For the mountain gorilla, detailed information has been provided by Harcourt et al. (1980). All sixty-nine copulations recorded by Harcourt and his colleagues in the Virungas National Park involved dorso-ventral postures, 'although twice one pair briefly assumed the ventro-ventral position before turning to copulate normally.' Same-sex mounting has also been described in male mountain gorillas and Yamagiwa (2006) includes illustrations of both dorso-ventral and ventro-ventral postures adopted by these males. Overall, 16 per cent of male–male mounts were of the ventro-ventral type.

The phylogenetic distribution of various types of copulatory postures among the extant prosimians,

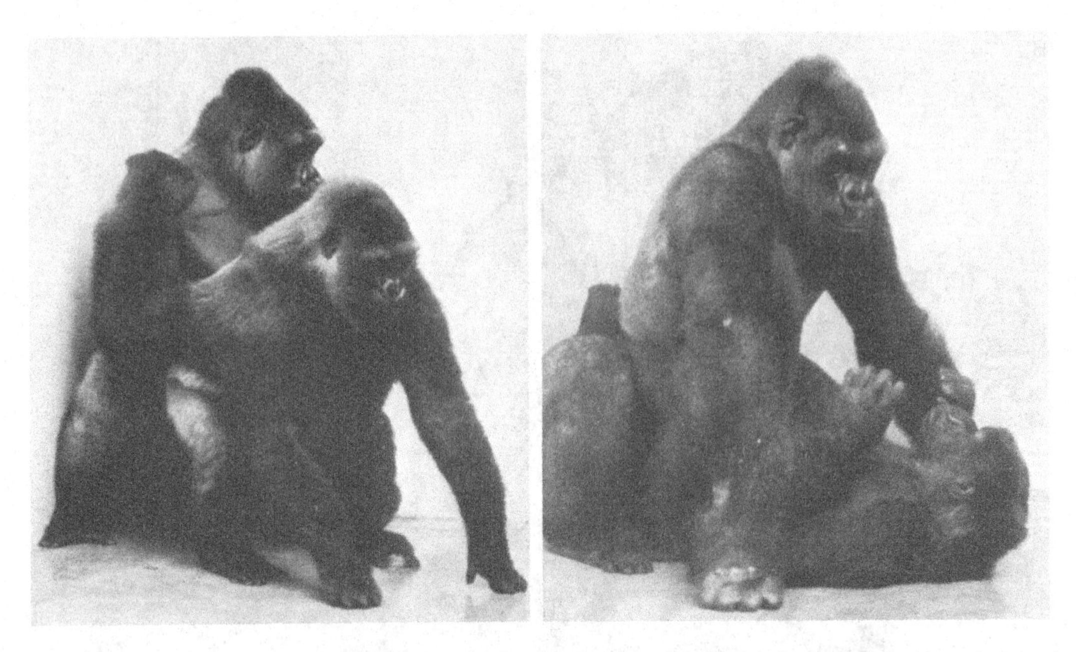

Figure 5.6 Copulatory postures in captive western lowland gorillas (*Gorilla g. gorilla*). Left: A female invites mating by backing towards a male, and the male mounts in a dorso-ventral position. Right: The less-frequently used ventro-ventral position.

Source: Photos by Dr. R.D. Nadler.

monkeys, apes, and humans is shown in Figure 5.7. The greater versatility of copulatory positioning, and use of ventro-ventral positions by the apes (and especially by the great apes) and humans probably springs from several causes. Firstly, as was suggested long ago by Bingham (1928), the increased brain development and intelligence of apes may be associated with a reduced reliance upon stereotyped copulatory patterns and a closer resemblance to human patterns of behaviour. The progressive enlargement of the brain which has occurred throughout hominid evolution has already been discussed in Chapter 1 (see Figure 1.10). A reduced reliance upon stereotyped copulatory patterns would likely have occurred in the australopithecines and in forms such as *Homo ergaster*. Secondly, it may also be important to consider that face-to-face patterns of copulation facilitate eye contact and continued facial communication between partners during mating. Eye contact plays an important role during pre-copulatory behaviour in many anthropoids. Female monkeys of various species frequently attempt to look back at males during dorso-ventral copulations (Dixson 1998*a*). Such communication is greatly facilitated during ventro-ventral copulation, however, as has been noted for orangutans, bonobos, and gorillas (Nadler 1980; 1988; Schurmann 1982; Kano 1992).

Finally, I suggest that the propensity for apes and humans to engage in face-to-face copulatory postures has been greatly facilitated by anatomical specializations which originated in arboreal ancestors. The apes have a long evolutionary history of using suspensory postures, as well as arm-over-arm swinging forms of locomotion (brachiation) in arboreal environments. Although brachiation is best developed in the gibbons, the orangutan also brachiates slowly and the African apes occasionally do so, although they are specialized for terrestrial patterns of locomotion (knuckle walking). *Homo sapiens* descended from arboreal, ape-like ancestors, and shares with apes many anatomical features of the shoulder joints, rib cage, and fore-limbs. *Australopithecus afarensis*, for example, had long arms and curved fingers; this ape-like hominid probably climbed and suspended itself below the branches efficiently, as well as walking bipedally on the ground. Humans, like the apes, have broad, flat chests and mobile shoulder joints allowing a greater range of arm movement than is possible for quadrupeds, such

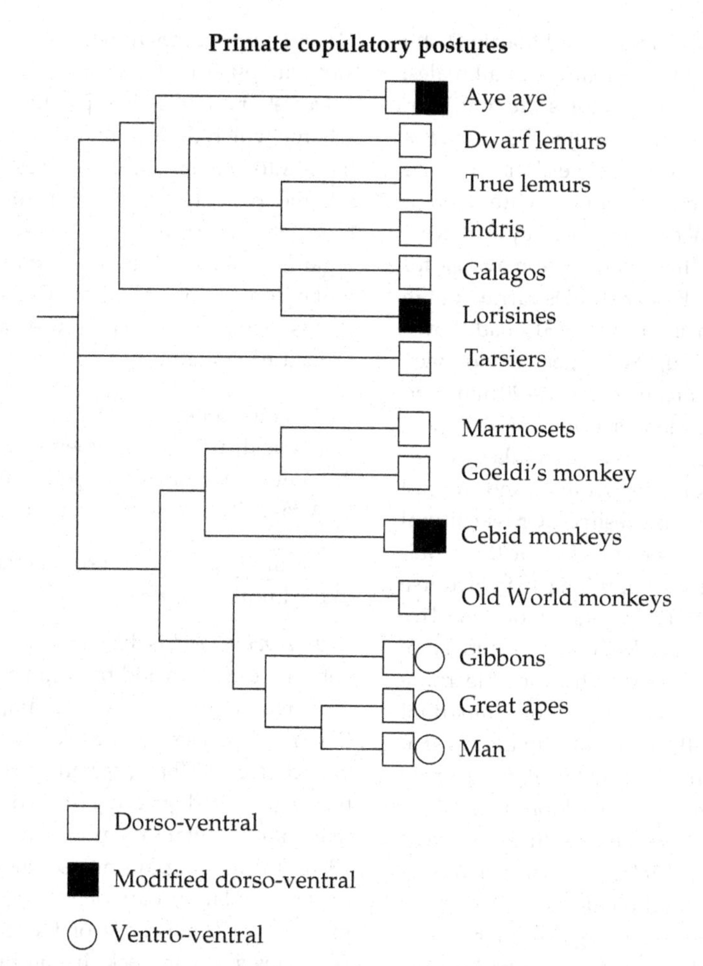

Figure 5.7 Phylogenetic relationships between the six extant superfamilies of the Order Primates, together with the distribution of various copulatory postures throughout the order. The basic mammalian dorso-ventral copulatory posture occurs in all six superfamilies. Specialized variants of this position occur in the lorisines, and in the aye-aye (inverted postures) and the atelines among the cebid monkeys (leg-lock postures). Ventro-ventral postures have been recorded in the great apes, in humans, and (much less frequently) in gibbons.

as monkeys. The use of flexible, suspensory postures may have facilitated the emergence of ventro-ventral and female superior copulatory positions during hominoid evolution. Animals which habitually hang, or swing by their arms whilst foraging might also suspend themselves in this way during mating. This is frequently the case in orangutans, and may explain the retention of ventro-ventral postures in apes which are now more terrestrial, such as the bonobo. Human beings may therefore have inherited a propensity to mate in face-to-face positions from remote ape-like ancestors. Add to this predisposition the marked lengthening of the lower limbs which occurred during human evolution, and a shift from dorso-ventral to more frequent use of ventro-ventral positions during intercourse might then have become established.

Copulatory durations

Many men are able to reach orgasm and to ejaculate quickly during sexual intercourse. Quantitative information on this subject is quite sparse, however, and particularly for non-westernized societies. In

North America, Kinsey, Pomeroy, and Martin (1948) found that 'for perhaps three-quarters of all males, orgasm is reached within two minutes after initiation of the sexual relation.' They also noted that 'for a not inconsiderable number of males the climax may be reached in less than a minute or even within ten or twenty seconds of coital entrance.' Kinsey et al. were well aware, however, that intercourse may last for much longer than this. Earlier studies carried out in the USA, by Dickinson and Beam (1931), had shown that in 362 married couples 17 per cent of men engaged in sexual intercourse for 15–20 min, and 9 per cent for 30 min or longer prior to ejaculation. In a small number of cases (3%), husbands reported that they could prolong intercourse and control attainment of orgasm as desired. Cross-cultural studies show that some societies favour the acquisition of such techniques of prolonged intercourse and delayed orgasm by males (e.g. among the Trobriand Islanders, Balinese, Marquesans, and Hopi Indians: Ford and Beach 1951). The role of learning and experiential factors is probably very important in such cases. Suggs (1966), for example, says that among adolescent Marquesans, who often engage in clandestine liaisons, 'the sex act seldom takes more than five minutes, most often two or three minutes.' Interestingly, he reports that 'Marquesan girls apparently have practically no difficulty in experiencing orgasms; they seem to attain orgasm after a relatively small number of sexual experiences, learning to control it quite quickly.' Among married couples, however, 'the duration of intercourse increases up to ten minutes or so in many cases. Prolongation of the act by *coitus interruptus* is attempted.... Obviously, the prolongation of the act is possible in a recognized union, as the fear of being apprehended is removed and the couple is completely at leisure.'

The question to be examined here concerns the evolutionary origin of human copulatory patterns. Were prolonged patterns of intercourse the norm for our species, or were matings typically brief? Is prolongation of intercourse due primarily to cultural factors, rather than reflecting the outcome of sexual or natural selection in remote ancestral populations in Africa? In an attempt to produce meaningful answers to these questions, it is helpful to place human copulatory behaviour within a comparative framework. How does the copulatory behaviour of *Homo sapiens* compare with that of other primates, and with other mammals in general?

Dewsbury (1972) has produced a classification scheme which divides mammalian copulatory patterns into sixteen possible types (Figure 5.8). This scheme provides a useful taxonomy of copulatory patterns. However, it has some drawbacks as regards its application to evolutionary questions, as will become apparent below. Dewsbury's classification is based upon the presence, or absence, of four traits during mating:

1. A genital lock
2. Pelvic thrusting movements by the male
3. Multiple intromissions prior to ejaculation
4. A capacity to exhibit multiple ejaculations

Let us firstly examine these four traits, as they relate to the human situation.

1. *A genital lock* is the mechanical tie which occurs between the penis and the vagina during mating in some mammals. In dogs, for example, the distal portion of the penis swells markedly once intromission has occurred, so that it becomes firmly lodged within the vagina. If dogs are castrated, their capacity to maintain a genital lock gradually diminishes (Beach 1970). In the hopping mouse, by contrast, it is the presence of large, androgen-dependent *penile spines*, rather than tumescence of the glans penis which facilitates a genital lock during mating (Dewsbury and Hodges 1987). Genital locks are rare among the primates. Some of the nocturnal prosimians, such as the galagos, have greatly enlarged penile spines and it is likely that a genital lock occurs in certain species (Dixson 1989). Among the anthropoids, the stump-tail macaque is unusual in having a greatly elongated, lanceolate glans penis; this forms a complementary fit with the narrow vaginal opening of the female. Stump-tails often remain together in a *post-ejaculatory pair-sit* and they are the only monkeys in which a partial form of genital lock is known (Lemmon and Oakes 1967). Humans, in common with the monkeys and apes in general, do not exhibit a mechanical tie between the sexes during copulation.

2. The next decision to be made when applying Dewsbury's classification of mammalian copulatory patterns concerns the presence or absence of *pelvic*

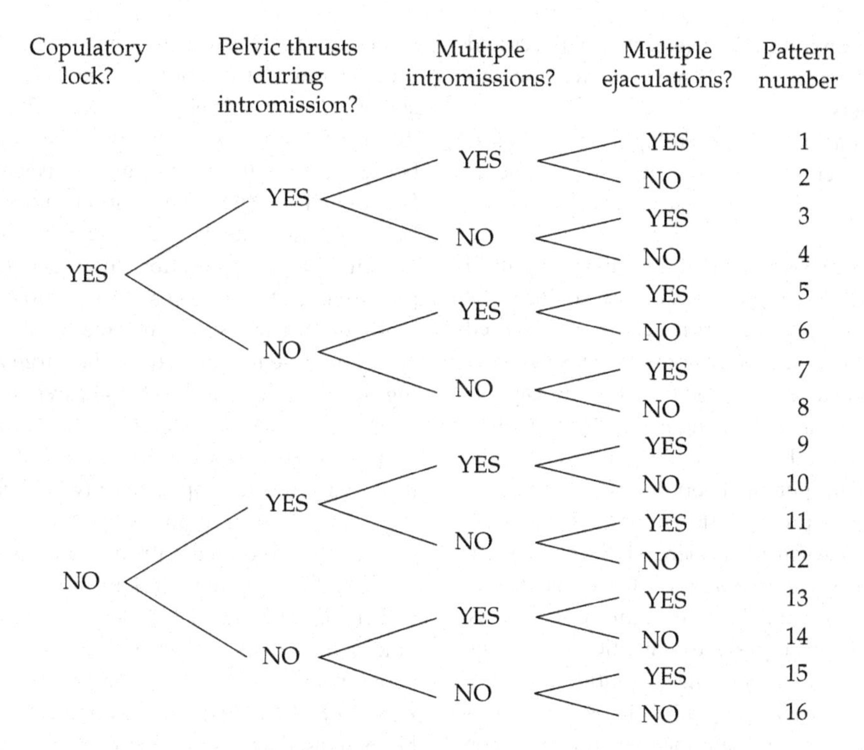

Figure 5.8 Dewsbury's (1972) classification scheme for masculine patterns of mammalian copulatory behaviour.

thrusting by males during intromission (Figure 5.8). Where the primates are concerned, this decision is straightforward, because pelvic thrusting occurs in all species of prosimians and anthropoids studied thus far. It represents a primitive condition for the Order Primates as a whole. This is not the case for all mammals, however. In the laboratory rat and the rabbit, for example, males make rapid pelvic thrusts only during the pre-intromission phase of the mount; once intromission is attained, there is a single thrust only, and not the repetitive thrusting patterns which occur in the primates.

3. *Multiple intromissions* characterize the copulatory behaviour of some mammals, such as the laboratory rat, although the number of intromissions a male makes prior to ejaculation is influenced by many factors, including age, and experiential and genetic variations (Larsson 1956). In the rat, the male's intromissions evoke neuroendocrine responses in the female which are important for the secretion of prolactin and for support of the corpus luteum during the ensuing pregnancy (Adler 1978; Freeman 1994). They also serve to dislodge copulatory plugs deposited by other males during previous matings (Wallach and Hart 1983). The role played by sexual selection in the evolution of multiple intromission copulatory patterns is also apparent in some primate species. In various macaques and baboons, red colobus monkeys, and red uakaris, males make a series of mounts with intromission and pelvic thrusting prior to ejaculation. All these monkeys have multi-male/multi-female mating systems, in which males have large testes and engage in sperm competition (Dixson 1998*a*). However, human beings, like male gorillas, chimpanzees, mandrills, talapoins, howlers, and many others, usually ejaculate during a single intromission with pelvic thrusting. This is a relatively simple copulatory pattern which is much more widespread among the primates than the multiple intromission pattern. Single intromission copulatory patterns may equally occur in species which commonly

engage in sperm competition (e.g. chimpanzees and bonobos) and those which do not (e.g. owl monkeys and marmosets).

4. The capacity to have *multiple ejaculations* is the fourth criterion used in Dewsbury's scheme (Figure 5.8).

It is usual for males to exhibit a period of reduced sexual arousal once orgasm has occurred; this is the *post-ejaculatory refractory period*. Dewsbury pointed out that whereas males of some species cease to copulate for extended periods, others are capable of mating again and of having a second ejaculation within 60 min. Yet this criterion, which results in the division of eight potential copulatory patterns into sixteen, is highly problematic. Among the primates, for example, Dewsbury and Pierce (1989) could identify only four species, out of thirty-three considered, which were not capable of multiple ejaculations. Even these four examples are contentious, however, and in two cases (*Otolemur crassicaudatus* and *Cercopithecus* (*Chlorocebus*) *aethiops*) it is likely that ejaculation can occur more than once per hour (Dixson 1998*a*). The problem is that almost all primate species which have been studied adequately are capable of ejaculating more than once per hour, at least when they are observed in captivity. The post-ejaculatory refractory period is more labile in monkeys and apes than in many rodents and other mammals commonly studied in the laboratory. Thus, classifying humans, chimpanzees, marmosets, and rhesus monkeys as multiple ejaculators serves no useful purpose, as it obscures very real differences which exist between them as regards copulatory frequencies in relation to sperm competition. The evolution of copulatory frequencies in humans and in other primates will be considered in the next section.

Unfortunately, none of the four criteria defined above are adequate to address the question of *copulatory durations* and their evolutionary basis. Whilst it is true that species which exhibit a genital lock tend also to mate for extended periods, it is the case that extended matings also occur in many mammals which lack a genital lock. Intromission duration is an important trait and cannot be ignored if we wish to understand the evolution of mammalian copulatory patterns. It has been argued that long intromission durations originated in nocturnal, ancestral mammals and may represent a conservative trait for the Class Mammalia as a whole (Ewer 1968; Eisenberg 1977; 1981). Traditionally, mammals have also been classified as being either *induced ovulators* or *spontaneous ovulators* (Asdell 1964). One notion uniting the evolution of copulatory and ovulatory mechanisms is that induced ovulation may also represent a primitive trait among mammals. Thus, prolonged matings, in early nocturnal mammals, may have been adaptive in stimulating the surge of luteinizing hormone required to trigger ovulation (Eisenberg 1981). It has to be kept in mind, however, that among extant mammals which are induced ovulators, some species copulate only for brief periods (e.g. the rabbit and domestic cat), whilst others engage in prolonged intromissions (e.g. the ferret and other mustelids). Some authors stress that rather than a rigid dichotomy a continuum exists among mammals in the expression of induced and spontaneous ovulation (Conaway 1971; Weir and Rowlands 1973), so that both mechanisms may be expressed within a single species (e.g. among hystricomorph rodents: Weir 1974).

I emphasize that these interesting hypotheses still remain largely untested, due in part to the limited information on relationships between copulatory stimuli and ovarian physiology in the 4,500 species of extant mammals. For an authoritative discussion of mating-induced ovulation and its evolution in mammals, the reader is referred to the review by Kauffman and Rissman (2006). In order to make Dewsbury's scheme more applicable to assessments of evolutionary questions, it is important to include the presence (or absence) of prolonged intromission as a copulatory trait. Therefore, in Figure 5.9 prolonged intromission (defined as an intromission lasting for more than 3 min) replaces the largely redundant category of 'multiple ejaculations'. Sixteen copulatory patterns are thus defined, the same number as in Dewsbury's (1972) original classification. Just four of these copulatory patterns (nos. 3, 10, 11, and 12) are represented among extant primates, as listed in Table 5.1.

Prolonged single intromission patterns of mating, including a genital lock (pattern no. 3) or lacking a lock (pattern no. 11) occur in various nocturnal

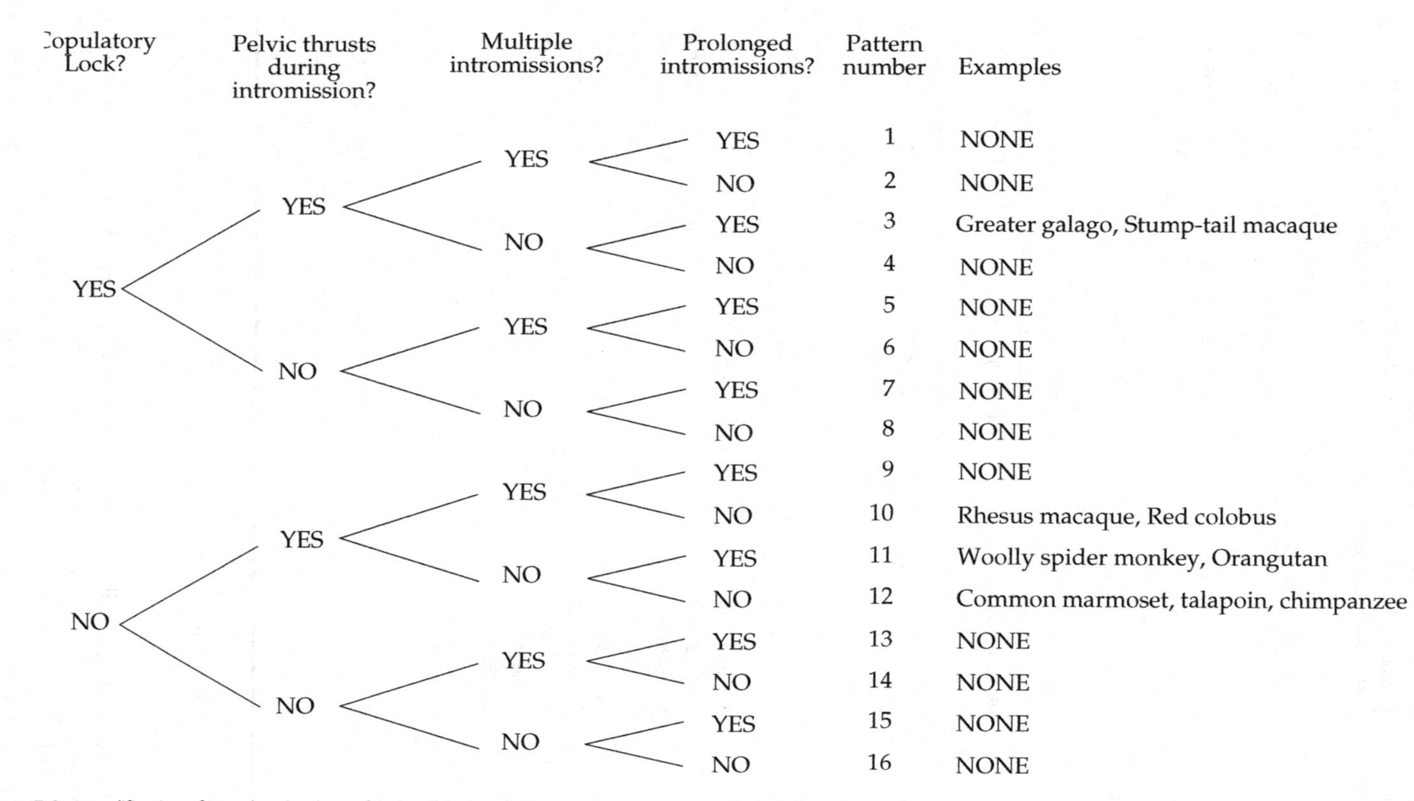

Figure 5.9 A modification of Dewsbury's scheme for classifying masculine copulatory patterns, to include intromission duration as an important evolutionary trait.
Source: After Dixson (1998a).

Table 5.1 Classification of primate copulatory patterns, according to the criteria laid down in Figure 5.9

Pattern no. 3	10	11	12
Possible genital lock: a single prolonged intromission and pelvic thrusts	No genital lock: multiple brief intromissions and pelvic thrusts	No genital lock: a single prolonged intromission and pelvic thrusts	No genital lock: a single brief intromission and pelvic thrusts
Prosimians			
Otolemur garnettii	None described	*Microcebus murinus*	*Lemur catta*
O. crassicaudatus		*Galago moholi*	*Varecia variegata*
Galagoides demidoff		*Arctocebus calabarensis*	*Cheirogaleus major*
Daubentonia madagascariensis		*Loris tardigradus*	*Tarsius bancanus*
		Nycticebus coucang	
New World monkeys			
None described	*Cacajao calvus*	*Lagothrix lagotricha*	*Callithrix jacchus*
	*Saimiri sciureus**	*Ateles belzebuth*	*Saguinus oedipus*
	Leontopithecus rosalia	*A. geoffroyi*	*Cebuella pygmaea*
		A. fusciceps	*Callimico goeldii*
		Brachyteles arachnoides	*Callicebus moloch*
			Aotus lemurinus
			Cebus nigrivittatus
			Alouatta palliata
Old World monkeys			
Macaca arctoides	*Colobus badius*	None described	*Nasalis larvatus*
	Macaca silenus		*Miopithecus talapoin*
	M. nemestrina		*Erythrocebus patas*
	M. nigra		*Chlorocebus aethiops*
	M. maurus		*Mandrillus sphinx*
	*M. fascicularis**		*Cercocebus albigena*
	M. mulatta		*Theropithecus gelada*
	M. fuscata		*Papio anubis*
	M. thibetana		*Macaca radiata*
	*Papio ursinus**		*M. sylvanus*
	P. hamadryas		
The apes and man			
None described	None described	*Pongo pygmaeus*	*Pan trogolodytes*
			P. paniscus
			Gorilla gorilla
			Homo sapiens

*Pattern no. 12 also occurs in these species.

Source: Data are from Dixson (1998a) and references cited in the text.

prosimians (such as galagos, angwantibos, lorises, and mouse lemurs), and in a few large-bodied diurnal anthropoids, most of which are highly arboreal (e.g. spider monkeys, woolly monkeys, and orangutans). The presence of prolonged patterns of intromission in nocturnal prosimians may indeed represent a primitive trait, retained from nocturnal, arboreal stem forms of the Order Primates. Many of the extant species are non-gregarious, and the sexes associate only infrequently, during the female's oestrus. Prolonged patterns of mating may be advantageous on several levels: as a mate-guarding tactic by males; as a means of inseminating maximum numbers of sperm and gaining an advantage in sperm competition; or to impart tactile stimuli which influence female fertility (Dixson 1995*a*; 1998*a*). Natural selection as well as sexual selection may have played a role in the evolution of prolonged matings (patterns 3 and 11) in the nocturnal primates. Such prolonged matings are absent in all small-bodied *diurnal* primates, and virtually absent in terrestrial species (the exception being the stump-tail macaque: Table 5.1). It is thought that predation risk may have selected against prolonged matings in such cases (Dixson 1991).

Among the anthropoids, the occurrence of prolonged patterns of intromission almost certainly represents a derived condition, rather than a primitive retention from nocturnal ancestors. In the atelines (spider monkeys, muriquis, and woolly monkeys), it is also associated with the occurrence of the specialized 'leg lock' copulatory pattern described earlier, and with multiple partner matings and sperm competition in social groups. The case of the orangutan is intriguing, as one wonders whether the dispersed nature of their social system, and infrequent contact between the sexes, may have facilitated the evolution of prolonged copulations during their consortships. All the primates are considered to be spontaneous ovulators. However, given the possibility of gradations between induced and spontaneous ovulation in mammals, it would be premature to deny that this might occur in orangutans and some other species. In greater galagos and ringtailed lemurs, for example, copulation is associated with a shortening of the female's receptive period (Dixson 1995*a*). In both these prosimians, males have very large penile spines. Stockley (2002) has produced evidence that the possession of such spines in male primates is associated with shorter durations of sexual receptivity in females.

Pattern number 10, which involves multiple intromissions with pelvic thrusting by males, occurs in a number of monkeys in which the sexes live in large social groups and in which large, relative testes sizes and sperm competition pressures are pronounced. Examples include many of the macaques and baboons (Table 5.1). This is *not intended to imply* that multiple intromission patterns of copulation are the norm in multi-male/multi-female primate groups, however. The reason why this pattern is present in some species, but not in others where females engage in multiple-partner matings, has not been explained. It is the case, however, that where multiple intromission copulations (pattern no. 10) and prolonged single intromissions patterns (nos. 3 and 11) do occur, they are much more likely to be found in species which engage in sperm competition (i.e. in multi-male/multi-female and dispersed mating systems) than in monogamous and polygynous species (Figure 5.10).

In the fourth and final primate copulatory pattern, males make relatively brief single intromissions with pelvic thrusting to attain ejaculation. This pattern (no. 12 in Figure 5.9 and Table 5.1) is shared by a large number of primate species, including some which have multi-male/multi-female mating systems (e.g. *Pan*, *Mandrillus*, and some of the macaques) as well as polygynous forms (e.g. *Gorilla*, *Theropithecus*) and monogamous New World monkeys, such as *Callithrix jacchus* and *Saguinus oedipus*. This pattern is therefore widespread among *diurnal primates* and appears to have been relatively little affected by sexual selection. It is more likely that natural selection and the requirement to avoid predators has led to shortening of intromission durations among the diurnal primates as a whole (Dixson 1991). Intromission durations in the pattern number 12 group of species vary from less than 10 s, (e.g. in *C. jacchus* and *P. troglodytes*), up to 1.5 min in the gorilla, and 2 min or less in human males. Further examples of brief intromission durations are given in Table 5.2, together with those exhibited by primates in which copulations are usually much more prolonged.

There are convincing reasons to include *H. sapiens* in the pattern number 12 category, as the fundamental human copulatory pattern involves relatively brief intromission times. This conclusion emerges from a consideration of the data collected by Kinsey, Pomeroy, and Martin (1948), as well as from cross-cultural studies that indicate that a large degree of cultural conditioning is involved in those societies where men copulate for extended periods.

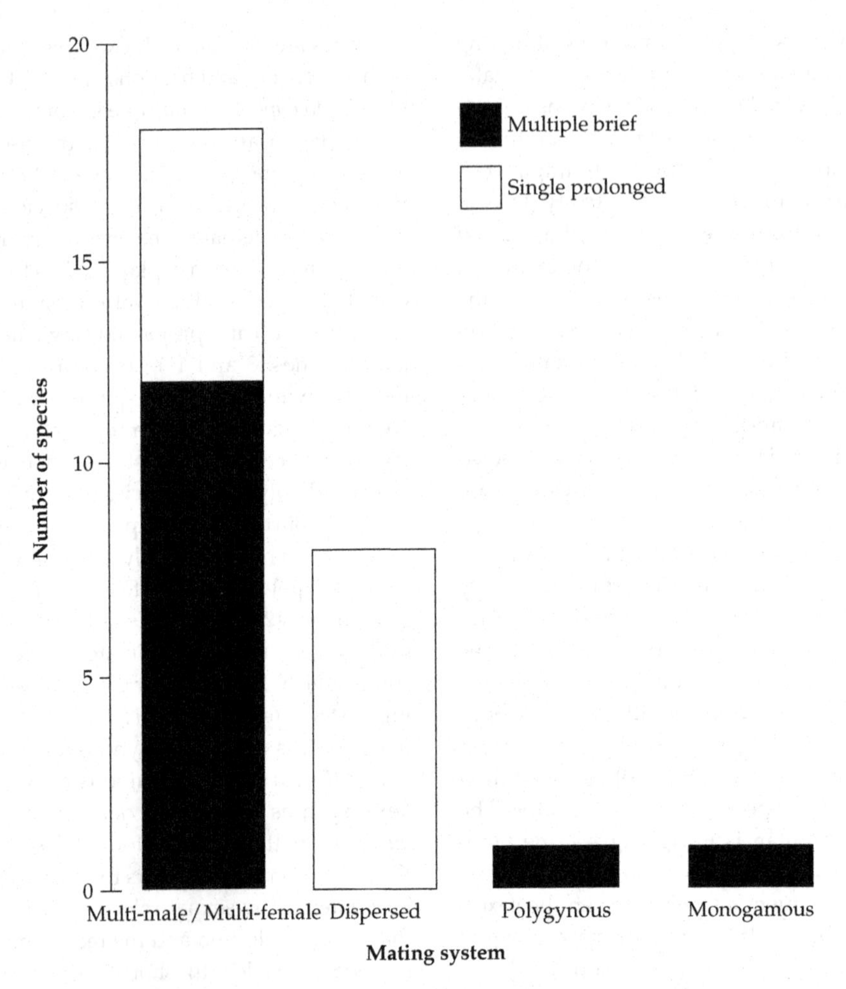

Figure 5.10 Relationships between the occurrence of complex copulatory patterns (involving multiple brief intromissions, or single prolonged intromissions) and mating systems in primates.

There is also the interesting fact that the propensity for males to ejaculate quickly is identified as a common problem in clinical studies, where it is usually referred to as *premature ejaculation*. This phenomenon is more likely to occur in younger men. Masters and Johnson (1966), for example, comment that

> Problems of premature ejaculation disturbed the younger members of the study-subject population. These fears of performance were not associated with problems of erection; rather they were directed towards the culturally imposed fear of inability to control the ejaculatory process to a degree sufficient to satisfy the female partner.

That these fears are partly culturally induced is supported by the observation that they tend to be greater in men who are more educated and aware that rapid ejaculation should be viewed as a problem (Masters and Johnson 1966; 1970). Bancroft (1989) comments that 'The impulsive, inexperienced young man may ejaculate rapidly and enjoy the experience.' Control develops as a result of greater experience, and 'because of the dampening effects of the ageing process and also because of the lessening of novelty that comes particularly with a stable relationship.' Bancroft also points out, however, that 'It may be that for some men this

Table 5.2 Intromission durations in primates: distinguishing between single brief intromission (SBI) and single prolonged intromission (SPI) copulatory patterns

Species	Intromission duration (s)	Copulatory pattern
Tarsius bancanus	60–90	SBI
Cebuella pygmaea	4–10	SBI
Callithrix jacchus	Mean 5 or 6	SBI
Alouatta palliata	Mean 32	SBI
*Saimiri sciureus**	<30	SBI*
Nasalis larvatus	Mean 41.5	SBI
Miopithecus talapoin	<60	SBI
Cercopithecus aethiops	<60	SBI
Cercocebus albigena	Mean 13.9	SBI
Mandrillus sphinx	<60	SBI
Macaca radiata	<60 Mean 10	SBI
Pan troglodytes	Mean 7.0 Mean 8.21	SBI
Pan paniscus	Mean 15.3	SBI
Gorilla gorilla	med 96	SBI
Homo sapiens	<120	SBI
Microcebus murinus	>180	SPI
Daubentonia madagascariensis	Mean 3720	SPI
Otolemur garnettii	780–15600	SPI
G. moholi	420–720	SPI
Galagoides demidoff	3600	SPI
Arctocebus calabarensis	240†	SPI
Loris tardigradus	600–1020	SPI
Nycticebus coucang	180–420	SPI
Brachyteles arachnoides	100–496	SPI
Ateles belzebuth	480–1500	SPI
A. paniscus	Mean 600	SPI
Lagothrix lagotricha	180–840	SPI
Macaca arctoides	>180	SPI
Pongo pygmaeus	med 840 max 2760	SPI

*Multiple intromissions also occur in *Saimiri sciureus*.
†One observed copulation only.

Source: From Dixson (1998*a*), and based upon multiple sources cited there.
Abbreviations: med = median duration; max = maximum duration.

aspect of autonomic function is peculiarly difficult to control—they have over-excitable and too easily triggered "ejaculation centres."'

Let us assume for a moment that the propensity of many human males to ejaculate quickly, once intromission occurs, is not abnormal but is indicative of an evolutionary background of relatively brief sexual intercourse in human ancestors. Given that the australopithecines, as the precursors of humans, were relatively small bodied, largely terrestrial, and diurnal in habit, it is possible that natural selection (predation pressure) also selected against prolonged matings. However, with the evolution of the genus *Homo*, the tendency for intercourse to occur in greater seclusion, at night, and in more secure surroundings might then have facilitated the cultural emergence of lengthier patterns of intercourse, and associated patterns of foreplay between the sexes. Prolonged sexual intercourse is probably not the result of sexual selection during human evolution, but represents instead a culturally based phenomenon. Many aspects of human behaviour have been modified by culture; it should not surprise us that the duration of sexual intercourse might be determined by cultural factors. Among the non-human primates, by contrast, the prolonged type of copulatory pattern has been greatly influenced by both sexual and natural selection. None of the non-human primates which copulate for extended periods have monogamous or polygynous mating systems, however, and this fact distinguishes them from the human condition.

Copulatory frequencies

There is a popular misconception that human beings are in some way exceptional, as regards frequencies of sexual intercourse; after all, our species is sexually active all year round and throughout the menstrual cycle. Many other mammals, by contrast, are seasonal breeders and mating is largely confined to the peri-ovulatory phase of the oestrus cycle. The next chapter will examine the concept of 'loss of oestrus' during human evolution. Here, we are concerned primarily with frequencies of copulation in *Homo sapiens*, as compared to the non-human primates, and whether sexual selection might have played a role in the evolution of copulatory frequency.

Studies conducted in the USA, UK, and in China provide useful information about frequencies of intercourse, based upon large samples of subjects. In China, Liu et al. (1997) recorded that the average frequency of intercourse for 7,602 married men was 4.8 times per month. There was considerable individual variability, however, and rates were slightly higher in village dwellers (5.43 times per month), as compared to married couples living in the city (4.66 times per month). As expected, age was important, so that the highest copulatory frequencies occurred in men who were less than 25 years old. Kinsey, Pomeroy, and Martin's (1948) studies in North America are more informative in this respect and frequencies of intercourse were higher than those recorded in China. Thus, married North Americans in the age band from 16–20 years had intercourse 3.75 times *per week*, on average, but this declined progressively, to an average frequency of 0.83 per week in men aged 56–60 years (Figure 5.11). For each age group, individual variations in frequencies of sexual activity were marked. The maximum weekly frequency of intercourse reported was twenty-nine times (for a man in the 21–25 year age group), declining to fourteen times (41–50 years), and five times per week for one man in the 61–65 year age group.

A survey conducted in the UK by Wellings and her colleagues (1994) also showed that frequencies of intercourse declined with age, from a median value of twice each week (in men aged 16–24 years) to 0.75 times per week in middle age (45–49 years). Although the majority of sexual activity took place between married partners, and those in long-term relationships, Wellings et al. also found that 'among single people, more than a quarter of men (28.1 per cent) and close to one-fifth of women (17.5 per cent) reported two or more partners in *the last year* while 13.1 per cent of men and 6.1 per cent of women reported more than two partners, a pattern that contrasts markedly with that of married individuals.' The tendency for males to have a larger number of partners over time is interesting and probably does reflect a biological predisposition for males to be more sexually active. As an aside, I mention here a most interesting hypothesis advanced by Symons (1979) regarding high frequencies of sexual activity with multiple partners among some male homosexuals. Symons suggests that this behaviour may reflect the masculine predisposition for high levels of sexual outlet, freed from the feminine constraints which regulate copulatory frequency within heterosexual relationships.

Although it is difficult to compare the results of large-scale surveys conducted in China, North

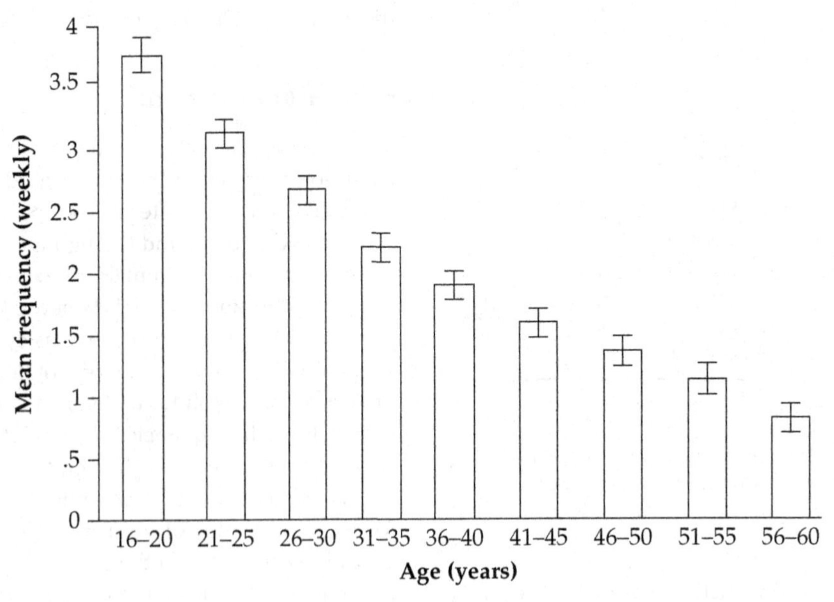

Figure 5.11 Age-related changes in weekly frequencies of intercourse in married men. Data are from the USA.
Source: From Kinsey, Pomeroy, and Martin (1948).

America, and Europe directly, because they used different methodologies, the general conclusion reached is that younger married men have intercourse an average of two to four times each week. Sexual activity may be higher in some individuals, including single men, but human beings exhibit a strong tendency to form lasting relationships, so that 'availability of a regular partner continues to be a dominating influence on coital frequency' (Wellings et al. 1994). In some indigenous cultures, intercourse is reputed to be more frequent than cited above, but the evidence is largely anecdotal, and is probably unreliable. Ford and Beach (1951) noted that 'in most of the societies on which evidence is available every adult normally engages in heterosexual intercourse once daily or nightly during periods when coitus is permitted.' Extreme examples cited by these authors included the Chagga of East Africa (ten times per night), the Aranda of Australia (three to five times per night), and the Lepcha of Eurasia, of whom it was said that 'when they were first married they would copulate with their wives five or six, even eight or nine times in the course of the night, though they would then be tired the next day.' In all these cases, the highest frequencies of sexual intercourse almost certainly refer to younger men. It is far from clear, however, whether these reports concern intercourse with orgasm and ejaculation. In terms of evolution and reproduction this is an important point. Sperm counts rapidly diminish if men engage in multiple copulations. Thus the reproductive system of the human male is not adapted to sustain normal sperm counts under conditions where multiple daily copulations occur.

How do the frequencies of sexual intercourse in human societies compare with those reported for the non-human primates? Is there any relationship between mating frequency, mating systems, and sperm competition? In many monkeys and apes, females may copulate at times other than when conception is likely and there is no restricted period of sexual receptivity or oestrus (this topic is discussed in the next chapter). This raises the question of how males might allocate their sexual activity to secure a reproductive advantage. In those species which are primarily monogamous, such the gibbons, owl monkeys, and marmosets, opportunities for females to mate with multiple males are more limited. The same is true of highly polygynous species, such as the gorilla and the hamadryas baboon, where a single male accounts for most of the copulations within a one-male unit. This is not say that monogamous or polygynous primate mating systems are sexually 'watertight' and exclusive. Females can, and sometimes do, mate with other partners. However, incidences of such additional copulations are much lower than in the multi-male/multi-female social groups of macaques, chacma baboons, chimpanzees, and bonobos. Under these conditions males engage in a variety of *alternative tactics* which may enhance their mating success and improve their chances of siring offspring. One option open to a male is to monopolize access to a female when she is most likely to ovulate, and to copulate more frequently during this potential fertile period. The consortships of chacma baboons with females at maximum sexual skin swelling provide an example of such behaviour (Hausfater 1975). However, two or three males sometimes form a coalition to oust a dominant male from consortship with an attractive female (Bercovitch 1988). Alpha male chimpanzees engage in 'possessive' tactics to monopolise sexual access to a particular female. Alliance formation between males also plays an important role in the determination of social rank (Goodall 1986). High-ranking male mandrills follow and mate-guard individual females as they develop their sexual skin swellings during the annual mating season (Dixson, Bossi, and Wickings 1993). It is the female which dictates many of the male's movements under these circumstances; the male follows her, sits close by, and chases away potential rivals. Female monkeys are not passive objects of inter-male competition and sexual interest, for they too engage in mate choice and may exhibit preferences for certain individuals such as younger males or novel partners. In Barbary macaques, subadult males sometimes try to gain proximity to females as dusk approaches; they are more likely to avoid competition with higher ranking adults at this time (Paul et al. 1993). Sub-adult male mandrills, which are smaller and more agile than fully developed adult males, sometimes associate with females in the tree-tops and as night falls. In this way they gain copulations which are not possible during the daylight hours.

Thus, in many primates which have multi-male/multi-female or dispersed mating systems, females may copulate with a succession of partners during the fertile period. Examples of such multiple-partner matings by females, which result in sperm competition between males, are provided in Table 5.3. We have seen that certain features of copulatory behaviour, such as prolonged single intromission and multiple brief intromission copulatory patterns, are more common in such cases, where sperm competition pressures are greatest. It is also logical to suggest that, in these circumstances, sexual selection might favour males that can sustain high ejaculatory frequencies, as well as high sperm counts. That this is indeed likely to be the case is shown in Table 5.4, which lists frequencies of copulation in primates which have multi-male/multi-female mating systems, as compared to those which are primarily monogamous or polygynous (Dixson 1995*b*). Twenty species are represented including *Homo sapiens*, so that the

Table 5.3 Evidence for multiple-partner matings by females in multi-male/multi-female primate groups

Species	Type of study	Numbers of males mated per female	Source
Prosimians			
Lemur catta	Field study	3–5	Koyama (1988)
Propithecus verreauxi	Field study	1–2	Richard (1976)
New World monkeys			
Alouatta palliata	Field study	>1	Jones (1985)
Ateles paniscus	Field study	>1	Van Roosmalen and Klein (1988)
Brachyteles arachnoides	Field study	4	Milton (1985)
Cebus apella	Field study	1–7	Janson (1984)
Old World monkeys			
Chlorocebus aethiops	Field study	>1	Andelman (1987)
Miopithecus talapoin	Field study	>1	Rowell and Dixson (1975)
Papio ursinus	Field study	1–3	Hall (1962)
Papio cynocephalus	Field study	>1	Hausfater (1975)
Papio anubis	Field study	>1	Scott (1984)
		1–3	Smuts (1985)
Cercocebus albigena	Field study	>1	Wallis (1983)
Mandrillus sphinx	Semi-free ranging	1–3	Dixson et al. (1993)
Macaca mulatta	Field study	1–11 (mean 3.2)	Conaway and Koford (1965)
	Field study	Mean 3.8	Loy (1971)
	Field study	1–9 (mean 3 or 4)	Manson (1992)
Macaca fuscata	Semi-free ranging	>1	Wolfe (1984)
	Captive group	3–19 (mean 10.7)*	Hanby et al. (1971)
Macaca fascicularis	Field study	>1	Van Noordwijk (1985)
Macaca radiata	Captive group	>1	Glick (1980)
Macaca sylvanus	Field study	3–11 (mean 6.0)	Taub (1980)
	Field study	5–11 (mean 7.12)	Ménard et al. (1992)
Macaca nemestrina	Captive group	1–5 (mean 2.5)	Tokuda, Simms, and Jensen (1968)
Apes			
Pan troglodytes	Field study	8+	Goodall (1986)
	Field study	>1	Hasegawa and Hiraiwa-Hasegawa (1990)
Pan paniscus	Field study	>1	Furuichi (1987)
	Field study	>1	Kano (1992)

*In this one case, data refer to numbers of male partners during a single mating season and not necessarily during a single ovarian cycle.

Notes: Cases where multipartner matings occur but the number of males involved is unknown are indicated by >1. Data in this table refer to matings during a single ovarian cycle, including the presumptive period of ovulation. Copulations occurring during pregnancy are not considered.

Source: Modified from Dixson (1997).

Table 5.4 Hourly frequencies of ejaculation in primates with multi-male/multi-female (MM), polygynous (PG), or monogamous (MG) mating systems

Species	Study type	No. of males	Ejaculation Range	Ejaculation Mean	Mating system
*Lemur catta**	F	5	1–1.75	1.35	MM
Callithrix jacchus	CG	6	0.095–0.202	0.155	MG
Aotus lemurinus	CG	5	–	0.07	MG
Papio ursinus	F	–	–	0.75	MM
P. cynocephalus	F	9	0.41–1.46	0.83	MM
*Mandrillus sphinx**	SF	6	0–3	0.85	MM
*Macaca mulatta**	CG	–	0–0.69	0.38	MM
M. fascicularis	CG	–	0–1.33	0.69	MM
*M. radiata**	CG	–	0.44–2.44	0.97	MM
M. arctoides	CG	18	0–11	1.91	MM
*M. sylvanus**	F	21	0–4.14	2.28	MM
M. nigra	CG	1	–	0.63	MM
*M. thibetana**	F	7	0–0.9	0.45	MM
*Erythrocebus patas**	F	5	0.19–0.74	0.43	MM†
Theropithecus gelada	F	–	–	0.18	PG
Hylobates syndactylus	F	–	–	0.005	MG
Pan paniscus	F	5	0.19–0.42	0.27	MM
P. troglodytes	F	–	0.03–1.14	0.52	MM
Gorilla gorilla	F	3	0.32–0.38	0.35	PG
Homo sapiens	NA	3342	0–0.119	0.025	PG/MG

*a species with pronounced mating seasonality.
†*E. patas* is classified as a MM species because in this study influxes of additional males during the mating season created a multimale–multifemale group.
Data on *Mandrillus sphinx* include the author's unpublished observations and the mean frequency relates to the highest-ranking male in the group. The siamang (*Symphalangus*) is listed here as *Hylobates syndactylus*. Data are from Dixson (1995*b*, 1998*a*).
Abbreviations: F = Field study; CG = captive group(s); SF = semi-free-ranging group; NA = North American population.

reader may compare hourly frequencies of ejaculation in healthy adult males with access to attractive and receptive females. In those cases where there is an annual mating season, for example in *Lemur catta*, then the data refer only to the mating period and not to the rest of the year when sexual interactions are infrequent. An obvious pattern emerges from consideration of these data. Primates which have multi-male/multi-female mating systems, in which sperm competition is likely to be high, also have the highest ejaculatory frequencies. In six such genera, the average hourly ejaculatory rate is 0.81, whilst in members of the six monogamous and polygynous genera it is only 0.13 ejaculations per hour. Human beings have an average rate of just 0.025 ejaculations per hour, based upon the Kinsey data, which is comparable to the situation in monogamous forms such as the siamang and owl monkey. Ejaculatory frequencies are also low in polygynous species, such as the gelada (0.18/h) and gorilla (0.35/h). In reality, male gorillas are far less sexually active throughout the year than most married men. This is because female gorillas only enter breeding condition every 4 or 5 years. For the rest of the time, they are pregnant or lactating and relatively unattractive sexually to the dominant silverback in their group. When a female is likely to ovulate, however, the male can achieve the frequencies of copulatory activity cited in Table 5.4. Such low frequencies in monogamous and polygynous primates contrast with exceptionally high rates of sexual activity in multi-male/multi-female species. The stump-tail macaque provides the most extreme example. In a video-taping study of pairs of stump-tails, one male ejaculated with his female partner

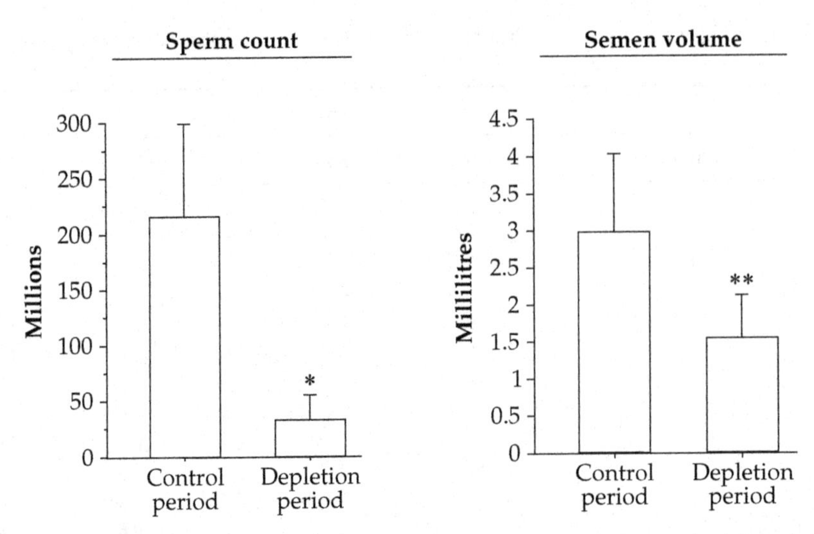

Figure 5.12 Effects of ejaculatory frequency upon sperm counts and semen volumes in men. During the *control period*, ejaculations occurred twice weekly. During the ensuing 10 day *depletion period*, ejaculatory frequencies increased to two to three times daily. Significant decreases in sperm counts and semen volumes were measured during the depletion phase. $^{*}P < .05$; $^{**}P < .01$. Based on data (means ± SEM) for six subjects.
Source: From Freund (1963).

thirty-eight times in 24 h. This stump-tail showed a higher frequency of copulatory activity in a single day than was recorded, on a weekly basis, for any of the 3,342 North American men interviewed by Kinsey and his colleagues.

Comparative studies of primate copulatory behaviour indicate that frequencies of human sexual intercourse are consistent with a monogamous or polygynous mating system. Although some individuals may have much higher frequencies of intercourse, they are not representative of humanity as a whole. 'Recreational sex' is an option in some human populations, for those fortunate individuals who are well off, well fed, and comfortably housed. However, for the earliest representatives of the genus *Homo*, living a challenging existence with minimal technological advantages, sex was almost certainly a less frequent activity. Moreover, it occurred primarily within established relationships, rather than with multiple partners, as in chimpanzees. It is for this reason also that the ability to maintain high sperm counts after repeated ejaculations is so much better developed among multi-male/multi-female primates, as compared to monogamous or polygynous forms. In a study conducted by Marson et al. (1989), semen samples were collected (by masturbation) from adult male chimpanzees, six times during a 5 h period. Total sperm counts in the initial sample averaged 1278 ± 872 million, decreasing to 587 ± 329 million in the sixth sample, collected 5 h later. Thus, even after multiple ejaculations, sperm counts in male chimpanzees greatly exceed those produced by most human males. The pronounced depletions in extra-gonadal sperm reserves and decreases in semen volume which occur in men as a result of ejaculating two or three times daily, over a ten day period, are shown in Figure 5.12. These data taken from the work of Freund (1963) show that sperm counts declined to an average of just 16.1 per cent of pre-depletion levels in the six men who took part in this study. These profound decreases in sperm counts confirm that the human reproductive system is not adapted to adjust and sustain sperm numbers in response to the kinds of sexual activity which are to be expected in primates which engage in sperm competition.

Conclusions

Variable as human copulatory positions may be (due to cultural factors) they are fundamentally the same as those which occur in non-human primates. The, ventro-ventral, or 'missionary position' and female superior copulatory positions of humans are homologous with patterns described for the great apes. Elsewhere among the primates, distinctive copulatory postures are restricted to particular phylogenetic groupings. Examples discussed in this chapter include the inverted copulatory positions of the lorisines of Africa and Asia, and the leg-lock mating postures of some of the ateline monkeys in South America.

The evolution of face-to-face copulatory positions in hominoids was probably facilitated by several factors, including increased brain size and intellectual capacities, as well as the communicatory advantages of continued eye contact between partners during mating. However, another possibility discussed in this chapter is that anatomical specializations of the shoulder joint and suspensory patterns of arboreal locomotion in ancestral apes may have facilitated the emergence of face-to-face copulatory postures. These patterns may then have been retained and modified among the progressively more terrestrial australopithecines which gave rise to the genus *Homo*. An alternative hypothesis, which ascribes the evolution of ventro-ventral copulation in humans to enhancement of clitoral stimulation and facilitation of female orgasm, is examined and rejected on the basis of information discussed in Chapter 4.

Sexual selection has affected patterns of intromission and pelvic thrusting during copulation in non-human primates. Specialized copulatory patterns, involving multiple brief intromissions or a single prolonged intromission, occur most frequently in primate species where males exhibit large relative testes sizes and females engage in multi-partner matings. Sperm competition and cryptic female choice may have moulded the evolution of these specialized copulatory patterns, as well as complex genital morphologies in such species. However, there is very little evidence that sexual selection has influenced the evolution of human copulatory patterns. Although prolonged intercourse may be facilitated by learning and encouraged by cultural preferences, the fundamental human pattern is mostly likely to have been relatively brief, involving a single intromission with pelvic thrusting to achieve ejaculation.

Frequencies of copulation are also greatest in those primate species in which females mate with multiple partners during the fertile period. Although data on ejaculatory frequencies in monkeys and apes are limited, they are sufficient to show that males in multi-male/multi-female groups copulate much more frequently than those which have polygynous/monogamous mating systems. Large-scale surveys of human sexual behaviour conducted in North America, Europe and China confirm that for the majority of couples, frequencies of intercourse are commensurate with those that occur in polygynous or monogamous primates. Moreover, experimental studies that require men to ejaculate at artificially high daily frequencies clearly show that human extra-gonadal sperm reserves rapidly become depleted under such conditions (see Figure 5.12). Men, unlike males of multi-male/multi-female species such as chimpanzees or macaques, are not physiologically adapted to sustain optimal sperm counts under conditions of sperm competition.

The comparative studies of sexual behaviour discussed in this chapter support the conclusions reached in Chapters 2, 3, and 4 on the basis of comparative analyses of mammalian reproductive anatomy and physiology. Humans have evolved from forms which had primarily polygynous or monogamous mating systems. Chapter 7 will address the relative importance of polygyny versus monogamy in ancestral hominids. First, however, the (often vexed) question of oestrus in primates, and the 'loss of oestrus' during human evolution will be examined in Chapter 6.

CHAPTER 6

The Oestrus That Never Was

'The Impetus, the Lyrical oestrus, is gone.'
Fitzgerald

A recurrent theme in discussions about the origins of human sexual behaviour concerns the concept of oestrus, as it applies to female mammals, and to the presumptive *loss of oestrus* during human evolution. Oestrus, as originally defined by Walter Heape (1900), denotes a period of heightened sexual activity in female mammals, in association with the likelihood that ovulation will occur and that mating will result in conception. Heape noted that 'it is during oestrus, and only at that time, that the female is willing to receive the male.' The term derives from the Greek word for the gadfly, whose bite drives animals into a frenzy; the implication being that females are in a state of sexual frenzy or *in heat* during oestrus. Ovarian hormones (and especially oestrogen and progesterone) control the onset and duration of oestrus, so that ovariectomy is usually associated with loss of the female capacity to display sexual receptivity in many mammals. This is the case, for example, in rodents such as the rat, in carnivores (e.g. the cat and ferret), in rabbits, ungulates (such as cattle and sheep), and in many others. It is not, however, a universal attribute, applicable to all mammalian groups or species.

The choice of terminology can have profound effects upon the progress of science, just as in other areas of human affairs. Clearly, women do not come into heat; nor do they display a restricted period of sexual interest and receptivity when ovulation is likely to occur. Human sexual behaviour differs in important ways from the behaviour that occurs during the restricted oestrous periods of rats, sheep, and dogs. In human beings, copulations may take place throughout the monthly menstrual cycle, after ovarian cycles cease at the menopause, or as a result of ovariectomy. This is not to imply that the menstrual cycle and ovarian hormones have no effects upon human sexual behaviour; I shall return to this subject later in this chapter. However, since women do not experience circumscribed periods of sexual receptivity, anthropologists have sought to explain how and why, there has been *loss of oestrus* during human evolution. Nor is it obvious when ovulation is likely to occur during the human menstrual cycle, as women do not exhibit external cues, such as the oestrogen-dependent sexual skin swellings found in chimpanzees and some Old World monkeys. Thus, in addition to searching for the reasons why oestrus was lost in human ancestors, scientists have sought to explain the origins of *concealed ovulation*. Symons (1979), for example, proposed that 'estrus was lost some time after humans last shared a common ancestor with any living non-human primate'. This would mean that oestrus occurred in the common ancestors of *Homo sapiens* and the modern day African apes, and that it was lost during the evolution of the australopithecines or during the emergence of the genus *Homo*. A number of selective forces have been proposed to explain these events. It has been argued that loss of a circumscribed period of oestrus, or signals of impending ovulation, would have promoted *pair bonding* in human ancestors. Increased opportunities for copulation within the pair bond could have benefited male partners in terms of greater certainty that they had sired any resulting offspring. Such pair-bonded males might also have increased their fitness by providing greater resources to their mates and offspring; an important selective strategy given the relative helplessness of human infants, slow rates of development, and extended inter-birth intervals in the human species (e.g. Morris 1967; Alexander and Noonan 1979; Lovejoy 1981; Fisher 1982). It has also been posited that loss of oestrus and concealment of ovulation during human evolution might have provided females with increased opportunities to mate with additional males (i.e. to engage in extra-pair copulations), and thus to *confuse paternity* and reduce the risks that males might engage in infanticide (Hrdy 1981). However, the most peculiar line of reasoning that I have encountered in relation to explaining the concealment of ovulation during human evolution is due to Burley (1979). She advanced the hypothesis that concealment of ovulation would have prevented women from avoiding matings during the fertile period; their motivation for such avoidance would have been to prevent unwanted pregnancies and to reduce the (certainly large) physiological costs involved in pregnancy, childbirth, and post-natal care. Women who developed concealed ovulation were unable to identify the fertile period, and so could not choose to avoid copulation at that time; therefore, they produced more offspring and passed on more of their genes to future generations.

The notion that sexual behaviour might play some cohesive role in primate social organization is not new. In *The Social Life of Monkeys and Apes*, published in 1932, Zuckerman advanced the view that 'reproductive physiology is the fundamental mechanism of society' and that 'the main factor that determines social grouping in sub-human primates is sexual attraction.' Zuckerman thought that monkeys and apes are sexually active all year round, or at least that the sexes are held together by sexual attraction throughout the year. This was postulated to be the physiological and behavioural basis of group life. Subsequently, it became clear that some monkey species exhibit distinct mating seasonality (as reviewed by Lancaster and Lee, as early as 1965). Yet the absence of sexual activity for large periods of the year did not preclude these monkeys from living in permanent social groups. It is peculiar that despite evidence amassed from a burgeoning program of field studies of the non-human primates conducted during the 1960s and 1970s, Zuckerman refused to change his viewpoint, which he reiterated and defended in the second edition of his book, published in 1981.

The ecological factors which might underpin the evolution of social groups in monkeys and apes were also beyond Zuckerman's scope; indeed, he stated that 'Lack of relevant information makes it impossible to discuss here the subject of the ecology of the primates' (Zuckerman 1932; 1981). Yet, research on primate behavioural ecology supports the view that the physical nature of the habitat, including the distribution of food (and seasonal fluctuations in its availability) has crucial effects upon the size and composition of social groups. Females, in particular, benefit from group membership if this enhances their ability to obtain resources for themselves and for their dependent offspring. Males, in turn, compete for access to females, which are a limiting resource in reproductive terms. As Wrangham (1979) encapsulated it, 'the entwined distribution of females and males yields the social system.' Thus, ecology and not just sexual attraction is likely to play a pivotal role in determining the sizes and compositions of primate groups. The same arguments apply to other mammals, as for example the social carnivores. As an example, lions form prides because the survival and reproductive success of individual pride members is enhanced by life within a social group. A group of lionesses is more successful than a single female could be, at holding territory, hunting larger prey, and rearing their cubs in communal 'crèches'. Males compete in order to associate with a group of females, but sexual behaviour per se is not the reason that lions live in prides (Schaller 1972; Packer, Scheel, and Pusey 1990).

Although Zuckerman greatly over-estimated the importance of sexual attractiveness and sexual behaviour in relation to the origins of primate social organization, he made valuable observations on the menstrual cycle and noted its relationship to patterns of behaviour. Adult females of all the Old World monkey and ape species exhibit menstrual cycles which are physiologically homologous with the human menstrual cycle. Zuckerman noted that baboons, macaques, and other monkeys tend to engage in sexual activity most frequently during the first half (follicular phase) of the monthly cycle, including the peri-ovulatory period, rather than during the second half (luteal phase) of the cycle. He was clear that these were not oestrous cycles, as females were 'ready to accept the advances of the male at all times, whereas the female lower mammal will mate, as a rule, only at those isolated intervals when she is in the physiological state of heat'.

Clearly, such observations should have constituted a major challenge to the development of theories about the origins of human sexuality, couched in terms of some unique loss of oestrus in the evolutionary lineage which produced *Homo sapiens*. It appears more likely that oestrus was lost in the common ancestors of the monkeys and apes, and not, as Symons (1979) proposed, after divergence of the hominid lineage from the common ancestors of human beings and the African apes. Indeed, a substantial number of publications has appeared over the years which critique the application of the concept of 'oestrus' to descriptions of sexual behaviour in the anthropoid primates, and some of these address the issue that absence of oestrus in women cannot be explained as a uniquely human trait (Rowell 1972; Hrdy 1981; Keverne 1981; Dixson 1983*a*; 1992; 1998*a*; 2001; Loy 1987; Goy 1992; Pawlowski 1999*a*; Martin 2003).

To emphasize the point about the importance of ecological factors as determinants of primate social organization and mating systems, and the absence of oestrus in anthropoids, I shall describe some field and laboratory observations of two African monkey species: the talapoin (*Miopithecus talapoin*) and the mandrill (*Mandrillus sphinx*). Both these species concentrate their copulatory activity within annual mating seasons, and both exhibit interesting shifts in social organization at such times. However, neither the talapoin nor the mandrill exhibits a peri-ovulatory oestrus of the type defined by Heape.

The talapoin is the smallest of the Old World monkeys; males weigh about 1.4 kg, and females just 1.0 kg when adult. Talapoins live in dense riverine rainforests in Cameroon and Gabon and occur in large multi-male/multi-female troops, commonly numbering more than eighty individuals (Rowell 1973; Rowell and Dixson 1975). Adult and sub-adult males form a distinct sub-group, and they do not mix extensively with the adult females except during the annual mating period. In Cameroon, the mating season occurs between the months of January to March, and occupies the longer of the two dry seasons which occur each year. Figure 6.1 shows monthly levels of rainfall in Cameroon at Mbalmayo, where these studies were conducted, together with data on the timing of talapoin mating and birth seasons. Interestingly, in northeast Gabon, where the dry and wet seasons are 6 months out of phase with those in South Cameroon, the mating season is also shifted by 6 months (June–August) with a birth season in December and January (Gautier-Hion 1968). Photoperiodic or temperature cues are most unlikely to account for these circannual rhythms of matings, conceptions, and births, which occur in equatorial environments. It is more likely that

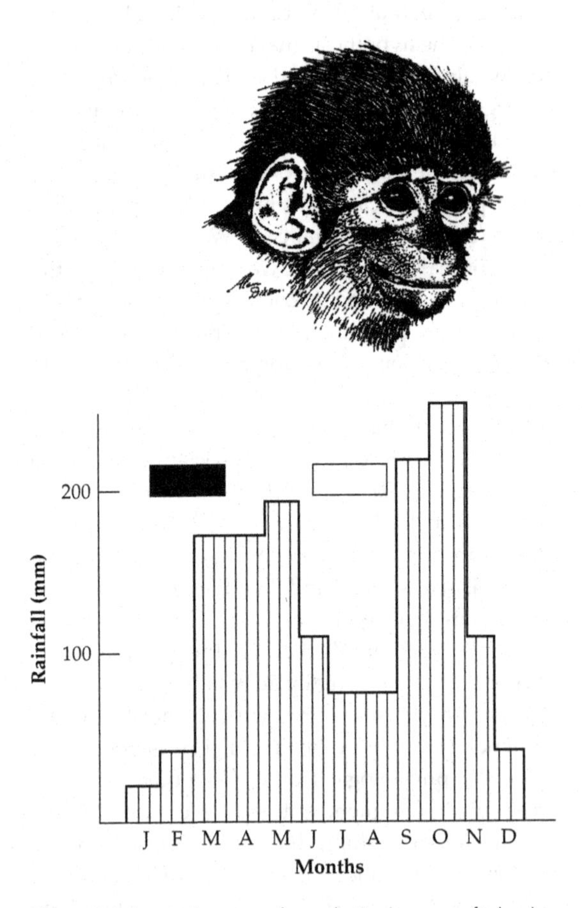

Figure 6.1 Seasonal patterns of reproduction in groups of talapoin monkeys (*Miopithecus talapoin*) in Cameroon. Monthly variations in rainfall correlate with annual cycles of mating behaviour (black bar) and births (open bar).

Source: Based on Rowell and Dixson (1975).

changes in nutritional or other factors secondary to seasonal shifts in rainfall trigger the neuroendocrine events which govern the start of the mating season. Female talapoins develop pink sexual skin swellings at this time, and it is suspected that social cues also play some role in the fine-tuning of their swelling cycles. Thus, in two troops of talapoins studied in Cameroon, females within each group showed much tighter co-ordination of their sexual skin swelling phases than was apparent between the troops, even though they occupied adjacent home ranges (Rowell and Dixson 1975).

During the annual mating season, males move from their subgroup to associate with females and to copulate; the extent of these movements and associations correlates strongly with the extent to which females have developed their sexual skin swellings (Figure 6.2). Females mate with multiple males, and there is a high level of aggression in the troop. Detailed observations of individual animals are exceedingly difficult to obtain because talapoins are small and live in dense forests. However, it is apparent that copulations occur over an extended period of the swelling cycle, and this conclusion is strengthened by the results of studies on captive groups of talapoins, which quantified changes in behaviour throughout the menstrual cycle (Scruton and Herbert 1970; Dixson et al. 1972). As can be seen from the results of these studies (Figure 6.3) matings occur throughout the follicular phase of the cycle, as the females' swellings enlarge due to oestrogenic stimulation, and are much less likely to occur during the luteal phase. There is no restriction of matings to the peri-ovulatory period, however, and no limitation of female sexual presentations or willingness to accept males in a period of heat or oestrus.

By contrast with the talapoin, the mandrill is a massive monkey and the adult males in particular are spectacular creatures, weighing more than 30 kg and adorned with red and blue sexual skin on the

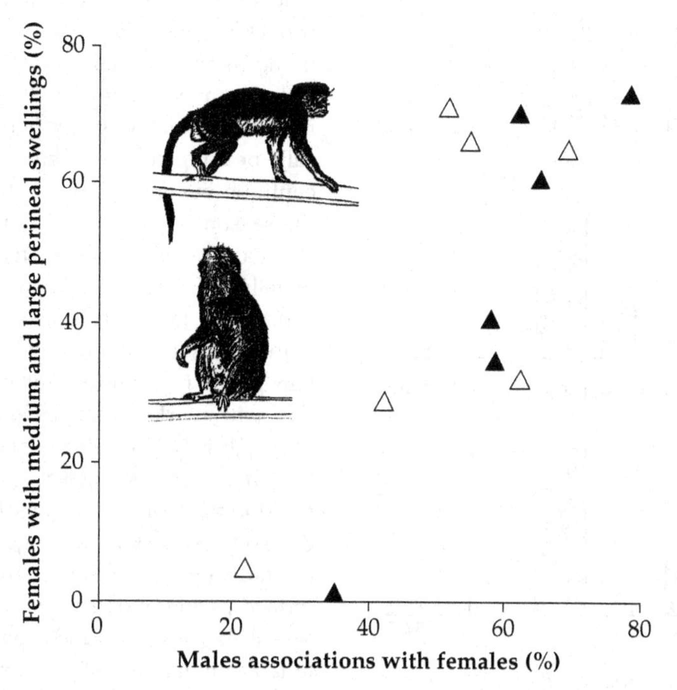

Figure 6.2 Relationships between numbers of females with medium/large sexual skin swellings and percentages of males' associations with females throughout the mating season in two groups of talapoin monkeys (*Miopithecus talapoin*) in Cameroon.

Source: From Dixson (1998a), after Rowell and Dixson (1975).

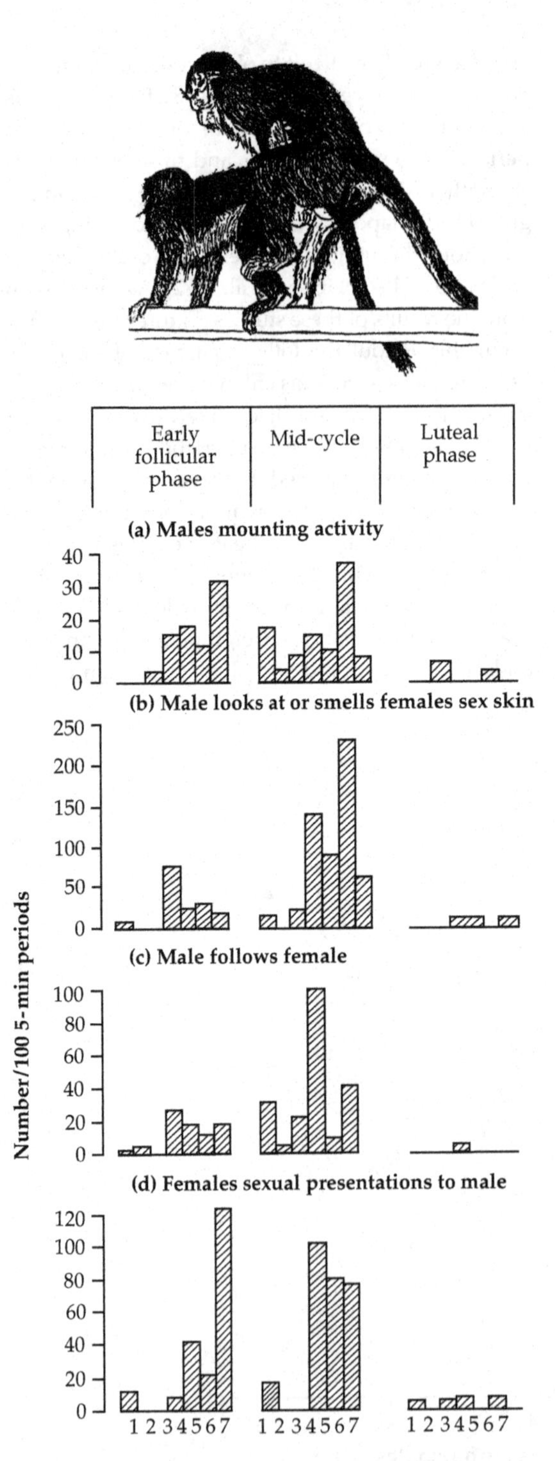

Figure 6.3 Sexual behaviour during the menstrual cycle in captive groups of talapoin monkeys (*Miopithecus talapoin*). Data are for seven females.

Source: From Dixson et al. (1972); after Scruton and Herbert (1970).

face, rump, and genitalia. Mandrills occur in the rainforests of Gabon, Rio Muni, and Congo, and they form multi-male/multi-female social groups. So extensive are mandrill home ranges and so elusive are these animals that much remains to be learned about the details of their social organization under natural conditions. However, it is known that very large groups, sometimes called 'hordes' may occur; more than 1,000 monkeys have been counted in one such horde in Gabon. Hordes sometimes split into smaller sub-groups, and it used to be thought that these might represent single male units, indicative of a polygynous mating system (Stammbach 1987; Barton 2000). However, observations of semi-free ranging and wild groups of mandrills in Gabon have shown that mandrills have a multi-male/multi-female mating system (Dixson et al. 1993; Abernethy et al. 2002).

Abernethy et al. were able to attach radio collars to free-ranging male mandrills, and this produced some interesting findings. Adult males spent substantial periods outside of the social group, living a 'solitary' existence, but entering the group to compete for copulations during the annual mating season. Detailed studies of a semi-free ranging group, living in an enclosed area of natural rainforest in Gabon, have shown that the mating season corresponds to the long dry period of the year, typically peaking between May and July. Matings may continue beyond these months, however, so that the season is longer and more flexible in its timing than that described for the talapoin. Mandrills usually give birth from January to March (Figure 6.4). Social factors can also impact the timing of mating seasons, so that moving animals to set up a new group in a rainforested enclosure was associated with marked changes in the onset and timing of females' cycles (Setchell and Dixson 2001).

Although the mating season tends to last for longer in mandrills than in talapoins, female mandrills also develop sexual skin swellings and are receptive for extended periods of time, so that there is no circumscribed period of oestrus. Figure 6.5 shows data on sexual skin swellings and copulatory behaviour collected from mandrills in Gabon, early in the annual mating season when the first four of fourteen adult females in a semi-free ranging troop had entered a reproductive condition. The pattern of sexual skin

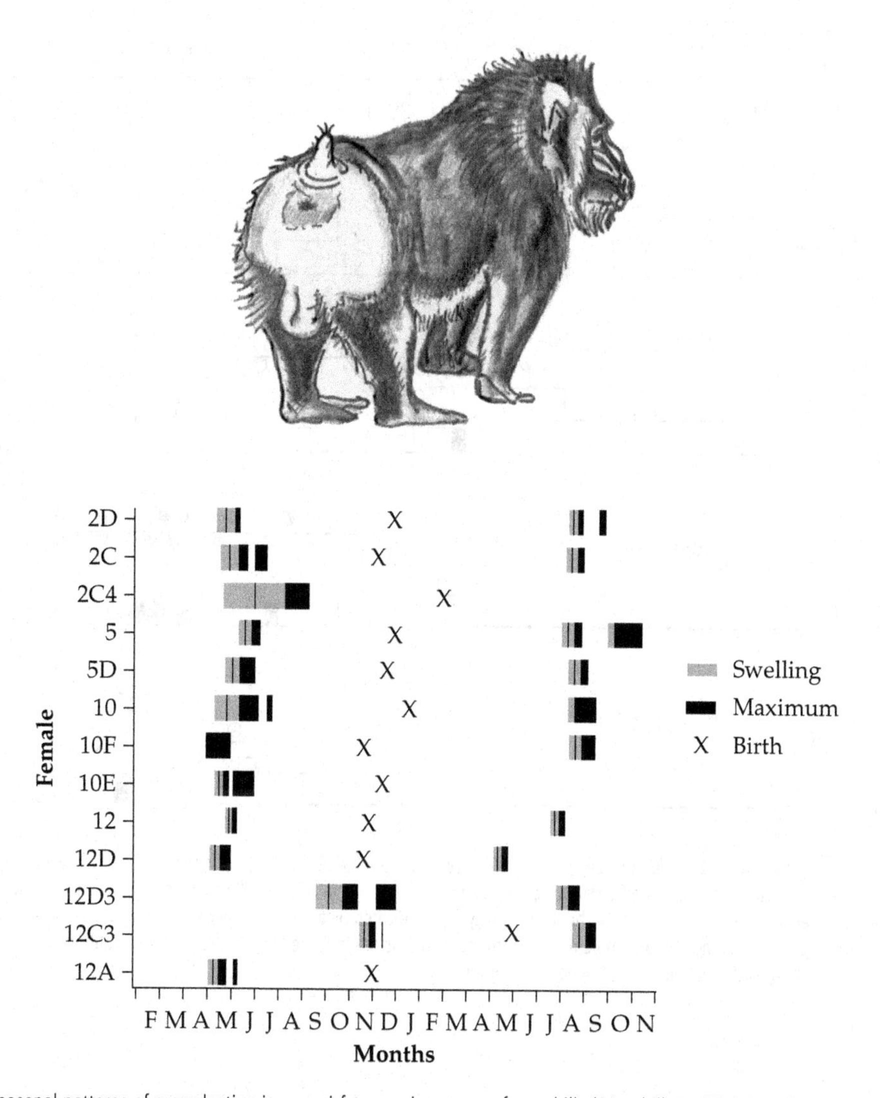

Figure 6.4 Seasonal patterns of reproduction in a semi-free ranging group of mandrills (*Mandrillus sphinx*) in Gabon. Females develop their sexual skin swellings and conceive during the long dry season, with most of the subsequent births (X) occurring between November and January.

Source: After Setchell and Dixson (2001).

changes in these females was related to social rank. The three lowest ranking females (no. 17A: rank 12; no. 16: rank 13; no. 6: rank 14) had very long follicular phases, as indicated by the lengthy periods (32–39 days) it took for their sexual skin swellings to enlarge and then to begin to detumesce (initial detumescence or breakdown of the sexual skin is readily observable). Although menstrual cycles last for one month, on average, in female Old World monkeys, apes, and in women, these female mandrills had greatly extended the first half of the cycle, perhaps due to some problem in secreting sufficient ovarian oestradiol during follicular development, or to some reduction in sensitivity of the sexual skin to oestrogenic stimulation. In marked contrast was the situation observed in the most dominant female in the troop (no. 2 in Figure 6.5). She exhibited the expected 15 day follicular (swelling) phase and her sexual skin was maximally swollen for just 4 days before breakdown occurred.

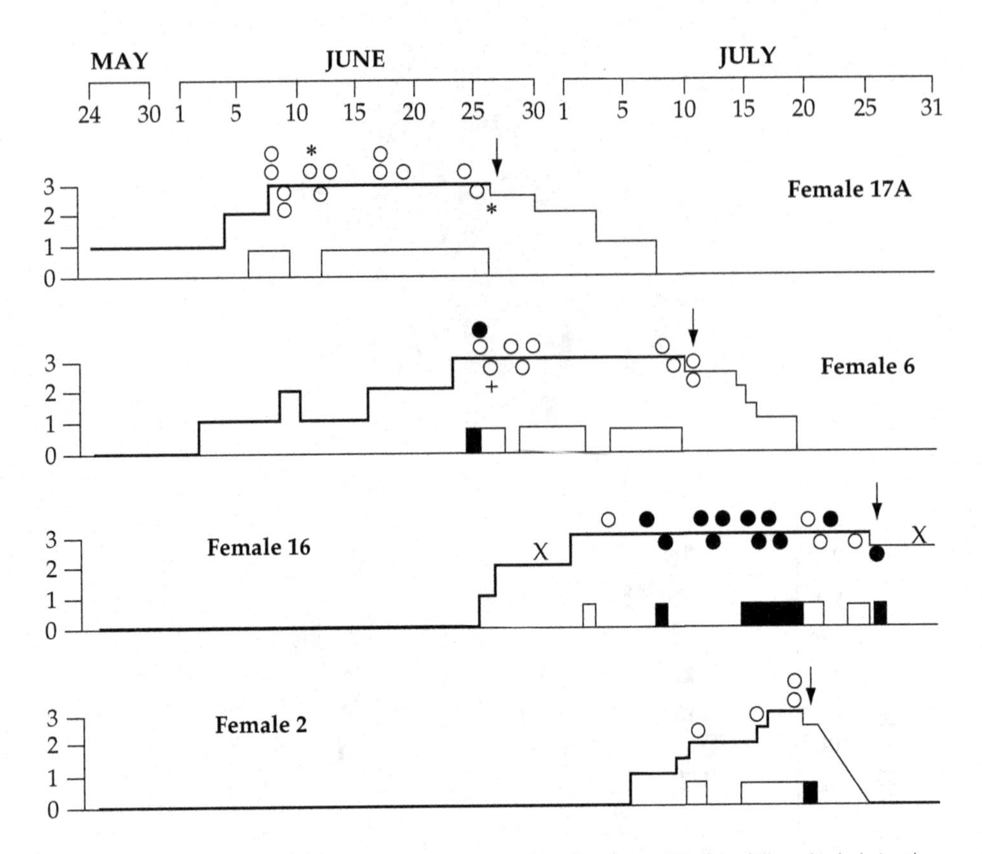

Figure 6.5 Details of sexual interactions during sexual skin swelling cycles in four female mandrills (*Mandrillus sphinx*), during the annual mating season in Gabon. Swelling sizes are rated on a four-point scale (0 = flat; 1 = small; 2 = medium; 3 = fully swollen). The day of initial sex skin breakdown, at the end of the follicular phase of the menstrual cycle, is indicated by a vertical arrow. Mate-guarding episodes by males are shown as horizontal bars (□ = male no. 14; ■ = male no. 7). Ejaculatory mounts are indicated by symbols for each adult male (○= no. 14; • = no. 7; X = no. 15; * = no. 13) and subadult male(+ = no. 2B). Author's data.

Patterns of sexual behaviour involving these females and the various adult and sub-adult males are also shown diagrammatically in Figure 6.5. There were three troop-associated adult males (alpha male 14; beta male 7; and theta male 15), three peripheral/solitary adults, and three troop associated sub-adult males. The two most dominant troop males (no. 14 and no. 7) engaged in mate-guarding episodes with individual females at the height of sexual skin swelling. Mate guarding thus focused upon 'single' females (males did not guard or herd units of females as hamadryas or gelada males do). Guarding must be energetically very costly for such males as they follow each female closely, often giving deep 'two phase grunts'. Activities such as feeding take second place to the defence of their mating opportunities. Male no. 14 (the most dominant male) showed guarding behaviour on 126 (82 per cent) days during the 156 day annual mating period. Such strategies are highly effective, as DNA typing studies show that dominant males have greater reproductive success (Dixson et al. 1993). It may be for this reason that such males develop large reserves of subcutaneous fat in the rump and flanks to sustain them in the mating season (Wickings and Dixson 1992*a*). The observation that male mandrills may leave troops after the mating season and forage alone (Abernethy et al. 2002) may be connected to the conflicting demands of reproduction and feeding ecology.

In female baboons, ovulation occurs at the end of the maximum phase of sexual skin swelling, in a time window of 2–5 days before sexual skin detumescence, or on the day of initial sexual skin breakdown (Wildt et al. 1977). This is likely to be the case in mandrills as well, although the necessary endocrine and laparoscopic studies have not been carried out on this species. It is clear, however, that dominant males terminate guarding abruptly once a female's sexual skin begins to detumesce. Up until then, copulations occur throughout the phase of maximal swelling. For example, it can be seen in Figure 6.5 that male no. 14 began to guard female no. 17A and to copulate with her as soon as she had developed a full swelling; he continued to do so for a 20 day period, until sexual skin breakdown brought an end to sexual interactions. Thus it seems likely that the female's sexual skin acts as a *graded signal* of female reproductive quality in mandrills, as has been proposed for other monkeys and chimpanzees which possess such swellings (Nunn 1999). Male mandrills are under considerable selection pressure to limit guarding and copulation to the ovulatory period; in reality female swellings provide only approximate information concerning likelihood of ovulation, and females are attractive and receptive sexually for extended time periods (especially so in the case of lower-ranking troop members). Thus there is no restricted period of female sexual receptivity or oestrus in mandrills.

These findings on the sexual behaviour of mandrills and talapoins are consistent with the results of field and laboratory studies on a wide range of Old World monkeys and the great apes. Copulations are not distributed randomly throughout the menstrual cycle; mounts and ejaculations are more frequent during the follicular phase or at mid-cycle than during the luteal phase in macaques, baboons, mangabeys, chimpanzees, gorillas, and orangutans (Dixson 1998*a*). Students of primate sexuality have long been aware that ovarian hormones have powerful effects upon female sexual attractiveness in some species, so that increases in male sexual activity may be due, in part, to enhanced female attractiveness during the follicular phase or at mid-cycle (Herbert 1970). Females may also actively invite males to copulate; Beach (1976*b*) coined the term *proceptivity* to define this aspect of female sexuality. The distinction between proceptivity (female sexual initiation) and *receptivity* (the female's willingness to accept the male) is an important one, because the neuroendocrine mechanisms underlying the two kinds of behaviour differ in significant ways. At the hypothalamic level, for example, lesions which cause marked deficits in proceptivity in female common marmosets (a New World monkey species) have little or no effect on sexual receptivity (Kendrick and Dixson 1986; Dixson and Hastings 1992). The temporal patterning of female proceptivity and receptivity may also differ markedly during the menstrual cycle. Thus, in the chacma baboon, females present the rump as a sexual invitation to males, but they may also seek to make eye-contact at the same time (Figure 6.6). Such *eye-contact proceptivity* increases in frequency during the follicular phase, peaks during the peri-ovulatory period, and then declines markedly during the luteal phase of the cycle. Yet, as Figure 6.6 also shows, female chacma baboons continue to accept males' mounting attempts and remain sexually receptive throughout the menstrual cycle. The decline in copulatory frequencies during the luteal phase is due, in large measure, to decreased female attractiveness and loss of sexual skin swelling at this time.

Alternatives to the use of simplistic oestrous terminology to define female sexuality have thus been available for more than 30 years; the definitions of female attractiveness, proceptivity, and receptivity (Beach 1976*b*) are broadly applicable in studies of sexual behaviour, and are especially useful in research on primate (including human) sexuality. The propensity for female monkeys and apes to display sexual receptivity at times other than when ovulation is likely is widespread, and should not be dismissed as an artefact of studying behaviour under captive conditions. In free-ranging rhesus monkeys, for example, Loy (1970) observed that females showed 'dual estrus periods', at mid-cycle and peri-menstrually. When male and female rhesus monkeys are pair tested in the laboratory, some pairings show this same pattern, so that frequencies of mounts and ejaculations peak at mid-cycle and prior to menstruation (Figure 6.7A). However, other pairings show either an increase in copulations throughout the follicular phase (Figure 6.7B) or they may mate only infrequently and without any obvious cyclical changes in behaviour (Figure 6.7C).

Detailed studies of rhesus monkey sexual behaviour have shown that these patterns result, in part, from variations in female sexual attractiveness, as well as from differences in the relationships between partners during free-access pair tests (Goy 1979; 1992). Robert Goy noted that *mutual attractiveness* is an important determinant of copulatory frequency and its cyclical patterning during pair tests. The fact that copulation is situation dependent, and not simply dependent upon female hormonal status, has been amply demonstrated by studies of monkeys and apes. In many species males are larger and physically more powerful than females. Males may attempt to initiate sexual interactions more frequently than females under laboratory conditions where space is restricted, just as in the wild males sometimes attempt to coerce females into mating with them. Within the confines of a laboratory cage, or a small enclosure, it is often difficult for females to refuse males when they attempt to mate, and females are also less likely to exhibit their normal cyclical patterns of proceptivity. However, when testing conditions are manipulated so as to provide females with greater control over mating interactions, then mid-cycle increases in proceptivity are better defined and copulations are more frequent at that time (e.g. in the rhesus monkey: Wallen 1982; Wallen and Winston 1984).

When orangutans are pair tested to observe sexual behaviour in captivity, the huge males may force their female partners to mate on almost every test. However, if a vertical dropgate divides the testing area, so that only the (much smaller) female has room to slide underneath and reach the male, then she is more proceptive at mid-cycle and copulations are more frequent at this time (Nadler 1977; 1988). Such observations indicate that cyclical changes in the effects of ovarian hormones upon the brain and sexual behaviour certainly occur in anthropoids, but they are subtle, situation dependent, and more influenced by variations in mutual attractiveness between the sexes than is the case

Figure 6.6 Sexual behaviour during laboratory pair tests in chacma baboons (*Papio ursinus*). Data (collected during four cycles in eight females) are aligned to the day of sexual skin breakdown (BD). Female proceptivity (presentations and presentations accompanied by eye-contact) is shown in the lower graph. Acceptance ratios to invitations made by the opposite sex are shown in the upper graph. Note that females accept males' attempts to mount throughout the cycle, whereas males accept far more of the females' invitations when the latter have large swellings and are sexually attractive.

Source: From Dixson (1998a); after Bielert (1986).

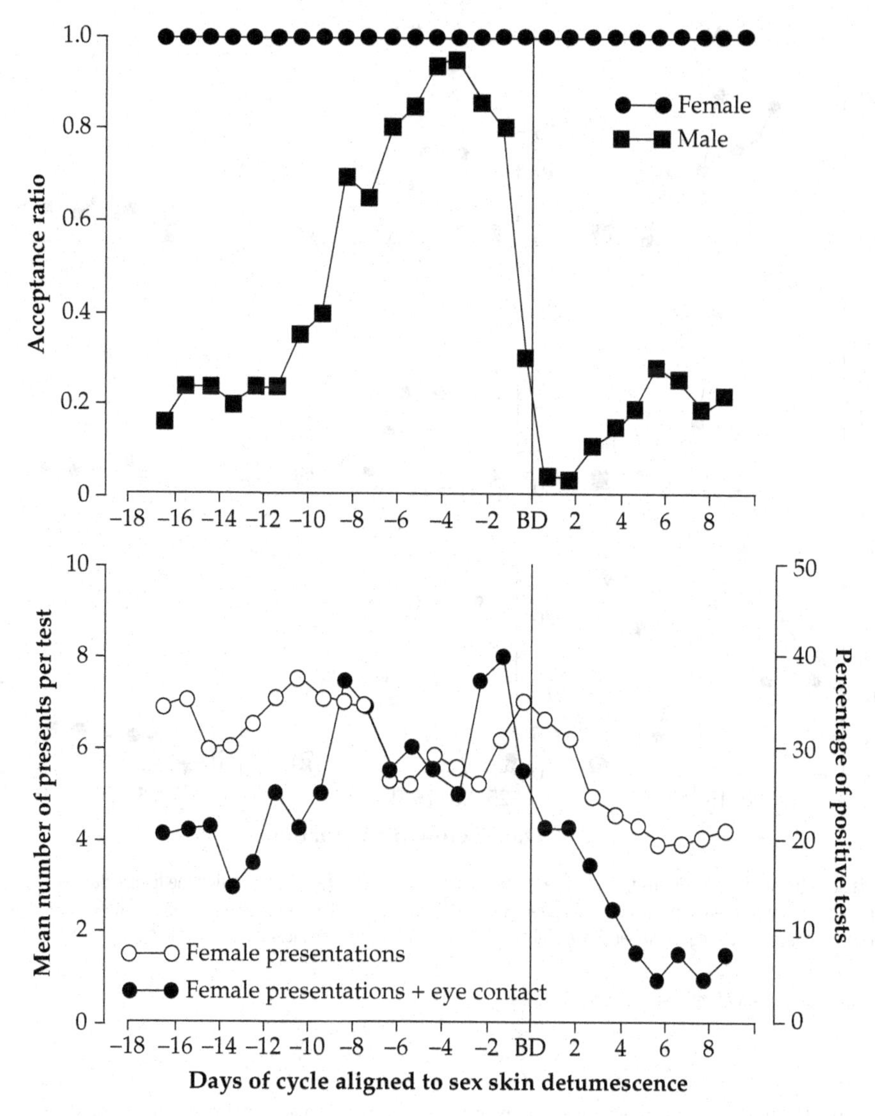

Figure 6.6 *continued*

for many mammals. Hence the term 'oestrus' is no more appropriate when applied to descriptions of sexual behaviour in female monkeys or apes, than it would be in descriptions of human sexuality.

Given that female Old World monkeys and apes do not exhibit oestrus, is the same true for other non-human primates? Prosimians such as the lemurs, galagos, and lorises exhibit circumscribed periods of sexual receptivity, which are tightly controlled by ovarian hormones (Dixson 1995*a*; 1998*a*). It is more acceptable to retain the term 'oestrus' in relation to studies of sexual behaviour in prosimian primates. However, among the New World monkeys, females have been observed to permit copulation outside of the peri-ovulatory period and during pregnancy in a number of species. In the common marmoset (*Callithrix jacchus*), for example, females remain sexually receptive and proceptive after removal of the ovaries and adrenal glands (Dixson 1987*c*). The adrenal cortex represents a significant secondary source of sex steroids, but it

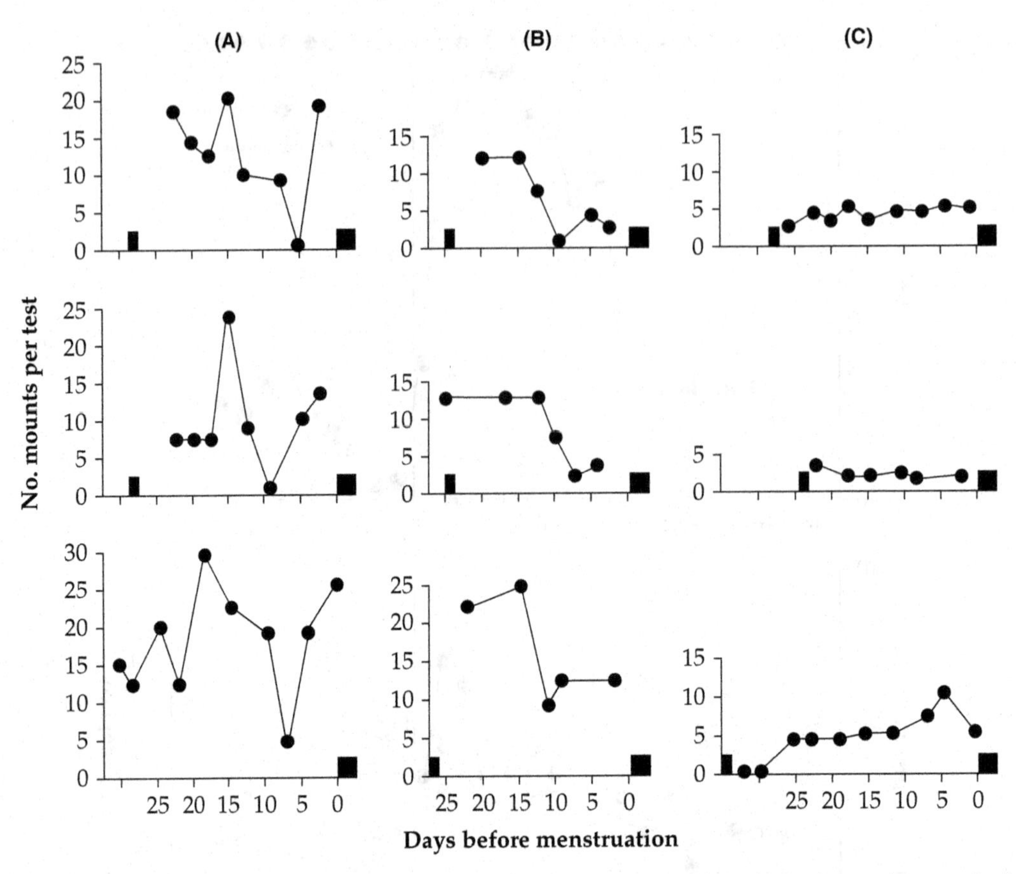

Figure 6.7 Variable patterns of copulatory behaviour in pairs of rhesus monkeys (*Macaca mulatta*) during the female partner's menstrual cycle. (A) In these pairs, mounts peak in frequency at mid-cycle and pre-menstrually (the black bars = menstruations). (B) Mounts are most frequent during the follicular phase, and decrease during the luteal phase of the cycle. (C) Sexual activity is at low levels in these pairs, and there is little evidence of cyclical changes.

Source: From Dixson (1998a); after Everitt and Herbert (1972).

plays no essential role in the maintenance of sexual behaviour, at least in marmosets (Figure 6.8). Thus, although female marmosets certainly display cyclical changes in their sexual activity (Kendrick and Dixson 1983; Dixson and Lunn 1987), they are not dependent upon oestrogen, progesterone, or other steroid hormones for the expression of proceptivity or receptivity.

In evolutionary terms, it appears that a relaxation of rigid hormonal control of female sexual behaviour is a trait shared by the anthropoids in general, and would have been present in the common ancestors of the New World and Old World representatives of the sub-order (Figure 6.9). Loss of oestrus, if that is the appropriate term, is an ancient phenomenon, and represents part of our evolutionary inheritance from non-human ancestors. Precursors of *Homo sapiens*, among the australopithecines, would not have exhibited oestrus, nor is it necessary to postulate that they possessed external signals of reproductive status, such as sexual skin swellings. The evolution of sexual skin swellings in the chimpanzee and bonobo (Genus *Pan*) may have occurred after the common ancestors of these apes diverged from the line which ultimately gave rise to *Homo*. This possibility should not surprise us, as sexual skin swellings have arisen and have been lost a number of times during anthropoid evolution (Dixson 1983*b*; 1998*a*; Sillen-Tulberg and Møller 1993).

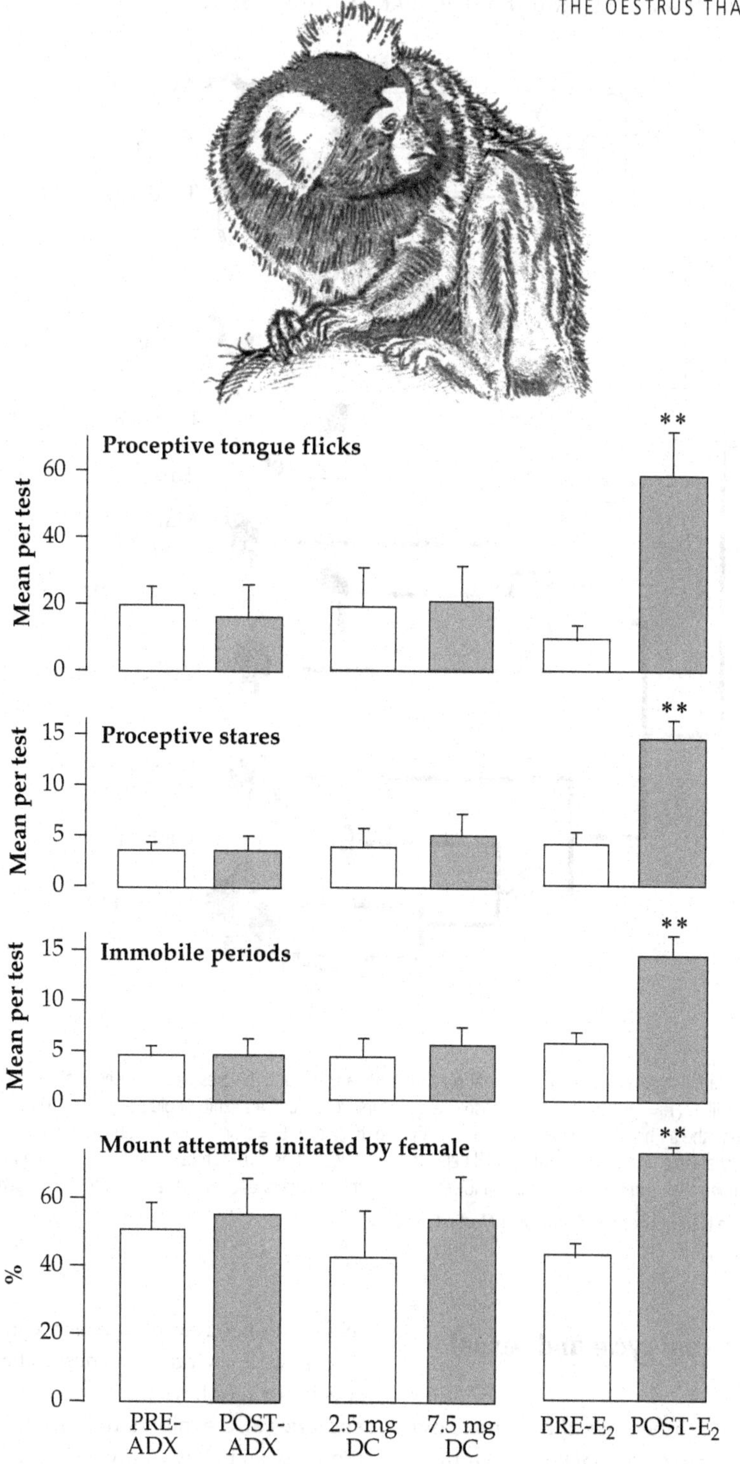

Figure 6.8 Effects of adrenalectomy (ADX), deoxycorticosterone replacement (DC), and oestradiol 17β (E2) upon sexual behaviour in ovariectomized common marmosets (*Callithrix jacchus*). Females remain proceptive and continue to initiate copulations with their male partners after removal of the ovaries and adrenal glands. Nonetheless, proceptivity increases after oestradiol treatment, indicating that central mechanisms underlying female sexual behaviour are sensitive to (but not dependent upon) oestrogen.

Source: From Dixson (1998a); after Dixson (1987c).

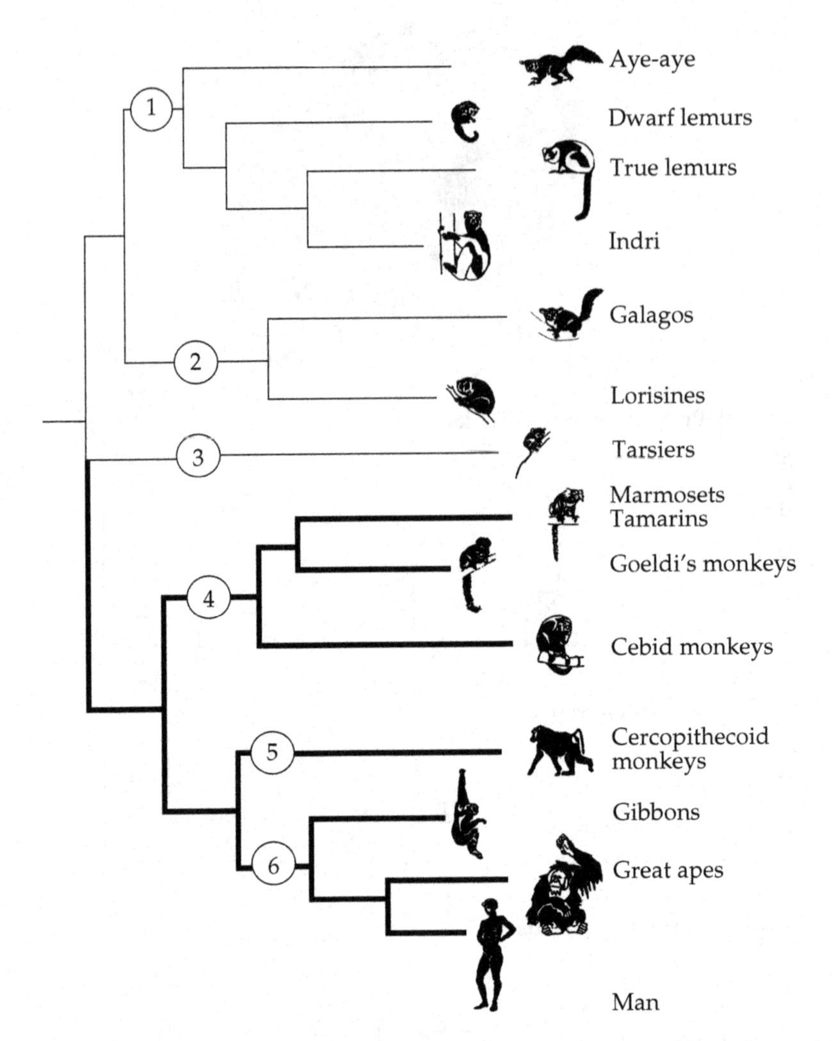

Figure 6.9 Phylogenetic relationships between the six extant primate superfamilies, and the presence or absence of oestrus in females. In the three prosimian superfamilies (1. = Lemuroidea; 2. = Lorisoidea; 3. = Tarsioidea) females exhibit rigid control of female receptivity during a peri-ovulatory oestrus. Among the anthropoids (4. = Ceboidea; 5. = Cercopithecoidea; 6. = Hominoidea) ovarian hormones have much less rigid effects upon the central mechanisms underlying female sexual behaviour. Thus 'loss of oestrus' is most likely to represent an ancient trait among anthropoids, originating among their common ancestors, as indicated by the thick, black branches of the anthropoid phylogeny.

Source: The phylogenetic tree is from Dixson (1998a); after Martin (1990).

The human menstrual cycle and sexual behaviour

Human beings have inherited the anthropoid capacity to dissociate female sexual behaviour from a rigid control by ovarian hormones. However, this statement should not be misinterpreted and taken to imply that the menstrual cycle has no influence upon human sexual behaviour. Rhythmic changes in female behaviour, and especially in proceptivity as well as attractiveness, characterize many species of monkeys and apes. Given that monkeys and apes also exhibit situation-dependent shifts in their sexual behaviour, we should expect the same to apply to human beings. Women and their male partners (no less than rhesus monkeys or chimpanzees) may show variations in mutual attraction and in sensitivity to hormones. The menstrual cycle is associated with changes in feelings of well-being and distress (physical and psychological) which are pronounced in some women. Thus, the follicular and peri-ovulatory phases of the cycle, when oestrogen levels are highest, are associated with feelings of increased well-being, whilst feelings of irritability, aggression, and physical distress are greatest at around the time of menstruation (peri-menstrual syndrome: Bancroft 1989). Such changes might secondarily impact the effects of hormones upon women's sexual thoughts and feelings. In early human populations, lacking contraceptive technology, women spent large portions of their adult lives pregnant, or lactating. Multiple non-conception cycles and deleterious peri-menstrual symptoms may have been much less frequent under those conditions.

Given such factors and the extreme complexity of situation-dependent variables in the lives of modern humans, it should not surprise the reader that scores of studies have failed to reveal simple relationships between the menstrual cycle and sexual behaviour. Failure to measure hormonal changes, to record behaviour with precision, and to control for the effects of mood has limited the scope of many studies. Almost 30 years ago, Schreiner-Engel (1980; 1981) reviewed the results of thirty-two studies which reported peaks of sexual activity at mid-cycle ($N=8$), pre-menstrually ($N=17$), during menses ($N=4$), and post-menstrually ($N=18$). Pawlowski's (1999*a*) review contains a majority of publications reporting increases in female sexual interest and proceptivity in the follicular and peri-ovulatory phases of the menstrual cycle (with pre-menstrual increases also in five out of sixteen reports). The results of a more recent study, by Wilcox et al. (2004), are shown in Figure 6.10. This focused upon women ($N=68$) in North Carolina who were sexually active, using non-steroidal contraceptive methods (intra-uterine devices or tubal ligation), and who kept records of sexual activity as well as collecting urine samples for hormonal monitoring. These women provided data on 171 (putatively ovulatory) menstrual cycles; the fertile phase of the cycle was defined as the 6 day period culminating in the occurrence of ovulation (Wilcox et al. 1995). Figure 6.10 shows that the proportion of women having sexual intercourse peaked during this 6 day fertile phase and that such proportions were generally higher during the follicular phase than in the luteal phase of the menstrual cycle. It is not known whether increased activity during the fertile period might hasten the onset of ovulation. Wilcox et al. (2004) examined this question in a larger sample of women (285 subjects and 867 cycles) and found preliminary evidence for such an effect. It should be noted, in relation to situation-dependent effects upon sexual behaviour, that Wilcox et al. found that their subjects 'were most likely to have intercourse on the weekend (Friday, Saturday, or Sunday). Ovulation, in contrast, presumably occurs without regard to day of the week.'

Unfortunately, these authors did not record details of female sexual initiation (proceptivity), autosexual (self-directed) behaviour, or frequencies of orgasm during menstrual cycles. Nor were the days on which menstruation occurred included in the analyses. Other studies have reported increases in female-initiated heterosexual activities and female sexual interest during the days leading up to ovulation (Adams, Gold, and Burt 1978; Dennerstein et al. 1994). Significant numbers of women experience initial increases in sexual interest during those days of the cycle which precede and include the day of ovulation (Stanislaw and Rice 1988; Wallen 1995).

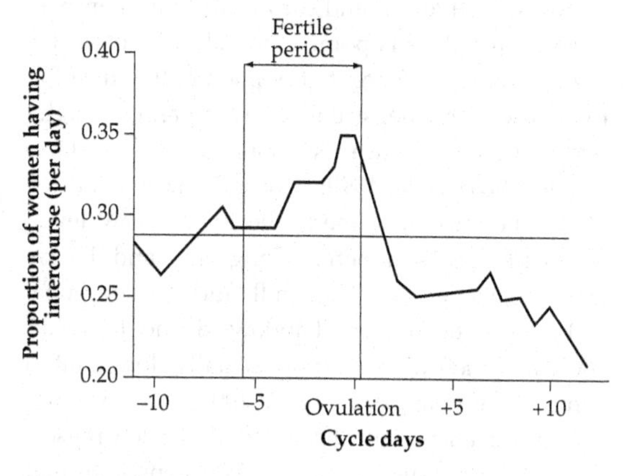

Figure 6.10 Variations in the proportion of women who report having intercourse at different phases of the menstrual cycle. Increases occur during the 'fertile phase' of the cycle (defined as the 6 days culminating in the probable day of ovulation).

Source: Redrawn and modified from Wilcox et al. (2004).

Sanders and Bancroft (1982) reported peri-menstrual increases in women's sexual interest, in addition to a mid-follicular (not a peri-ovulatory) peak in sexual interest. The increases in sexual interest during the peri-menstrual phase occurred despite heightened symptoms of physical distress and peri-menstrual syndrome in some subjects. Sanders and Bancroft noted that 'it is not unusual to find women whose sexual interest or preparedness to engage in sexual intercourse is restricted to a brief period of the cycle...in such women the peak is nearly always peri-menstrual. It is difficult to account for that pattern on the basis of abstinence in view of the long periods of unenforced abstinence for the rest of the cycle.' Therefore, it appears that cultural taboos concerning sexual abstinence during menstruation do not entirely explain the pre-menstrual increases in sexual interest experienced by some women. It will be recalled that pre-menstrual increases in copulations also occur in some rhesus monkeys, both in the wild (Loy 1970) and during pair-tests between captive animals (see Figure 6.7A).

Given that significant numbers of women experience follicular and (especially) mid-cycle increases in their sexual interest, it would not be surprising if they also reported greater responsiveness to sexually attractive traits of male partners (or potential partners) at such times. In the next chapter I shall consider the evolution of sex differences in human physique and secondary sexual traits, in relation to sexual attractiveness and the origin of human mating systems. At this point, it should be noted that there is some evidence that women at the most fertile phase of the menstrual cycle may report heightened attraction towards more masculine faces (Penton-Voak et al. 1999; Penton-Voak and Perrett 2001), the odours of men who exhibit low levels of fluctuating asymmetry (Gangestad and Thornhill 1998; Gangestad, Thornhill, and Garver-Apgar 2005*a*), and taller men (Pawlowski and Jasienska 2005) and men who are more socially dominant or competitive (Gangestad et al. 2004). The proposition has been widely discussed that such masculine traits are indicative of superior genetic fitness. Women who are likely to ovulate may thus be more attracted to such men, because their advantageous genetic traits will be passed on to any resulting offspring. However, it has also been proposed that women in the fertile phase of the menstrual cycle are more likely to find such heightened cues of masculinity most attractive when engaging in *short-term relationships* involving extra-pair matings with men who are not their long-term or marriage partners (reviews by Gangestad and Cousins 2001; Okami and Shackelford 2001; Gangestad, Thornhill, and Garver-Apgar 2005*b*; Pillsworth and Haselton 2006; Gangestad 2007). Indeed, standard works in the field such as *The Handbook of Evolutionary Psychology* (edited by Buss 2005) and the *Oxford Handbook of Evolutionary Psychology* (edited by Dunbar and Barrett 2007) contain chapters promoting these ideas. The suggestion is that evolution has resulted in *dual-mating strategies* in human females; that is to say, women have strategies which function as part of a long-term relationship with an established partner, and different strategies built around peri-ovulatory, extra-pair copulations with males having 'good genes', especially if these traits are not possessed by the long-term partner. Such arguments are often bolstered by references to the relevance of sperm competition and cryptic female choice as drivers of the evolution of human reproduction (Shackelford et al. 2005; Platek and Shackelford 2006). Dual-mating strategies and extra-pair copulations during the fertile period are thought to have resulted in selection via sperm competition between males.

I remain sceptical about such hypotheses concerning the evolution of female dual-mating strategies, the pursuit of 'good genes' via extra-pair copulations by women, and associated selection via sperm competition between their long-term and short-term male partners. The reader is referred to the arguments advanced in earlier chapters which show (convincingly in my view) that sexual selection via sperm competition and cryptic female choice is unlikely to have played any significant role in the evolution of human reproductive anatomy, physiology, and copulatory behaviour. Although some evolutionary psychologists argue that only the dual-mating and extra-pair copulation (EPC) theory can account for the reported shifts in female sexual preferences during their menstrual cycles (Gangestad et al. 2005), I do not think that this can be correct. Firstly, there is the question of how strong these shifts in female sexual preference (for masculine traits) really are in real-life situations affecting sexual decisions. This is not to deny the existence of increased female sexual interest at certain times; after all it has been established that many women do show increases in sexual interest during the follicular and peri-ovulatory phases of the cycle (and peri-menstrually). However, we have seen that in human beings, as in the monkeys and apes, such shifts are situation dependent and subtle by comparison with the effects of the ovarian cycle upon sexual behaviour in most mammals. Just because women (or at least some women) display greater interest in masculinized faces (or other masculine traits) during their fertile periods, it does not follow that these shifts are large in absolute terms or that they result in EPCs if a permanent partner has a less masculinized face. I suggest instead that there has been a regrettable tendency for some workers in this field to over-interpret the results of studies where women are asked to express preferences for masculine traits in relation to hypothetical long-term versus short-term mating strategies. Such over-interpretation has fuelled the view that women have evolved psychological mechanisms to 'cuckold' their long-term partners, in order to obtain 'better genes' for offspring that will be raised with those same long-term partners.

This brings me to a second important drawback to the dual-mating hypothesis. It is widely acknowledged that men and women have an evolved propensity to form long-term partnerships for reproductive purposes and that such relationships are indeed crucial determinants of infant survival and hence of reproductive success for human parents. Marriage is virtually universal among human cultures around the world, more than 80 per cent of which permit polygyny, although the majority of men have a single wife at any given time, so that monogamy is the norm (Ford and Beach 1951; Borgerhoff Mulder and Caro 1983; Low 2007). This strong predisposition to form long-term partnerships is facilitated, in part, by the human propensity to 'fall in love'; romantic love is a universal human attribute (Pillsworth and Haselton 2006). These authors cite a doctoral thesis by Harris (1995), who reviewed information from 100 cultures worldwide and found evidence for the occurrence of romantic love in every case. Jankowiak and Fischer (1992) confirmed its occurrence in 89 per cent of cultures for which data exist in the *Human Relations Area Files*. The formation of long-term relationships between the sexes for reproductive purposes is vital, given that human offspring are born in a relatively helpless state and require many years of parental support. Men benefit from forming lasting partnerships with women, because there is much greater assurance of paternity of the offspring which result, and male investment in offspring significantly increases the chances that they will survive. This is the case among the few hunter–gatherer cultures which still exist (e.g. the Ache of Paraguay: Hurtado and Hill 1992; the Hadza of Tanzania: Marlowe 1999*a*; 1999*b*; 2003) and would likely have been the case among the earliest modern humans, or their precursors in Africa. In addition to his superlative studies of the reproductive ecology of the Hadza, Frank Marlowe (2000) has conducted a meta-analysis of 186 human societies which showed that paternal investment and provisioning of offspring is a generalized trait that significantly assists their survival.

Throughout the earliest phases of human evolution, and during the evolution of the precursors of modern *Homo sapiens*, the choice of a long-term partner for successful reproduction (not just for sexual behaviour) was likely to have been a crucial decision. This statement applies to females in particular, as they invest so much in reproduction via pregnancies, lactation, and extended periods of parental care. For young women, engaged in the process of acquiring a long-term partner, preferences for an amalgam of masculine traits, such as facial and bodily cues, might have had selective advantages. A heightened preference for such features during the fertile phase of the cycle may simply reflect the results of such an ancient selective process during pair formation. It may be the case that too much has been made of poorly defined modern-day distinctions between female preferences for traits in short-term versus long-term partners. In evolutionary perspective, it seems most likely that female preferences for masculine traits such as height, body build, and facial cues would have been expressed primarily in relation to seeking a long-term mate, if those traits indeed signal 'good genes'.

It must be kept in mind that a comparison of thirty-seven human cultures led Buss (1989) to conclude that both men and women value traits such as kindness, generosity, and a good sense of humour very highly in prospective long-term partners: more so than physical attractiveness. If that is the case, it is difficult to accept the notion that women would seek dual matings during their fertile period to gain different genes (from men with more masculinized faces, for example) from those favoured as beneficial for long-term reproduction. A potent force in the evolution of human sexual psychology has been *jealousy* in response to threats (real or imagined) to a sexual relationship (Buss 2000). Both men and women are prone to jealousy if their long-term relationships are threatened by a third party. In some cultures women especially are subject to severe punishments if they engage in adultery. Of course, one interpretation of this observation might be that such mechanisms evolved to deter extra-pair copulations as they pose a significant risk to paternity in long-term relationships. However, it does not follow from such a line of reasoning that women are actively pre-disposed to seek such EPCs with men having more masculinized physical traits, or that women show significant increases in such behaviour when they are likely to ovulate and to conceive. On the contrary, the risks of engaging in such EPCs would be enormous, due to the potential damage to the long-term relationships and likely survival of a woman's existing children. It should be seriously questioned whether ancestral human female preferences for masculine traits of hypothetical genetic advantage would have fuelled the evolution of psychological mechanisms serving dual mating strategies and extra-pair copulations around the time of ovulation.

A third issue worthy of consideration concerns physiological factors which favour long-term sexual relationships and paternity certainty in human reproduction. Some women suffer from pre-eclampsia; a form of hypertension which afflicts 10 per cent of human pregnancies and which results from immunological incompatibility between mother and foetus (Robillard, Dekker, and Hulsey 2002; Martin 2003). The likelihood of pre-eclampsia decreases with the duration of a woman's sexual relationship with her partner, so that her immune system becomes habituated via repeated exposure to his seminal products, and better adjusted to tolerating a foetus which carries 50 per cent if his genes. However, if a woman mates with a new partner, such as might occur as a result of dual matings to obtain better genes during the fertile period, then the risk of pre-eclampsia is greatly increased should she conceive as a result.

For all of the above reasons, I regard the hypothesis that women have evolved a dual-mating strategy and that extra-pair copulations involving men with more masculinized facial (or other) traits are more likely to occur during the fertile period as weak and poorly substantiated by current evidence. The likelihood that any such preferences for masculine traits form part of selective mechanisms for primary (i.e. long-term) mate choices is more credible.

The whole notion that women might be predisposed by hormonal shifts to alter their sexual behaviour in such striking ways smacks of the continued existence of oestrus in human beings, rather than its loss. Indeed, some evolutionary psychologists now seek to resurrect the use of the term oestrus in relation to human sexuality (Thornhill 2006; Miller, Tybur, and Jordan 2007; Gangestad and Thornhill 2008). Gangestad and Thornhill (2008) conclude that 'the discovery of women's oestrus has penetrating and potentially revolutionary implications for a proper conceptualization of human mating.' On the contrary, the evidence reviewed in this chapter leads to the conclusion that oestrus was lost in the early anthropoids and was not present in human ancestors. Where *Homo sapiens* is concerned we encounter only misconceptions concerning *the oestrus that never was.*

Conclusions

The concept of *loss of oestrus* as it has been applied to the evolution of human sexuality is flawed, and its use should be discontinued. Heape's original definitions of oestrus in mammals are of very little value in discussions of primate sexual behaviour, and especially so where the monkeys and apes are concerned. Female anthropoids are not dependent upon the secretion of ovarian hormones for the expression of sexual behaviour. Beach's three-pronged approach to defining female sexuality (i.e. female proceptivity, receptivity, and attractiveness) has much greater explanatory power than ill-defined statements concerning the presence or absence of oestrus. Comparative studies indicate that a rigid neuroendocrine control of female receptivity and proceptivity during the peri-ovulatory period was lost in ancestral anthropoids. Thus, extant monkeys and apes, as well as women, lack oestrus, whilst it is still present among the prosimian primates.

These observations do not mean that ovarian hormones have no effects upon sexual behaviour or attractiveness in anthropoid primates. On the contrary, numerous studies have established that such effects exist in monkeys and apes. Examples are presented and discussed in this chapter. Effects of ovarian hormones upon female proceptivity (sexual invitations to males) and attractiveness (e.g. sexual skin swellings) are more pronounced than hormonal effects upon receptivity (willingness to accept the male and allow copulation).

Cyclical changes in sexual behaviour also occur during the human menstrual cycle. However, they tend to be even more subtle, and situation dependent, than in many monkeys or apes. The hypothesis that some women alter their mating strategies during the fertile phase of the menstrual cycle in order to conceive with a short-term male partner having 'better genes' than a long-term (e.g. marriage) partner is evaluated and rejected. Instead, it is proposed here that evolutionary psychology has created an artificial dichotomy between long-term versus short-term patterns of female mate choice, and over-interpreted relatively small cyclic fluctuations in women's preferences for certain masculine traits.

The notion that 'pair-bonding' was facilitated by loss of oestrus during human evolution, or that 'concealed ovulation' evolved in order to confuse paternity and reduce the likelihood of male infanticide, is likewise rejected here. There is currently no convincing explanation as to why the neuroendocrine mechanisms controlling proceptivity and receptivity are less dependent on ovarian hormones in anthropoids than most mammals. The explanation is likely to involve selection for female choice, and alternative mating tactics in non-human primate ancestors of *Homo sapiens* that had complex social organizations and relatively large brains. Such selection occurred long before the hominids appeared in Africa, so that absence of a peri-ovulatory oestrus would have been part of the evolutionary inheritance of the australopithecines, and of the genus *Homo*.

CHAPTER 7

Human Sexual Dimorphism: Opposites Attract

'There is love at first sight, not love at first discussion.'
Anon.

In the *Descent of Man*, Darwin observed that among mammals

> The greater size, strength and pugnacity of the male, his special organs of offence, as well as his special means of defence, have been acquired or modified through the special form of selection I have called sexual.

He also stressed that, as well as inter-male competition: 'there is another and more peaceful kind of contest, in which the males endeavour to excite and allure the females by various charms.'

Darwin's emphasis upon the effects of sexual selection on masculine body size, weaponry, attractive adornments and displays is consistent with the fact that females usually represent a limiting resource for males. Females invest a huge amount physiologically in reproduction and in the care of their offspring, so they are expected to be more 'choosy', whilst males may be expected to compete among themselves and to invest more in attempts to attract and control prospective mates. This is not to imply, however, that sexual selection does not act upon females. In the last chapter, the importance of female sexual skin swellings in mandrills, talapoins, chimpanzees, and other Old World anthropoids was discussed in relation to changes in female sexual attractiveness during the menstrual cycle. These extraordinary oestrogen-dependent swellings are prime examples of the effects of sexual selection acting upon females, to enhance their 'charms' during the 'more peaceful kind of contest' required to attract members of the opposite sex (Figure 6.6). Darwin (1876) was intrigued by these structures, and he recalled that 'in my *Descent of Man* no case interested and perplexed me so much as the brightly-coloured hinder ends and adjoining parts of certain monkeys'. Even in this context, however, he emphasized the functions of brightly coloured areas of skin in male monkeys (notably in the male mandrill) rather than exploring the effects of sexual selection in relation to female sexual attractiveness.

In this chapter, I shall examine the question of human sexual dimorphism, and explore the extent to which sexual selection may have influenced the evolution of the physique, facial traits and secondary sexual adornments of men and women. Given the limited information available to him in the nineteenth century, Darwin made important observations and advanced new hypotheses concerning the effects of sexual selection upon the evolution of human morphology. Although the huge potential to build upon Darwin's insights was neglected for many years, this field has been revitalized by the advent of modern research in the fields of evolutionary psychology and anthropology. A critical appraisal of some of these recent advances is included here; the goal being to understand how far current sex differences in our morphology might reflect the effects of sexual selection in the remote past, during the course of human (or pre-human) evolution. How far are the morphological traits of men and women consistent with an evolutionary history which involved sexual selection within monogamous, polygynous, or multi-male/multi-female mating systems? I am not concerned here with sex differences in olfactory cues. Although body odours have been implicated in mate choice and attractiveness in human beings (Wedekind 2007), given the current state of knowledge, I consider that studies of sexual dimorphism in bodily and facial cues provide greater insights into the likely origins of human mating systems.

Sexual dimorphism, body weight, and mating systems

Sexual dimorphism in adult body weight is most pronounced in those primates, such as geladas, hamadryas baboons and gorillas, which have polygynous mating systems (Figure 7.1). Competition between adult males for access to females is intense in such species, as is the case in other polygynous mammals (as, for example, elephant seals or red deer). Selection for increased male body size has also occurred in anthropoids such as macaques, chacma baboons and mangabeys, which have multi-male/multi-female mating systems. In many cases sex differences are not as pronounced in multi-male/multi-female forms as in polygynous primates, but there are exceptions, as will be discussed below. In monogamous primates, such as the owl monkeys, marmosets, and gibbons, the two sexes are usually very similar or equal in size.

Figure 7.1 also shows the degrees of sexual dimorphism in canine length in the various mating systems, and these broadly correlate with the data on body size. The largest canines occur in males of polygynous primate species, whereas there is little or no sexual dimorphism in the monogamous species. However, as discussed in Chapter 1, reduction in sexual dimorphism in canine size is an ancient hominid trait, occurring throughout the australopithecine lineage as well as in the genus *Homo*. Comparative studies of sexual dimorphism in canine size among extant primates are of limited value in attempts to reconstruct the origins of human mating systems.

Human beings are sexually dimorphic in body size, with the ratio of adult male-to-female body weight being approximately 1.1–1.2. Considerable individual variation is apparent, both within and between human populations, and some examples are provided in Figure 7.2. In the fourteen populations represented here, men average 63.5 kg and women 52.3 kg in body weight. Men are also taller than women on average, and the ratio of adult male/adult female height is 1.06 for the thirteen examples included in Figure 7.2. These sex differences are quite modest; slightly larger than those which typify monogamous primates, but much less than in many polygynous forms. Consider, as examples,

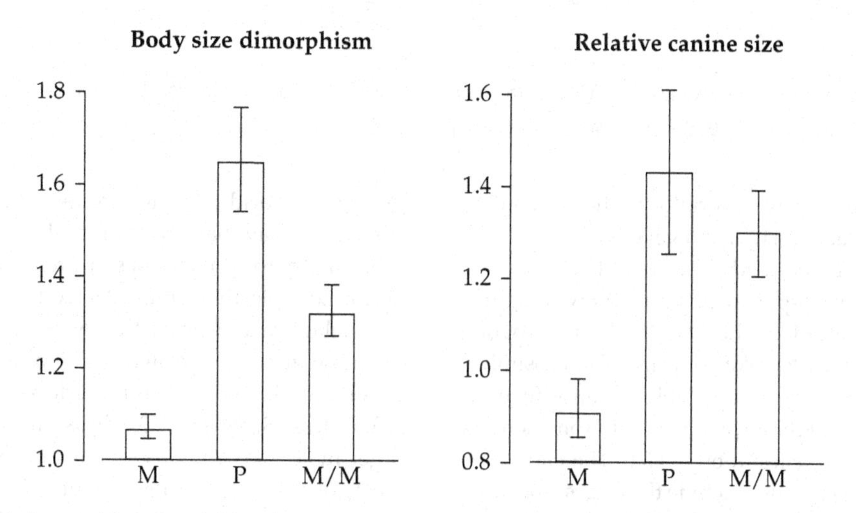

Figure 7.1 Body size sexual dimorphism (adult male body weight divided by adult female body weight) and relative canine sizes in primate genera which have monogamous (M), polygynous (P), or multi-male/multi-female (M/M) mating systems. Data are means ± SEM.
Source: From Dixson (1998a); after Clutton-Brock (1991).

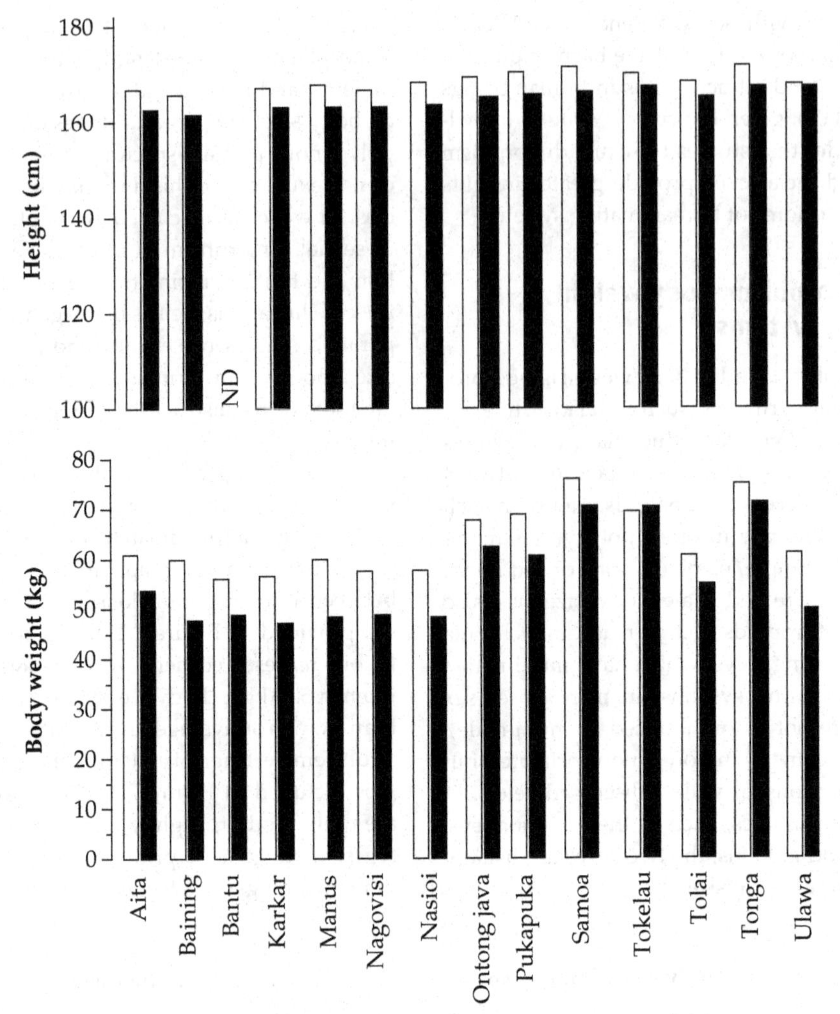

Figure 7.2 Sex differences in body weight and height in various human populations. Open bars = men; Black bars = women.
Source: Based on data in Collard (2002) and derived mainly from Houghton (1996).

male/female body weight ratios in the gorilla (2.4), hamadryas baboon (1.7), and gelada (1.6).

Martin, Willner, and Dettling (1994) showed that sexual dimorphism in adult body weight is more pronounced in the Old World anthropoids than among the New World forms. The prosimians should be considered separately, because many of them lack sexual dimorphism, and in some lemurs, females exceed males in body size. The Old World monkeys and apes also include the largest species of primates, and there is a tendency for sexual dimorphism to scale with increases in body size (Rensch's rule). More importantly, the fact that many of the larger Old World primates are terrestrial, or semi-terrestrial, may have favoured the evolution of larger males to combat risks of predation. In gorillas, for example, the dominant silverback male plays a crucial role in defending the females and offspring when the group is threatened in any way. Without the silverback, the group is leaderless and relatively defenceless (Schaller 1963). Thus, the evolution of larger male body size may be due to the entwined effects of natural selection (via predation risk) and sexual selection (inter-male competition for access to females). This should not surprise us. Moreover, the comparatively modest but consistent tendency for men to be heavier and taller than women may reflect ancient effects of natural and sexual selection during human evolution in Africa. One way of examining in more detail the degree of sexual selection in males of extant species is to compare ratios of male/female body weight with the *socionomic sex ratio* in groups of these species. The socionomic sex ratio of a primate group is the number of adult females divided by the number of adult males. The ratio tends to be highest in polygynous species (where only a single adult male may be present) and lowest in monogamous species (where there is often just one adult male and female). Previous analyses of this type have been criticized (Martin et al. 1994) on the basis of the limited sample sizes used and an over-representation of monogamous primate species (Clutton-Brock, Harvey, and Rudder 1977; Alexander et al. 1979). Figure 7.3 shows the results of an expanded analysis, made possible by virtue of the fact that field data on the socionomic sex-ratios of primate groups are now much more complete. One hundred and eleven species representing a total of thirty-nine genera of monkeys and apes are included in the analysis represented in Figure 7.3. There is, in general, a positive relationship between increasing sexual dimorphism in body weight and increasing numbers of adult females per adult male in social groups of anthropoids. It is important to note, however, that there is a wide scatter of data points on the graph. Some species with multi-male/multi-female mating systems exhibit greater body size dimorphism than more typically polygynous species. Some unusual socionomic sex ratios occur, as in the mandrill where adult females may outnumber adult and subadult males by a ratio of 6:1, despite the occurrence of a multi-male/multi-female mating system (Abernethy et al. 2002).

If we consider not only sex differences in human body size but also differences in muscle mass and strength between men and women (these will be discussed below), then these equate to a likely socionomic sex ratio between two and three. As regards human evolution, such a ratio might have occurred in an ancestor having either a polygynous or a multi-male/multi-female mating system. Given the strong evidence assembled in Chapters 2–4 indicating the absence of sexual selection within a multi-male/multi-female system during human evolution, it is much more likely that the data in Figure 7.3 are indicative of a polygynous background to human evolution. Two further lines of comparative evidence may be advanced in support of this conclusion.

Firstly, there is the question of reproductive bimaturism and its relationship to mating systems among the anthropoids. In many monkeys and apes, males take longer to reach sexual maturity than females; this is usually correlated with a later onset of puberty and slower progression through the stages of puberty and adolescence in males. These effects are most pronounced in species where sexual dimorphism in body size is apparent, so that males are larger than females and take longer to attain fully adult size. The ages at which males and females reach puberty and attain full body size in various Old World anthropoids including *Homo sapiens* are shown in Table 7.1. The age at which

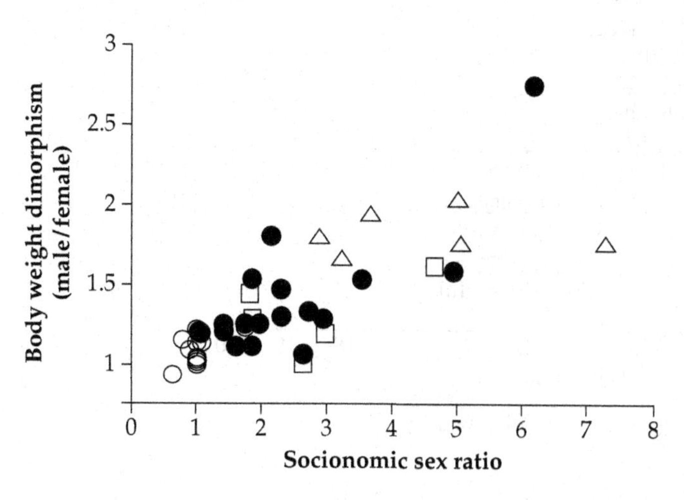

Figure 7.3 Relationships between socionomic sex ratios (*x* axis) and body weight sexual dimorphism (*y* axis) in New World monkeys, Old World monkeys, and apes. Sexual dimorphism is most pronounced in genera with large socionomic sex ratios (i.e. number of adult females per group/number of adult males). Mating systems are indicated by symbols: ○ = monogamous; Δ = polygynous; • = multi-male/multi-female; □ = mixed: polygynous or multi-male/multi-female groups both occur.

Source: Data are derived principally from Campbell et al. (2007) and also from Smuts et al. (1987); Davies and Oates (1994); Jablonski (1998), and Kappeler (2000).

puberty begins in males may be defined as the time when the androgenic functions of the testes increase markedly, and this is usually associated with overall increases in testicular growth, penile size and progressive secondary sexual development. There is considerable individual and inter-specific variation in the timing of these events. In boys, for example, increases in size of the testes and scrotum occur at about 11.5 years of age on average, and the penis begins to enlarge by 12.5 years. Despite individual differences in the onset of puberty, the general progression of physical changes follows an orderly sequence in members of both sexes (Figure 7.4). In girls, as in females of other Old World anthropoids, the occurrence of the first menstrual flow (menarche) provides a useful marker of puberty. However, this marker is not useful for the New World monkeys, because menstruation occurs in relatively few of these species. In girls, the earliest sign of puberty is usually growth of the breast-bud, and this may occur at anywhere between 8–13 years of age (average 11 years: Figure 7.4). Both sexes show increased growth in height; this is the *adolescent growth spurt*. However, it will be noted in Figure 7.4 that the growth spurt also begins earlier in girls than boys. This difference is the characteristic of other anthropoids in which the sexes are dimorphic in body size and in which sexual bimaturism occurs, in association with polygynous or multi-male/multi-female mating systems (Table 7.1).

One of the many contributions made to this field by J. M. Tanner was his discovery that since the

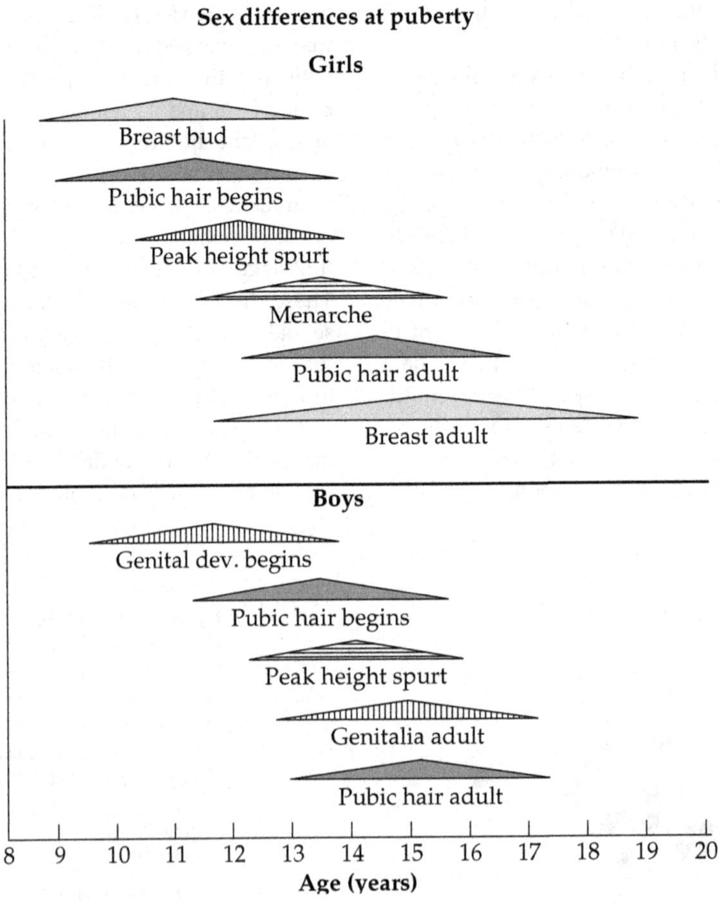

Figure 7.4 Timing of physical and associated changes throughout puberty in girls and boys.
Source: After Marshall (1970).

Table 7.1 Sexual bimaturism in Old World monkeys, apes, and humans

Species	Females (age in years)		Males (age in years)	
	At puberty	Full grown	At puberty	Full grown
Chlorocebus aethiops	2.8	–	5	–
Erythrocebus patas	–	3	4	5
Miopithecus talapoin	4	5	–	6
Lophocebus albigena	3.6	5	4.5	7
Cercocebus atys	3	5	5	6
Mandrillus	2.75–4.5	5	4–8	8–10
sphinx	(3.6)		(6)	
Papio cynocephalus	4.8	7	5.7	9
P. hamadryas	4.3	5.6	5.8	10
Macaca mulatta	2.5	6	3.5	8
M. sylvanus	3.5	5.5	4	7.5
Hylobates spp.	<6	8	<6	8
Pongo pygmaeus	7?	10–11	6–7	14+
Pan troglodytes	7.5–8.5	10–11	6–9	16.5
Gorilla gorilla	7.75	10	6–8	12–15
Homo sapiens: P	15–16	–	17	–
H. sapiens: M	12.8–13.2	16–18	13–14	20+

Source: Data from McCormack (1971); Tanner (1978); Harcourt et al. (1980); Dixson (1981); Dixson et al. (1982); Goodall (1986); Kingsley (1988); Wickings and Dixson (1992*b*); Bercovitch (2000); and Campbell et al. (2007). P = pre-industrial revolution; M = modern.

industrial revolution the age of puberty has declined steadily in Europe and in the USA. The average age at which girls reach menarche declined by about 4 months per decade, from 15–16 years in 1860, to 12.8–13.2 years during the 1960s (Tanner 1978). Menarche typically occurs in girls after the peak of the adolescent growth spurt in height. Girls increased in average height during the nineteenth and twentieth centuries, as well as entering puberty at a younger age. The same was true for boys, and the growth of the larynx, which takes place during puberty due to increased secretion of testosterone, now occurs at a significantly younger age than would have been the case historically. An ingenious demonstration of this was provided by Daw (1970) who examined the records of the ages at which 'breaking' and deepening of the voice occurred among boys in the J. S. Bach Choir in Leipzig. For the years 1729–49, such changes typically occurred at around 17 years of age. Nowadays, boys who are only 13 or 14 years old exhibit comparable development of the larynx resulting in deepening of the voice.

From the earliest phase of human evolution, the sexes would have exhibited bimaturism of reproductive development, with males passing through puberty, adolescence, and attaining sexual maturity later than females. Although the timing of these events occurs at younger ages in various modern human societies, the sequence of changes in growth, development of the reproductive systems and secondary sexual traits remains stable, as does the tendency for girls to mature earlier (Figure 7.4). As can be seen in Table 7.1, such bimaturism is typical of monkeys and apes which are sexually dimorphic in body weight. Indeed, Martin et al. (1994) were able to show that the age of attainment of sexual maturity is usually the same for males and females in those primates where the sexes are of equal size. However, for sexually dimorphic species (for which the ratio of adult male/adult female body weight exceeds 1.15) males reach maturity later than females. Human beings fall within the latter group, and the larger size and slower development of the human male is consistent with an evolutionary past involving a significant degree of inter-male competition for access to mating partners.

It is also relevant to mention here that males tend on average to have shorter lifespans than females in many vertebrates, including humans (Mealey 2000; Finch 2007). There is evidence that mating systems may have some impact upon this sex difference. Clutton-Brock and Isvaran (2007) reported that higher levels of inter-male competition for access to mates (such as occur in multi-male/multi-female groups and in polygynous one-male breeding units) are associated with reductions in male longevity (relative to females) in natural populations of mammals and birds. Table 7.2 summarizes the data from sixty-nine countries worldwide showing percentages of males and females in human populations at various ages. Initially, males tend to slightly outnumber females, probably due to a slight sex difference in birth sex ratios (105 male:100 female births). However, by 20–39 years of age percentages of males and females are almost equal and from 40 years of age onwards women increasingly outnumber men in the vast majority of countries that conduct population censuses. This sex difference in human longevity is intriguing, as it may be greater than expected for a monogamous mammal. Thus it could be relevant to judgments about the role of polygyny in human evolution. Unfortunately, no firm conclusion can be reached, as very few comparative data are available on longevity in monogamous mammals. Thus, Clutton-Brock and Isvaran's (2007) study relied mostly on information from socially monogamous birds and polygynous mammals.

Returning to the subject of changes in growth at adolescence in human beings, considerable differences occur in the development of muscle mass, fat depots and strength between the two sexes. Boys develop larger muscles than girls as well as having larger hearts and lungs, and they have greater numbers of erythrocytes (and a higher concentration of haemoglobin) in the blood. Harrison et al. (1988) summarize the evolutionary significance of these sex differences by stating that 'the male becomes at adolescence more adapted for the tasks of hunting, fighting and manipulating all sorts of heavy objects, as is necessary in some forms of food-gathering.' By contrast, females invest more physiological resources than males in the production and storage of fat, which is deposited particularly around the buttocks, hips, and thighs, as well as in the developing breast. These changes occur as a result of oestrogenic stimulation, as adolescent girls transition to womanhood and become reproductively mature (Pond 1998). The physiological demands of pregnancy and lactation are such that these fat reserves play a crucial role in determining the success of female reproduction and the viability of offspring.

Thus, a second point relating to the question of sexual dimorphism and the likely polygynous origin of human mating systems concerns sex differences in adult *body composition*. Sexual dimorphism in human body composition is much more pronounced than can be conveyed by the measurements of male and female body weight alone. Clarys, Martin, and Drinkwater (1984) dissected male and female cadavers to measure exactly how much various tissues contribute to body composition in adulthood. As can be seen in Figure 7.5, muscle constituted 38.8

Table 7.2 Sex differences in human longevity: Age-related changes in percentages of males and females in the total population, in sixty-nine countries worldwide

Percentage of total population	Age group (years)				
	0–19	**20–39**	**40–59**	**60–79**	**80+**
Males	19.5**	15.0	9.6	4.6	0.6
	±6.2	±1.4	±2.8	±2.2	±0.4
Females	18.7	15.0	9.8*	5.7**	1.2**
	±6.1	±1.3	±2.9	±3.1	±0.9

Note: Data are mean ± SEM percentages of males and females of each age group in the total populations of 69 countries, as listed in The Financial Times World Desk Reference, 2004. * $P < .05$; ** $P < .01$.

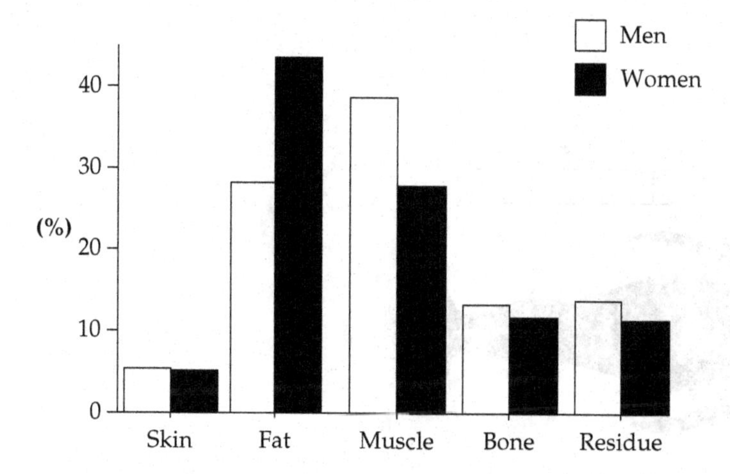

Figure 7.5 Human body composition. Men show higher percentages of muscle, and women show higher percentages of fat, in relation to their overall body composition.

Source: Based upon data from Clarys et al. (1984).

per cent on average of body mass in men, but only 27.7 per cent of overall mass in women. Conversely, women included an average of 43.6 per cent fat (adipose tissue) in their body make-up, as compared to 28.4 per cent in men. Unfortunately, I have been unable to locate body composition data of the kind summarized here for any of the non-human primates. Nor are the cross-cultural data on sex differences in human body composition adequate for comparative purposes. Nonetheless, it is interesting that the ratio of male muscle mass/female muscle mass in Clarys et al.'s (1984) sample is 27.4:17.89 kg, which yields a sexual dimorphism ratio of 1.53. This is considerably higher than the male/female body weight ratio (1.1) for these same subjects and strengthens the conclusion that selection has favoured increased development of muscular strength in men. Indeed, from puberty onwards, males begin to out-perform females in tests which assess motor performance (e.g. measures of hand-grip, arm-pull, and arm thrust strength: Harrison et al. 1988). Sex differences in human body composition thus also support the conclusion that polygyny, rather than monogamy alone, has played some role in the origin of human sexuality. However, to examine this question further, it will be helpful to enquire how far human body shape and composition might influence mate choice, and whether any such effects are cross-culturally consistent and likely to represent ancient traits for *H. sapiens*.

Human physique and sexual attractiveness

Masculine Somatotype

A number of studies have produced evidence that women find certain types of masculine physique more attractive sexually. A broad chest and shoulders, narrow hips, and a muscular torso are important traits which influence female assessments of masculine physical attractiveness (Lynch and Zellner 1999; Dixson et al. 2003). In order to quantify these preferences, some studies have made use of Sheldon's (1940) system of *somatotyping*. This classifies human physique according its degree of mesomorphy (muscular and powerful build), ectomorphy (slim and lean build), and endomorphy (heavy-set and fatter build). Examples of pronounced mesomorphic, ectomorphic, and endomorphic somatotypes are shown in Figure 7.6, along with an intermediate, or average, type of physique. Images of this kind, back-posed to avoid the distraction of facial cues and standardized for height and posture, have been used in cross-cultural studies to measure female preferences for masculine physique (Dixson et al. 2003; Dixson et al. 2007*a*; 2007*b*; in press). Results of studies conducted in China, New Zealand, USA, and Cameroon are summarized in Figure 7.7. In each case the images were modelled to include skin colour and hair appropriate for the population considered. In all these countries

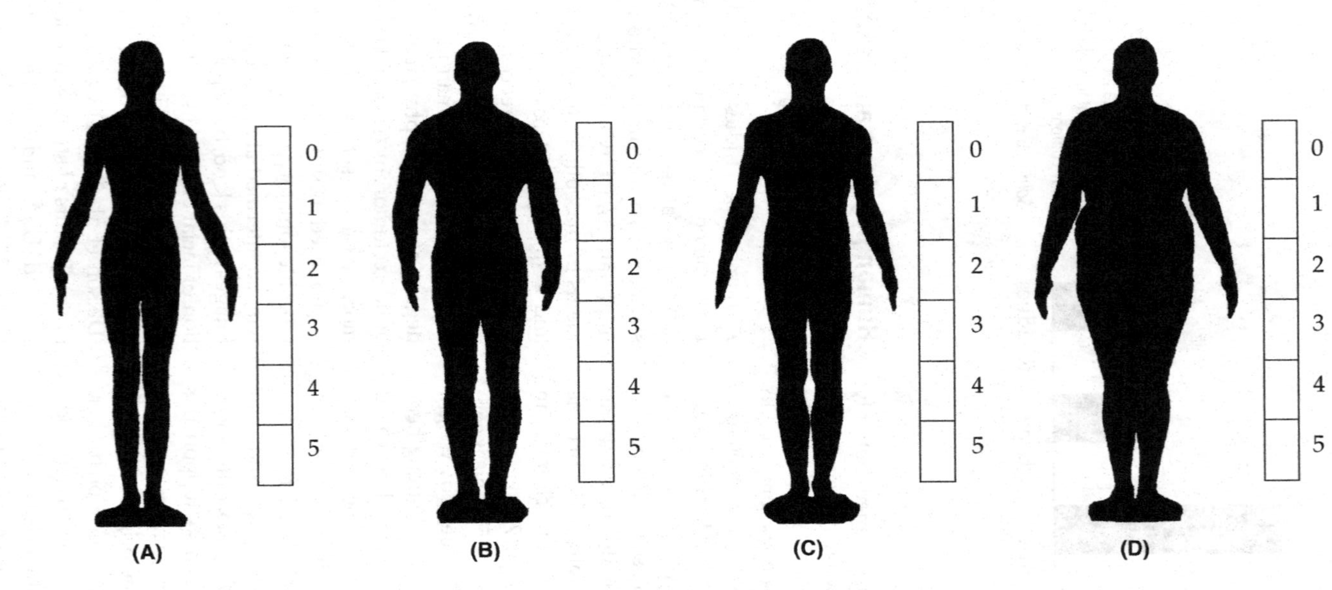

Figure 7.6 Back-posed images of male somatotypes ((A) ectomorphic; (B) mesomorphic; (C) average; (D) endomorphic) used in cross-cultural studies of masculine physique and sexual attractiveness. Results of some of these studies are shown in Figure 7.7.

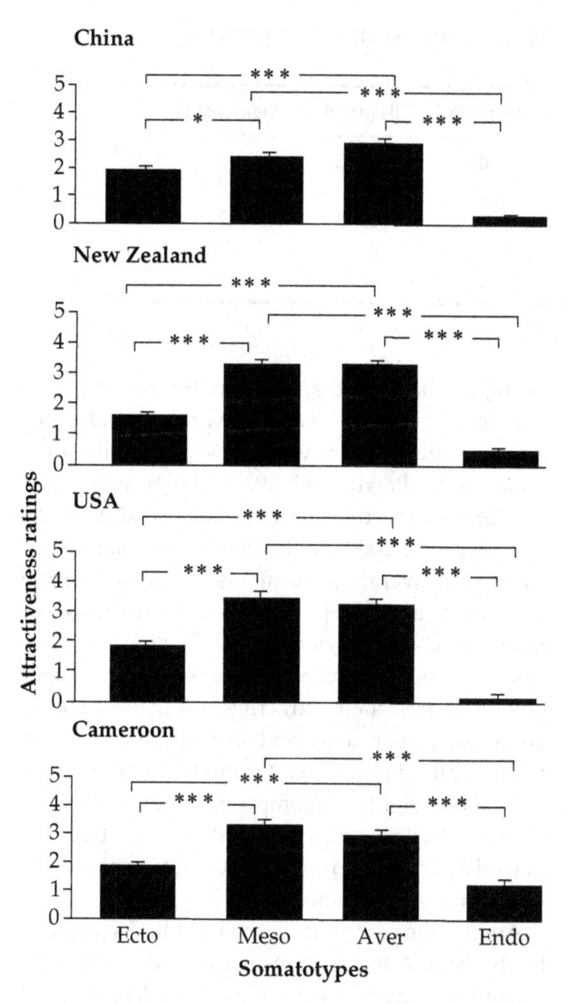

Figure 7.7 Women's ratings of the attractiveness of the four masculine somatotypes depicted in Figure 7.6. In China, New Zealand, USA, and Cameroon, women gave the highest ratings to mesomorphic and average somatotypes. *$P < .05$; ***$P < .001$.

Source: Data (means ± SEM) are from Dixson et al. (2007*a*; 2007*b*; and in press).

mesomorphic and average somatotypes received the highest ratings for sexual attractiveness, whilst the endomorphic physique was least preferred. The ectomorph did not receive the highest ratings in any of the populations studied, but this image did receive proportionately higher scores in China (Figure 7.7) and in Sri Lanka than in some other countries. Thus, some cultural variations probably exist in female preferences. As a second example of such differences, a mesomorphic somatotype was rated as significantly more attractive than an average male physique in UK, whereas the reverse was true in China.

Sheldon's original ideas concerning somatotyping, published in his *Varieties of Human Physique* (Sheldon, Dupertuis, and McDermott 1940) and the *Atlas of Men* (Sheldon, Dupertuis, and McDermott 1954), have been much criticized and refined over the years (Carter and Heath 1990). While measures of somatotype and body composition are not synonymous, it is fair to say that pronounced endomorphy equates to a significantly fatter body build, whilst mesomorphy, such as is represented in Figure 7.6, is indicative of a higher overall muscle mass. Mesomorphy is associated with greater physical strength and aptitude for pursuits requiring speed, strength, and stamina (Carter and Heath 1990). Ectomorphy, by contrast, is consistent with a less powerful physique, and endomorphy may be associated with greater risk of cardiovascular disease and diabetes (Katzmarzyk et al. 1998; Bolunchuk et al. 2000). Thus, female preferences for average or mesomorphic physiques may be related to the fact that such traits provide reliable signals of masculine health, strength, and fitness. These traits may have been of considerable importance in human evolution, given selective pressures upon males to compete with each other, protect their families and to display competence in hunting, scavenging, and other skills required for survival and successful reproduction (Ellis 1992; Buss and Schmidt 1993; Bramble and Lieberman 2004).

Bramble and Lieberman stress the likely importance of endurance running during evolution of the genus *Homo*. This might have been especially relevant to the success of hunting large game in open African savannah environments prior to the development of sophisticated weaponry. Persistence hunting survived until very recently among the !Xo and /Gwi bushmen of the central Kalahari in Botswana. Liebenberg (2006) has provided valuable information on the tracking and survival skills used by small groups of these hunters in pursuit of prey such as kudu, hartebeest, and gemsbok. The persistence method of hunting provided these bushmen with significant amounts of meat, although not as much as could be secured by hunting with dogs (Table 7.3). For the earliest precursors of human beings in Africa, such as *H. ergaster*,

Table 7.3 Persistence hunting: success rates as compared to other hunting methods used by the !Xo and /Gwi hunter–gatherers of the Kalahari

Hunting method	Days tried	Attempts	Successes (%)	Kills (per day)	Yield (kg/day)
Persistence	46	2	100	0.043	5
Dogs	5	5	60	0.6	24.8
Bow and arrow	46	41	5	0.043	2.9
Club and spear	46	11	45	0.109	1.6
Spring-hare probe	46	29	14	0.087	0.1

Source: After Liebenberg (2006).

equipped with only the most primitive technology, it is possible that endurance running and persistence hunting were crucial elements in the evolution of their hunter-gatherer cultures. Leibenberg concludes that 'before the domestication of dogs, persistence hunting may have been one of the most efficient forms of hunting. Endurance running and persistence hunting may therefore have been crucial factors in the evolution of humans.'

Human mate choice is an exceedingly complex subject, and cross-culturally the sexes exhibit an array of preferences for particular traits in long term partners, including aspects of personality, such as kindness and a good sense of humour (Buss 1989). However, in response to images offering a choice of masculine somatotypes, women consistently choose either men of average or muscular physique. Probably, the ideal lies somewhere between these two somatotypes. Honed by natural selection during human evolution, the more muscular physique of men conveyed advantages in hunting and survival. Inter-male competition for access to mating partners was probably also instrumental in moulding the evolution of masculine physique, as was female mate choice for the physically most healthy and able males. The preferences expressed by women in modern day human cultures (Figure 7.7) may thus reflect the results of natural and sexual selection acting upon our African ancestors.

Masculine stature

Like sexual dimorphism in body weight, sex differences in height are characteristic of human populations worldwide (Figure 7.2). Although stature varies considerably between human populations, men within each population tend on average to be taller than women. It is also the case, cross-culturally, that women express a preference for marriage partners who are slightly taller than themselves (Pawlowski 2003). Taller men have greater reproductive success (Pawlowski, Dunbar, and Lipowicz 2000; Nettle 2002). Leg length contributes to overall assessments of human stature and may thus affect judgements of attractiveness. Sorokowski and Pawlowski (2008) report that this is the case but increasing the leg lengths of images only enhances their attractiveness if the changes are small (5 per cent) and in true proportion to the overall height of the figures. As with masculine body weight and composition, it is likely that sexual selection as well as natural selection has played a significant role in the evolution of sex differences in human stature.

Among the extant apes, the hind limbs are relatively short. Although the australopithecines were capable of bipedal locomotion, their legs were of intermediate length, and it is with the advent of the genus *Homo* that stature increases markedly, perhaps in association with transitions to a hunter-gatherer mode of life. Thus, estimates of average stature in australopithecines such as *A. afarensis* (128 cm) and *A. africanus* (124.4 cm) are much less than for *H. ergaster* (185 cm), *H. erectus* (166.5 cm), *H. neanderthalensis* (165 cm), or for modern *H. sapiens* (Collard 2002). Carrier (2007) has advanced the interesting hypothesis that the short legs of great apes may be adaptive not only during climbing, but also during inter-male aggression. Carrier argues that males benefit from having a lower centre of gravity when they stand erect (as during the charging displays of male chimpanzees, or the chest-beating displays of silverback gorillas) or when they fight, by increasing the leverage exerted through the upper body and arms. Fights between males can be severe in chimpanzees (Goodall 1986), and although fights are infrequent between silverbacks, they can be intense (Harcourt 1978). The same is the case for aggression between mature male orangutans (Galdikas 1985*b*). Injuries due to fighting have been documented in skeletal studies of all these species and more so among males than females (Lovell 1990; Jurmain 1997). The australopithecines had relatively short legs, intermediate in length between those of the great apes and *Homo*, and Carrier points out that this feature persisted 'for over two million years, or approximately 100,000 generations'. This may have been connected to patterns of inter-male aggression which benefited from having shorter legs and a lower centre of gravity, as among the extant apes. We should keep in mind, however, that retention of somewhat shorter legs in the australopithecines may have been associated primarily with their locomotor specializations, which included climbing in arboreal environments, as well as terrestrial, bipedal locomotion. Lengthening of the legs in the genus *Homo*, may equally reflect shifts in ecological specialization towards an exclusively terrestrial existence and an increased ability to walk and run for long distances.

Current interpretations of the fossil evidence encourage the view that male australopithecines were larger than females; this is consistent with significant levels of inter-male competition such as may occur in polygynous or multi-male/multi-female mating systems (however, see Chapter 1, for a critique of evidence on sexual dimorphism in australopithecines). With the emergence of the genus *Homo* and the growth of inter-male competition involving tool use, the ability to run for long distances, and to combine physical strength with enhanced intellectual abilities, the advantages of short stature would have decreased. Indeed, sexual selection is likely to have favoured female preferences for taller partners who were more successful in these new kinds of inter-male competition, hunting, and survival skills. The preference which many women currently express for a taller long-term partner may thus also reflect ancient traits, developed in forms such as *H. ergaster* and incorporated into the mate-choice systems of *H. sapiens*.

The female hourglass figure: Singh's hypothesis

As well as significant sexual dimorphism in stature, body weight and body composition in human beings, there are marked sex differences in body shape between men and women. These differences emerge during puberty and adolescence. In women, oestrogenic stimulation results in greater deposition of fat in the buttocks, thighs, and breasts. In men, by contrast, testosterone promotes greater muscular development, and fat is laid down in the abdominal region, rather than in the buttocks, thighs, or breast area (Harrison et al. 1988; Rebuffe-Scrive 1991). Women therefore accumulate more fat than men in the lower part of their bodies (a *gynoid body shape*), with slimmer waists and broader hips, accentuating the skeletal sex difference in the pelvis. The *waist-to-hip ratio* (WHR) is a simple measure of this sexual dimorphism in adult body shape. The circumference of the waist, at its narrowest point, is divided by the circumference of the hips; in women the resulting WHR ranges from 0.67–0.80 (in healthy pre-menopausal subjects) whereas in men higher values (0.85–0.95) are the norm (data collected in Finland: Marti et al. 1991). Singh has conducted a series of studies in order to examine the role of female WHR in male mate choice. He proposes that a low WHR, such as occurs in young, non-pregnant women, is indicative of a healthy distribution of body fat and consistent with a fertile and reproductively advantageous physiology. Masculine preferences for women possessing a narrow waist and an hourglass figure may thus have been favoured by sexual selection during human evolution (Singh 2002, 2006).

There is evidence that the female WHR might provide an honest signal of health and reproductive potential. Possession of a narrow waist and large breasts correlates with the production of higher levels of oestrogen during the follicular phase of the menstrual cycle (Jasieńska et al. 2004), whereas women with high WHRs tend to have more irregular cycles or to fail to ovulate (Moran et al. 1999; Van Hooff et al. 2000). Women who suffer from polycystic ovarian syndrome, which is characterized by increased secretion of testosterone, reductions in oestrogen secretion and infertility problems, have higher WHRs than healthy women in the same age range (Pasquali et al. 1999; Velasquez et al. 2000). The hormonal changes and loss of fertility which occur at the menopause in women are also associated with shifts to higher WHRs (Arechiga et al. 2001).

Singh (2002) stresses that it is the 'interaction between WHR and body mass index (BMI) that affects health status and healthiness'. The body mass index (BMI) is a measure of weight, scaled for height (body weight in kilograms is divided by height in metres squared, to yield the BMI). However, measurements of the BMI alone do not capture the important sex differences in body shape and fat distribution which Singh implicates in men's ratings of female sexual attractiveness and health. Singh (1994) asked physicians of both sexes to rate the health and attractiveness of line drawings of female body shapes which varied in BMI (underweight, average, and overweight) and in four levels of WHR; examples of the drawings he used are shown in Figure 7.8. The results closely paralleled those obtained in an earlier study involving subjects who were not health professionals. Figures with a lower WHR (especially the 0.7 WHR images) were rated as more healthy, youthful, and attractive in both studies. The highest ratings were given to the average-weight female figure having a 0.7 WHR; underweight and overweight figures were not rated as highly for health or attractiveness, even when manipulated to have low WHRs.

A challenge for studies of the evolution of human morphology is the requirement to obtain data from a sufficiently wide range of cultures, given the variations in physique and sexual preferences which occur in different parts of the world. Some cross-cultural studies indicate that men might prefer a higher female WHR (Yu and Shepard 1998; Wetsman and Marlowe 1998) but I do not consider that these findings invalidate Singh's hypothesis. Yu and Shepard collected data in Peru from an indigenous population (the Matsigenka) living in a remote, protected area of the Manu National Park. Interestingly, men rated images of women with a 0.9 WHR as most attractive and healthy. However, it is important to note that Yu and Shepard used only images of women with either a 0.7 or a 0.9 WHR, rather than providing a range of images (e.g. no 0.8 WHR was available). The Matsigenka, like some other South American indigenous Indian cultures, tend to be quite thick-set in physique. The 0.7 WHR image may have appeared unusual, and less healthy for this reason; the only alternative choice being a 0.9 WHR female image. The line drawings used were the same as those in Singh's studies; front-posed, clothed in western-style bathing costumes and of Caucasian appearance (skin colour and hair style). This was also inappropriate, as it is important to use images which are relevant to the culture that is being sampled. Frank Marlowe's studies of the Hadza hunter–gatherers in Tanzania provide an

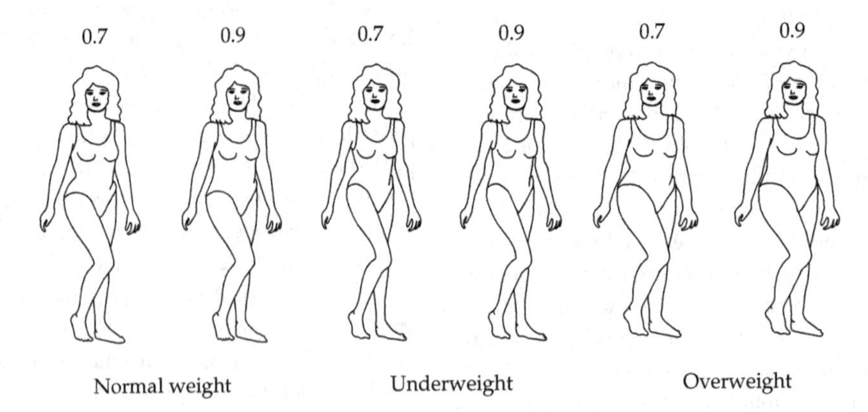

Figure 7.8 Examples of the line drawings of women, varying in waist-to-hip ratio (WHR) and body weight, used by Singh in studies of female WHR and attractiveness.

Source: From Dixson (1998*a*) and by courtesy of Professor D. Singh.

excellent example of this principle. Initially, Wetsman and Marlowe (1998) reported that Hadza men prefer images of women with higher WHRs. These investigators also used Singh's front-posed, line drawings of Caucasian women. However, Hadza men expressed their interest in viewing images of women which include the buttocks, as these are important in their judgements of attractiveness. Accordingly, a subsequent study (Marlowe, Apicella, and Reed 2005) used side-posed images of women; men's preferences for a 0.6 WHR emerged once they could assess images which included the buttocks (Figure 7.9A).

In our cross-cultural studies we have used back-posed images of women (Figure 7.9B). The results of this work confirm that men rate images of women having a slim waistline and a low WHR as most attractive sexually and as most desirable partners for a long term relationship. As examples, Figure 7.10 shows data on men's preferences collected in China, New Zealand, USA (California) and Cameroon. In each case, the same set of six back-posed female images was used, ranging in WHR from 0.5 to 1.0, but adapted to have skin and hair colours appropriate for each culture. It is clear from these studies, as well as others using different images, that men prefer female WHRs ranging from 0.6 to 0.8. Values within in this range are rated as more attractive and marriageable than are high WHRs (0.9–1.0). A WHR of 0.7 is not a universal preference, therefore, and we should not expect it to be so. Recall that, in Finland, healthy women of reproductive age have WHRs ranging from 0.67 to 0.8. It is reasonable to expect that sexual selection might have favoured male preferences within this range, and that the same might apply in different populations worldwide.

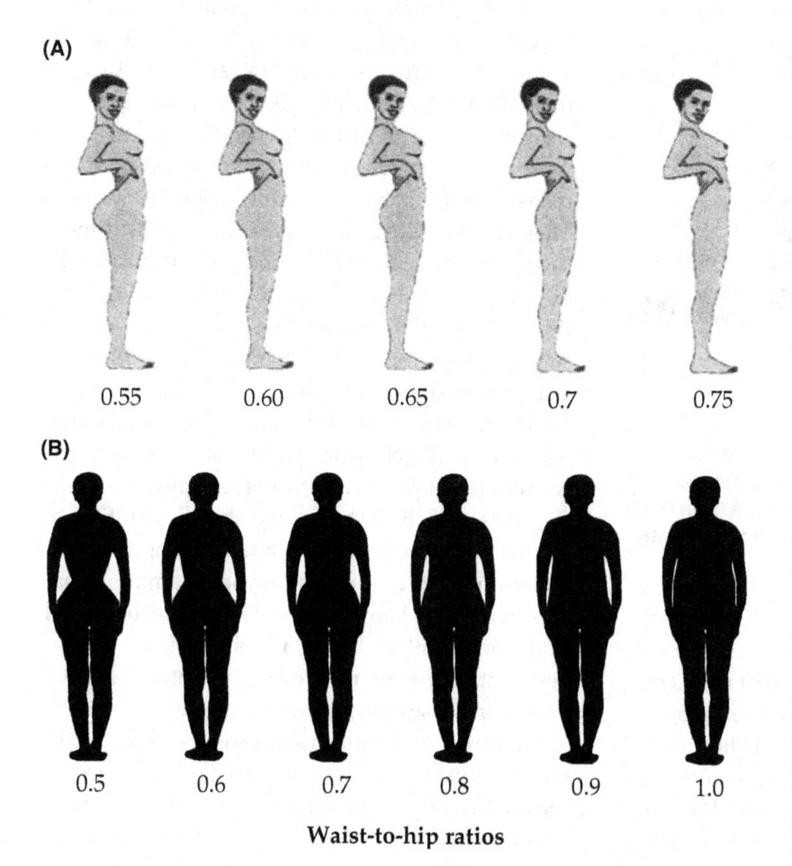

Figure 7.9 (A). Side-posed images of women, varying in waist-to-hip ratio (WHR), used by Marlow, Apicella, and Reed (2005) in their studies of female body shape and attractiveness in Hadza hunter–gatherers in Tanzania. (B). Back-posed images used in cross-cultural studies of female WHR and attractiveness. This version of the images was adapted for work in Cameroon. Some results of cross-cultural surveys are shown in Figure 7.10.

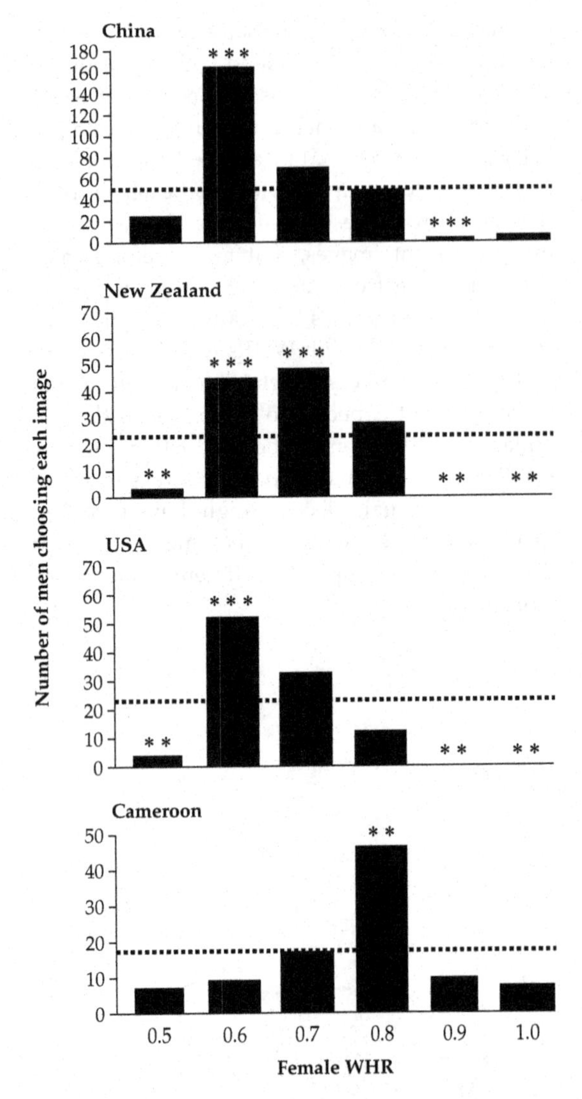

Figure 7.10 Men's ratings of the attractiveness of back-posed images of women, varying in WHR (as shown in Figure 7.9B). Histograms show the numbers of men who chose each image, compared to the numbers who would be expected to do so by chance (dashed line). χ^2 tests were used to determine levels of significance. $^{**}P < .01$; $^{***}P < .001$.

Source: Data are from Dixson et al. (2007*a*; 2007*b*; in press).

Although it is sometimes argued that men's preferences for female body shape have altered considerably down the ages, due to changes in fashion and other cultural constraints, there are good reasons to believe that a slim female waistline and a low WHR has always been rated as attractive. Singh (2002), for example, measured WHRs in 286 ancient sculptures from Africa, Egypt, Greece, and India. He found a sex difference in each case, with lower WHR values for the female images. In some cases, as in Indian sculpture, the female waists were very narrow; an artistic exaggeration of the fundamental sexual dimorphism. It is also the case that in literature and poetry, there has been a long-standing tradition to idealize the attractiveness of a slim female waistline (Singh, Renn, and Singh 2007).

There is, I suggest, at least one exception to the rule that a low female WHR is typical in ancient works of art. In the upper Palaeolithic period in Europe, small female figurines were sculpted, such as the 'Venus of Willendorf', which is shown, along with a similar example, in Figure 7.11. The fact that these *Venus figurines* have been unearthed at multiple sites across Europe, dated at 23,000–21,000 years ago, hints at a common cultural basis for their production (Gvozdover 1989; Stringer and Gamble 1993). Typically, the heads of such figures are rudimentary and lack facial details in most cases; likewise the arms tend to be thin and poorly modelled. The breasts, buttocks, thighs, and abdomen, by contrast, tend to be very large. Some of these figurines seem to represent mature women of late or post-reproductive status. Others have rounded abdomens and probably represent pregnant women. They do not appear to have a low WHR in most cases, but a formal study of their proportions would be most interesting. They may have symbolized the wish for a well-nourished and reproductively successful life rather than being representations of youthful, sexually attractive women. The period in which these figurines were sculpted was one of the cold and harsh climatic conditions in Europe. For scattered groups of human hunter-gatherers these figurines may have had some shared significance during ceremonies and the formation of alliances. However, they do not conform to the notion that slim waists and low female WHRs are universal traits of ancient sculpture, which leads me to suggest that they were not necessarily intended to be depictions of sexually attractive women.

A controversial issue in studies of the evolution of female physique and sexual attractiveness concerns the relative importance of a woman's WHR and her BMI as visual cues of her reproductive health and

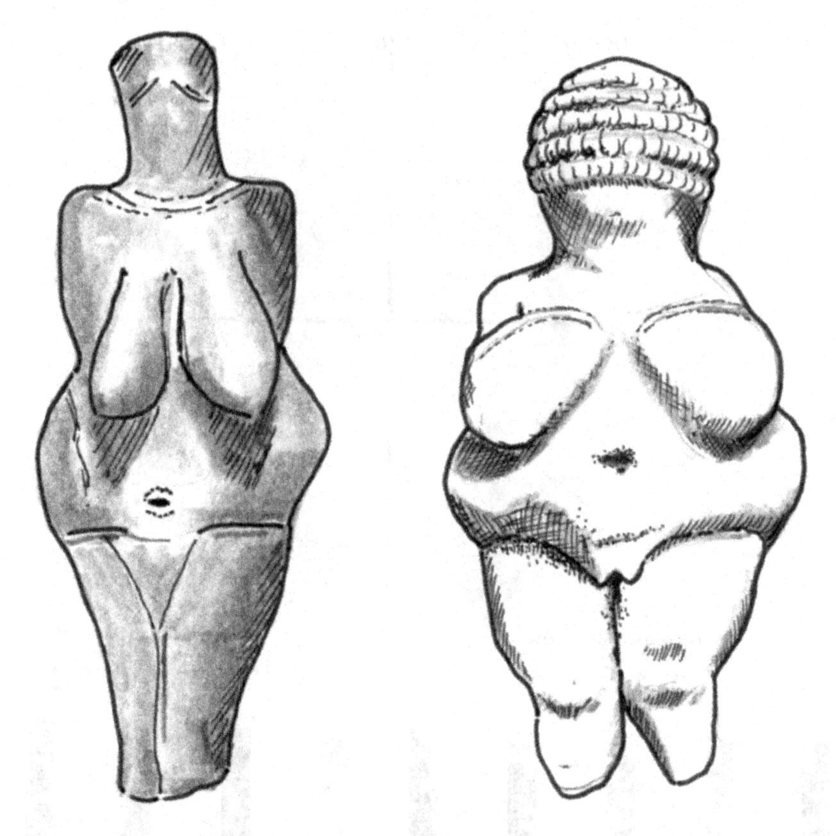

Figure 7.11 Examples of Palaeolithic 'Venus' figurines. (Left) A figurine from Dolni Vèstonice, made from clay and bone, fired at high temperature to produce one of the earliest known ceramic objects. (Right) The famous 'Venus of Willendorf'. This limestone figurine was found near the Austrian town of Willendorf and is thought to be approximately 25,000 years old.

Source: Author's drawings, from photographs.

attractiveness (Tovée and Cornellissen 1999; Tassinary and Hansen 1998; Manning et al. 1999; Henns 2000). Debates concerning this problem have been complicated by the fact that BMI and WHR are positively correlated; it has therefore proven difficult to tease apart these two variables in studies of female attractiveness. Figure 7.12 shows the results of a cross-cultural study which was conducted in order to address this question. Photographs were taken of women, before and after they underwent cosmetic micrograph surgery to reduce their waistlines and re-shape their buttocks. Men were asked to choose the most attractive of each pair of photographs (pre- and post-operative pictures were presented in random order). The study was conducted in New Zealand, Indonesia, Samoa, and Cameroon and included indigenous, as well as urban, populations (Singh et al. in press). In every case, men expressed a marked preference for the post-operative images of women, after reduction of their WHRs. Interestingly, of the ten women included in this study, five exhibited increases in BMI post-operatively, whilst five showed decreases. Yet these variations in female BMI had no significant impact upon men's judgements of female attractiveness (Figure 7.12).

The sexual dimorphism in body composition which develops from puberty onwards in human beings has doubtless resulted from natural selection, which has favoured a higher percentage of fat in the body composition of women for reproductive purposes. The distribution of female body fat has also been affected by other factors, given

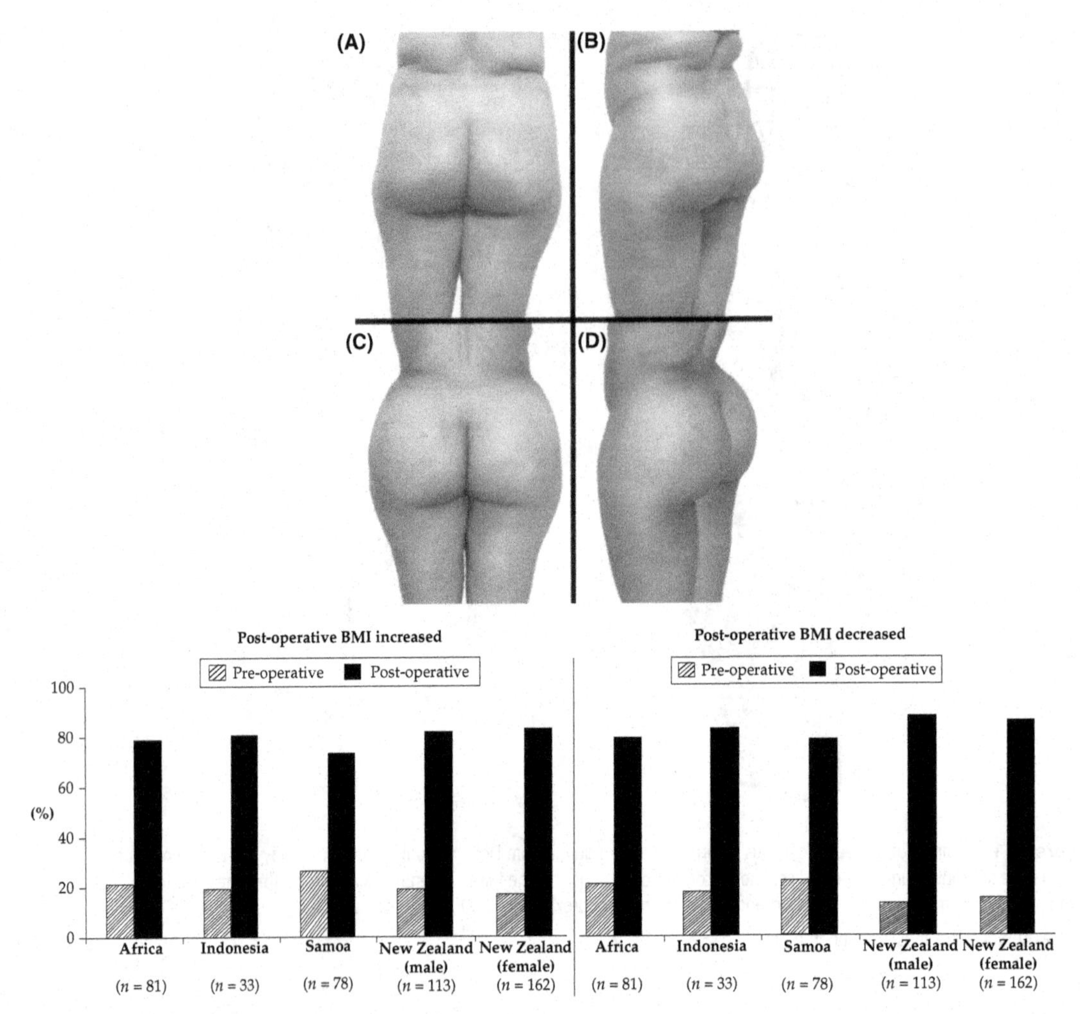

Figure 7.12 Upper: Pre-operative (A and B) and post-operative (C and D) photographs of a woman who underwent micrograft surgery in order to reduce her WHR and to enhance the shape of her buttocks. Lower: Results of a cross-cultural study to determine men's preferences for pre-versus post-operative images of the type depicted above. Photographs of ten patients were used. In five cases body mass indices (BMI) of patients increased post-operatively, whilst in five others BMI decreased. Irrespective of changes in BMI, men in Africa, Indonesia, Samoa, and New Zealand selected the post-operative images as being more attractive (as did women in New Zealand).

Source: From Singh et al., in press.

that storage in certain areas (buttocks, thighs, and breasts) may have been more advantageous for mechanical and thermoregulatory reasons. However, sexual selection has also influenced the evolution of the hourglass feminine shape, with rounded buttocks and a lower WHR than occurs in men, because these traits provide honest signals of a healthy distribution of body fat in women during their reproductive years (Singh 2002; 2006; Lassek and Gaulin 2008). Historically, men who reproduced with women possessing such traits would have secured greater reproductive success. Even if we argue that female WHR is of far less importance in some modern human populations, it is intriguing that men in a variety of countries and cultures consistently choose female images which have low WHRs (e.g. Figures 7.10 and 7.12). Thus, it is likely that men's choices of female images having a low

WHR reflect deeply ingrained psychological traits that have an ancient evolutionary origin.

Little has been said so far about sexual dimorphism in the human breast. Fat is deposited in greater quantities in the female breasts, as they enlarge markedly from puberty onwards. It has often, and correctly, been pointed out that *H. sapiens* is unusual among the primates, because the breasts enlarge in this way in young women who are neither pregnant nor lactating. (Short 1980). Indeed the size of the non-lactating breast is not indicative of its potential to produce milk, as it is adipose and stromal tissue, rather than milk-secreting tissue, which determines its size and shape. Symons (1995) has pointed out that the breast undergoes considerable changes in its external appearance in women throughout their lives, and that these changes may provide men with information about female age and reproductive status. Some of these variations in breast morphology, in pubescent, young adult and older women, are shown in Figure 7.13. It seems likely, therefore, that the contribution made by the female breasts to an hourglass body shape would have occurred in parallel with other changes in subcutaneous fat distribution which favoured the evolution of a lower WHR in women (Pawlowski 1999*b*). It is of great interest that Jasieńska et al. (2004) have shown that larger breast sizes and a low WHR correlate with higher levels of salivary oestradiol 17β during menstrual cycles (Figure 7.14). Progesterone levels are also higher in women with narrower waists, so that the WHR may provide a better indication of reproductive potential than breast size alone.

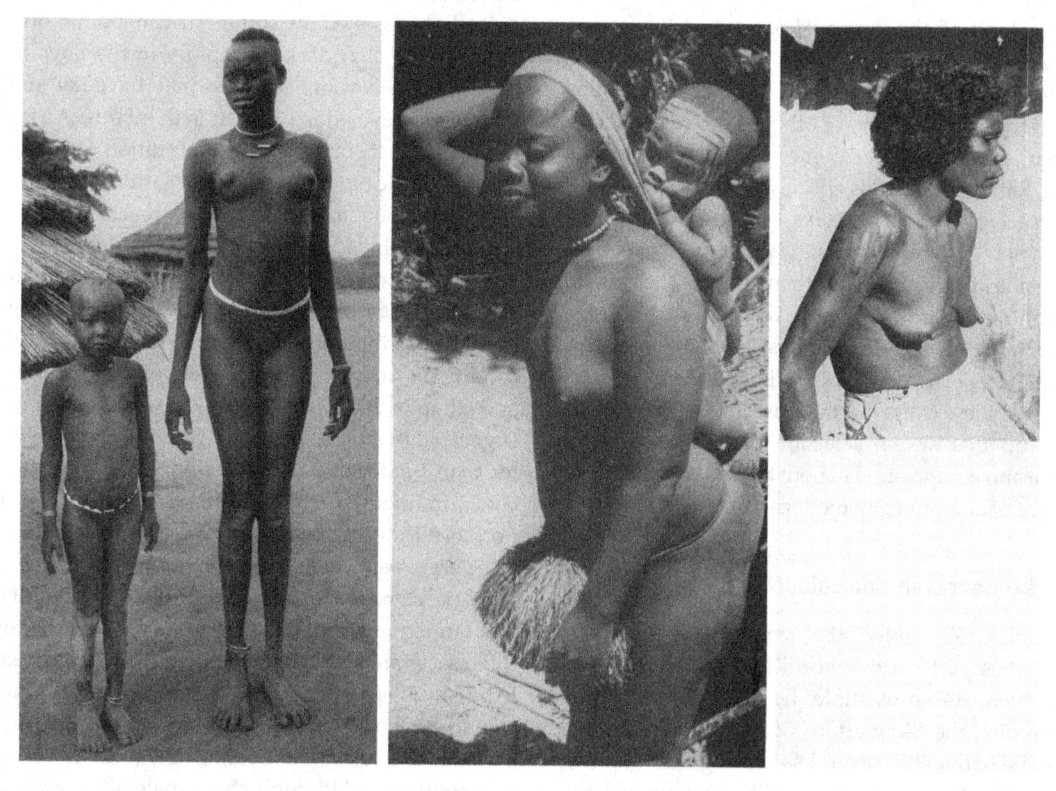

Figure 7.13 Changes in the appearance of the human female breast, with age and reproductive status. Left: Dinka girls from the Sudan. The taller girl exhibits adolescent breast development. Centre: A young Önge woman, from the Andaman Islands, who is lactating. She carries her baby on her back and also supports its weight on her buttocks (which protrude markedly due to extreme fat deposition: steatopygia). Right: An older (probably multiparous) Australian aboriginal woman.

Source: From Coon (1962).

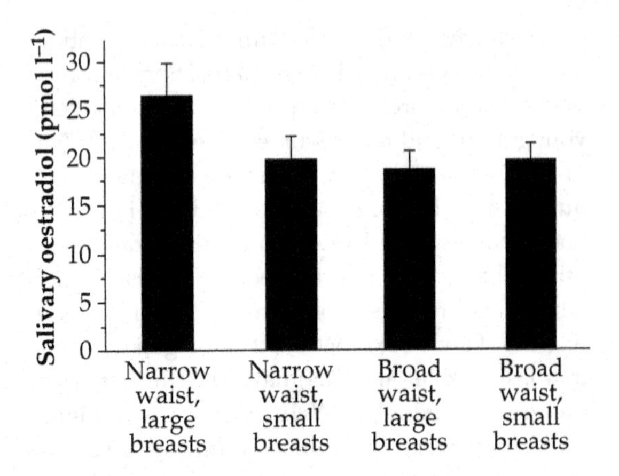

Figure 7.14 Relationships between waist size, breast size, and mid-cycle levels of salivary oestradiol in women. Women with narrow waists and large breasts have the highest oestrogen levels.

Source: From Jasieńska et al. (2004).

Members of the genus *Homo*, which preceded *H. sapiens* in Africa, had a very similar post-cranial anatomy, walked upright, and were probably subject to thermoregulatory or other selective pressures which resulted in reduction of body hair (Wheeler 1984; Rantala 2007). It is entirely possible, therefore, that *H. ergaster* or *H. erectus* females would have possessed narrower waists, fatter buttocks and thighs, and larger breasts than males. The males, by contrast, were likely to have been significantly taller, more muscular and physically stronger than the females. There are compelling reasons to suggest that these traits did not arise *de novo* in *H. sapiens*, and that sexual selection had affected body shape and composition in both the sexes prior to the advent of anatomically modern humans.

Sex differences in skin colour

Darwin (1871) noted the tendency for women in various parts of the world to have lighter skin than men. As an example, he cited Schweinfurth concerning the Monbuttoos of Africa: 'Like all her race she had a skin several shades lighter than her husband's, being something of the colour of half-roasted coffee.' Darwin also suggested that sex differences in skin colour might have been influenced by sexual selection during human evolution. Hormonal changes at puberty are now known to effect the expression of this sex difference. Thus, in young women, oestrogen promotes lightening of the skin, whilst in young men increased androgen secretion favours a darker complexion. This is due both to the deposition of melanin in the epidermis and the greater increases in red blood cells (and haemoglobin levels) in deeper layers of the skin, resulting in 'browner and ruddier complexions' in men than in women (Frost 1994).

Skin colour varies a great deal between human populations. Darker pigmentation, such as occurs in modern African populations, is most likely to represent the ancestral condition of *H. sapiens*. The layer of melanin in the epidermis helps to protect the skin against harmful effects of ultraviolet radiation, as the sun is especially powerful in equatorial regions. Protection against sunburn would obviously have been important for human ancestors in Africa. It is interesting that babies may initially be quite light-skinned in some African populations, but then the skin darkens rapidly in the days following birth. Natural selection may have favoured such rapid deposition of melanin in infants as a protection against sunburn. The emigration of *H. sapiens* from Africa to more-northerly latitudes, such as occurred when modern humans reached Europe approximately 40,000 years ago, was associated with the loss of melanin and the evolution of lighter skin. This may have been adaptive in maximizing the ability of the skin to synthesize vitamin D, under conditions of low ultraviolet exposure (Jablonski and Chaplin 2000; Jablonski 2006). It has also been suggested that lighter-skinned people may fare better than those who are dark-skinned in extremely cold conditions, such as those which pertained in northern Europe during the ice-ages of the Palaeolithic era (Post, Daniels, and Binford 1975).

Sexual dimorphism in skin tone has been reported in a large number of ethnic groups worldwide; the differences are subtle in many cases but they appear to be consistent, with women having slightly lighter skin than men (Frost 1988). Skin colour may provide a visual cue to female health, age, and reproductive condition. Although the female skin lightens due to oestrogenic effects with the onset of physical maturity (Tanner 1978), subsequent pregnancies and periods of lactation often bring about localized changes in pigmentation. These may affect the skin of the face, breasts, and abdomen, producing lasting changes and visible signs of aging (Wong and Ellis 1984; Sodhi and Sausker 1988). Symons (1995) has suggested that masculine preferences for lighter female skin tones might have undergone sexual selection, because such preferences would have enhanced the reproductive success of the men concerned. The skin, free of its covering of hair, becomes a canvas on which visual cues of health and reproductive condition may be displayed. Ethnographic comparisons confirm that male preferences for lighter skin in women exist in various human populations (van den Berghe and Frost 1986). As an example of recent work on this question, Figure 7.15 shows the results of studies carried out in Africa, New Zealand, China, and USA to examine male preferences for female skin colour. In each case, men were asked to choose which of five, back-posed female images they found most sexually attractive. The five images were identical except for skin colour. One image represented the colour typical for the culture being studied; two images were made progressively lighter in skin tone and two were darker. A computerized technique was used to control exactly the degree to which images were lighter or darker than usual (Dixson et al. 2007*a*; 2007*b*; in press). The results provide some support for the hypothesis that lighter female skin colours are more attractive to men, but they are not conclusive. Thus, in China, the lightest skin colour was clearly the most attractive, followed by the slightly lighter than average skin tone. Average and darker skin colours were rated as much less attractive by Chinese men (Figure 7.15). In both New Zealand and USA (California), darkening of the skin also rendered female figures less attractive, but only in California was lighter skin (one shade above average) rated as significantly more attractive. In the African (Cameroonian) sample, none of the five female images was rated as more attractive than would be expected by chance alone. The results of these studies are promising, but they should be extended to include other populations before conclusions can be drawn about the role of sexual selection in the evolution of human skin colour. In particular, it would be important to determine whether men are responsive to the subtle sex differences in skin colour which occur naturally in human populations. The images used in computerized questionnaires represent more pronounced variations in female skin tone than are typical of natural populations. Computerized images may represent *supernormal* stimuli (to borrow a term from ethology) in the same way that depictions of excessively slim female waists, or large breasts, might exaggerate features which men find attractive.

Sexual dimorphism of the human face

Hormonal changes during puberty and adolescence also affect growth and sexual dimorphism of craniofacial traits in human beings. Men tend to develop a larger jaw, chin, and cheekbones, a somewhat heavier brow-ridge and narrower, deeper-set eyes than women. These facial changes, which emerge gradually from puberty onwards in young men, reflect in part the effects of higher levels of testosterone. In young women, by contrast, a higher level of oestrogen, coupled with a much lower tendency to secrete testosterone, favours continued growth of more juvenile facial traits; a smaller chin and brow, larger eyes in relation to the size of the face and, often, fuller lips. As discussed above, the skin also exhibits differences between the sexes, so that the female complexion tends to lighten under the action of oestrogen. In men, testosterone stimulates the development of a darker complexion, as well as the growth of facial hair; this last trait will be considered in the next section.

Among the anthropoid primates, facial communication plays an integral role in many aspects of social and sexual behaviour (Van Hooff 1967; Dixson 1998*a*). Human non-verbal communication is complex, and we employ a rich repertoire of facial displays, many of which developed in ancestral forms prior to the evolution of language. *H. sapiens* is the only primate species in which the sclera of the eye is white, rather than being pigmented. This makes the iris and pupil of the human eye stand out in contrast to the surrounding 'white' of the eye. Judgments of gaze direction and eye-contact are most important during facial communication (Conway et al. 2008), so that these functions may be facilitated by the white sclera of the human eye (Kobayashi and Kohshima 2001). Facial cues are likely to play a fundamental role in human assessments of sexual attractiveness and in mate choice. Cross-cultural studies have shown that facial

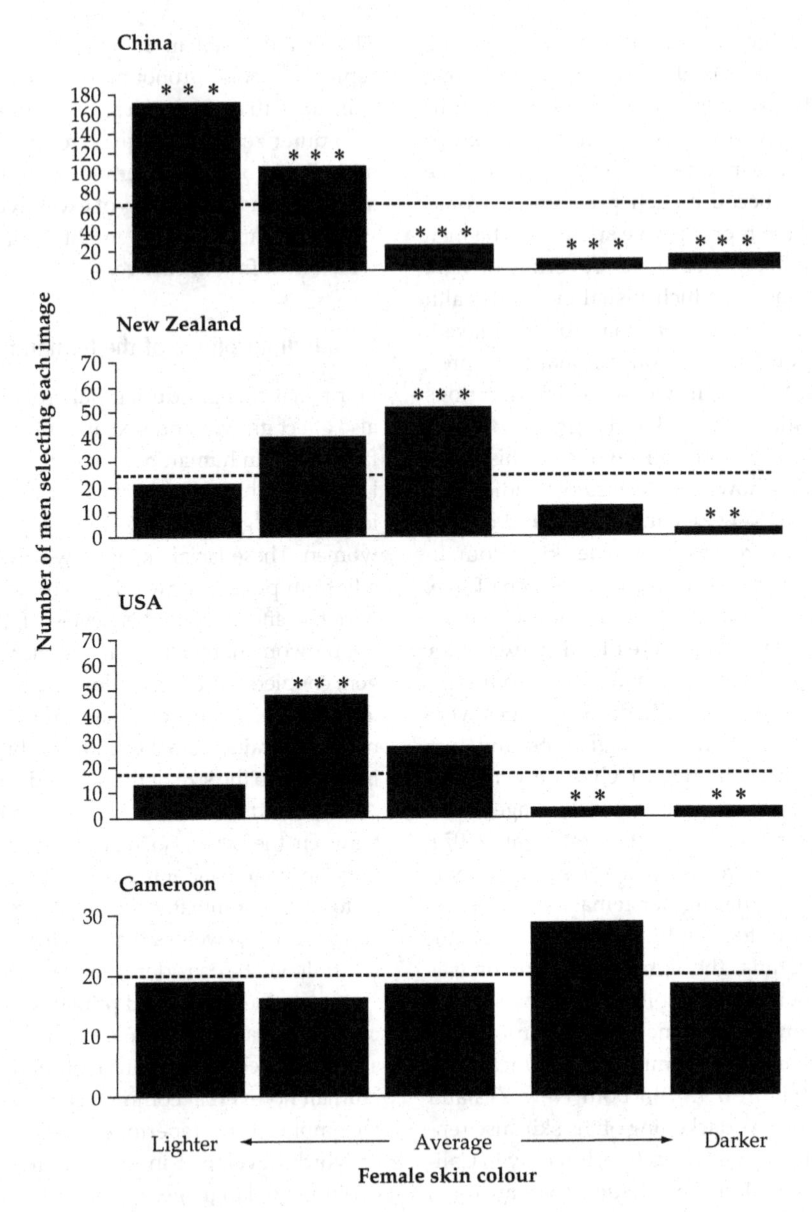

Figure 7.15 Results of cross-cultural studies to examine men's preferences for female skin colour. Data show the numbers of men who rated each of six female images as most attractive. The images differed only in their skin tone. The dashed line indicates the numbers of men who would be expected to select each image by chance alone. $^{**}P < .01$; $^{***}P < .001$ (χ^2 tests).

Source: Data are from Dixson et al. (2007*a*; 2007*b*; in press).

expressions reflecting certain fundamental emotions, such as joy, anger, surprise, and fear, are universal in *H. sapiens* (Ekman and Friesen 1971; Ekman 1973). That the human proclivity to exhibit such expressions is genetically determined is indicated by the finding that individuals who have been blind from birth show them, and even display distinctive familial traits of facial expression (Peleg et al. 2006). We should expect, therefore, that sexually dimorphic traits of the human face, whether in repose or when used for subtle non-verbal displays between the sexes, might be highly significant in assessing another individual's motivation and personality, as well as their health and attractiveness. A considerable volume of research has sought to define those qualities of the human face which determine beauty and attractiveness to the opposite sex. The roles of averageness and of symmetry have been mooted to be of importance in this regard (e.g. reviews by Penton-Voak and Perrett 2000; Rhodes 2006). The notion that averageness in human faces might be most attractive (Langlois and Roggman 1990) is based upon the finding that composite faces, formed by blending images of a number of subjects, are often considered beautiful. However, a careful study of this problem led Perrett, May, and Yoshikawa (1994) to conclude that subjects (in UK and Japan) find composites of the most attractive faces in a selection of images to be more attractive than the overall average for this larger set of images. They propose that certain fundamental determinants of facial attractiveness may be shared across cultures and that preferences for these traits might represent 'a directional selection pressure on the evolution of human face shape'. Facial symmetry might be an example of such a cross-culturally robust cue, as it is a signal of genetic quality and developmental stability. As an example, Little, Apicella, and Marlowe (2007) have shown that Hadza hunter-gatherers show preferences for facial symmetry that are more pronounced than those of people in the UK. In Figure 7.16, the reader may observe the very subtle changes required to produce symmetrical faces and also the degree of sexual dimorphism displayed by Hadza faces.

Here I am concerned primarily with the question of whether sexually dimorphic human facial traits might influence attractiveness and mate choice in men and women. If consistent effects can be identified, can these preferences tell us anything about the origins and evolution of sexual behaviour in the genus *Homo*? There is increasing evidence to support the view that, like the WHR, the female face provides visual signals of underlying health and fertility (Johnston and Franklin 1993; Perrett et al. 1994, 1998). Law-Smith et al. (2006) have demonstrated a positive correlation between circulating levels of oestrogen during the follicular phase of the menstrual cycle and ratings of women's faces for femininity, attractiveness, and health (Figure 7.17). Thus, women who have higher oestrogen levels are consistently rated as having more attractive (and healthy) faces. The reader may examine in Figure 7.17 the composite faces produced by Law-Smith et al; in my own experience of showing these to some hundreds of young men and women, they overwhelmingly rate the higher-oestrogen composite as more attractive in the absence of any prior information about possible differences between the photographs. It is evident that these composite faces differ subtly in a number of aspects of morphology; there are possible differences in complexion to consider as well. Indeed, there is some evidence that photographs of women's faces taken during the follicular phase of the menstrual cycle are rated as more attractive than photographs of the same individuals during the luteal phase (Roberts et al. 2004). There is the possibility, therefore, that women's facial attractiveness might convey information about short-term as well as longer-term cues to reproductive status and health.

The situation is more complex where masculine facial traits, health, hormonal status and attractiveness are concerned. Testosterone affects the degree to which male facial traits develop during puberty and adolescence, and there is evidence that more masculinized faces are perceived as more dominant, but not necessarily as more attractive, by women (Swaddle and Reierson 2002). Male teenagers with more dominant, masculinized facial traits report coital activity at younger ages than others (Mazur, Halpern, and Udry 1994). In a study of military officers who were graduates of the West Point Military Academy in the USA, dominant facial appearance was a reliable predictor of the ranks ultimately attained by men (Mueller and Mazur 1997). Those who became generals also had more children; an

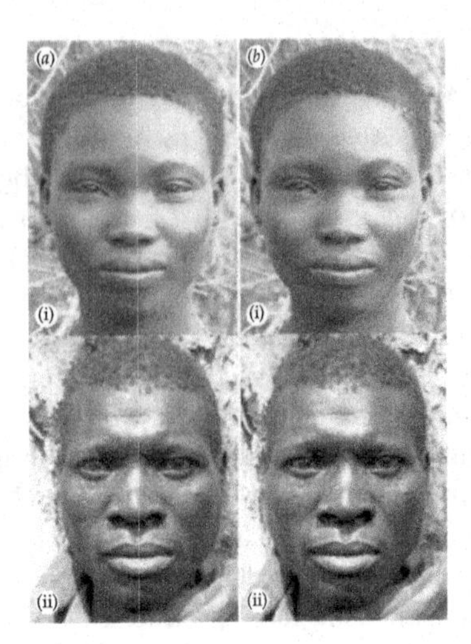

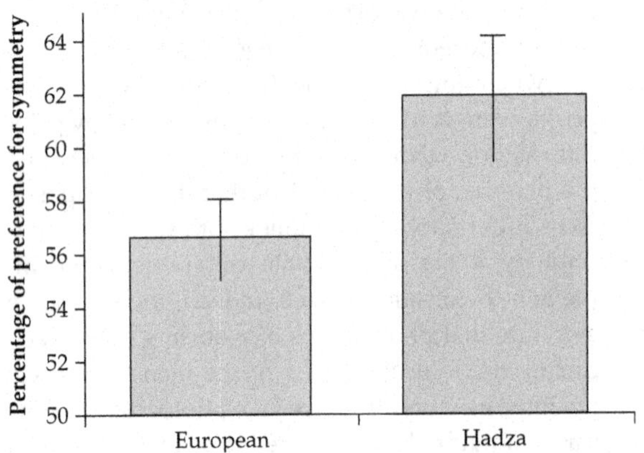

Figure 7.16 Symmetrical (ai and aii) and unaltered (bi and bii) photographs of the faces of a Hadza woman and man. The Hadza hunter–gatherers of Tanzania show greater preferences for facial symmetry than is the case for people tested in the UK.
Source: From Little, Apicella, and Marlowe (2007).

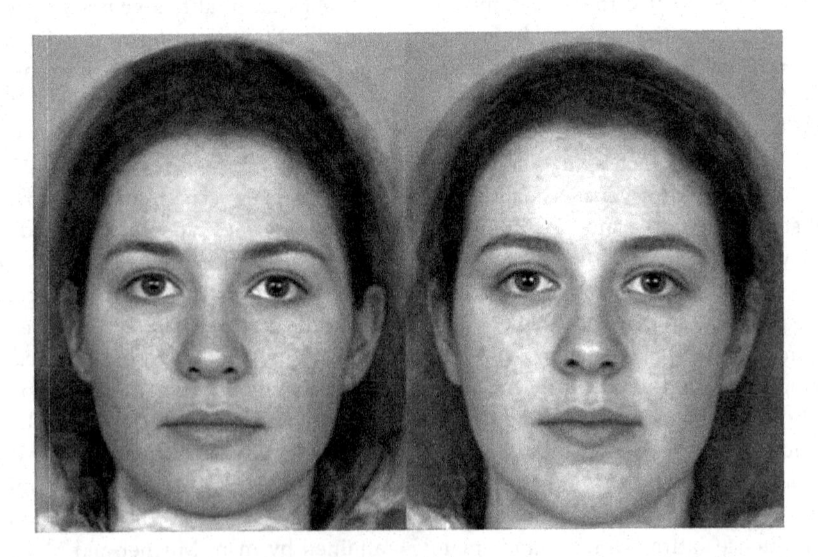

Figure 7.17 Which face is most attractive? Both photographs are composite images of a number of women. On the left, is a composite of faces of women who have high follicular phase levels of oestradiol 17β. The right hand composite image is of women who have lower oestrogen levels.
Source: After Law-Smith et al. (2006).

average of 3.67 vs. 3.02 for other officers ($P<.001$). Yet it is not the case that women always find masculinized faces more attractive; preferences for larger jaws and cheekbones in male faces have been reported in some studies (Grammer and Thornhill 1994; Scheib, Gangestad, and Thornhill 1999), whilst Perrett et al. (1998) found that women preferred more 'feminized' or 'baby-faced' males. As discussed in the last chapter, women's hormonal status can affect their judgments of the attractiveness of a variety of masculine traits, including more masculinized faces, which receive higher attractiveness ratings during the fertile phase of the menstrual cycle (Penton-Voak et al. 1999; Penton-Voak and Perrett 2000).

Women's preferences for facial masculinity may be predicted from their ratings of their own partner's masculinity and from stated preferences in a theoretical 'ideal partner' (DeBruine et al. 2006). Women who are facially very attractive and who have a low WHR are more likely to prefer more masculine faces (Penton-Voak et al. 2003). Rapidity of sexual development also affects female preferences; women who have intercourse at younger ages tend to prefer masculinized faces (Cornwell et al. 2006).

It is possible that such results may reflect different female strategies in relation to qualities signalled by masculine facial traits. Thus a highly masculine face may signal health and greater ability to withstand the immunosuppressive effects of testosterone (Rhodes et al. 2003). This line of reasoning is based upon Folstad and Karter's (1992) *immunocompetence–handicap hypothesis*, developed on the basis of animal studies and supported by research which demonstrates a correlation between masculine secondary sexual traits and immune function in various species (Møller, Christe, and Lux 1999). Well-developed masculine facial cues may also be reflective of other physical and behavioural qualities which are attractive to women. It has even been reported by Soler et al. (2003) that facial attractiveness in men correlates with measurements of semen quality, such as sperm motility and morphology. Although these correlations were not highly significant ($P<.05$), they are consistent with the results of experiments showing that masculine secondary sexual traits are positively correlated with semen quality in animals (e.g. in zebra finches: Birkhead and Fletcher 1995; red deer stags: Malo et al. 2005). Yet, if men's health, strength, and reproductive potential are reflected by facial masculinization, it is surprising that some women prefer more feminized faces (Perrett et al. 1998). In discussing this quandary Penton-Voak and Perrett (2000) point out that highly masculinized faces may also 'elicit negative personality attributions; coldness, dominance and dishonesty, for example'. Given the importance of personality assessments in human mate choice (Buss 1989), there may be advantages for women in choosing men whom they believe are likely to invest more in relationships and care of offspring. Roney et al. (2006) reported that women rated the images of men with more masculinized faces (and higher salivary testosterone levels) as more physically attractive, but assessed men with softer facial features as being more interested in children and more desirable for long-term relationships. One current limitation of all these studies is the paucity of cross-cultural data on human facial traits and attractiveness. With the exception of work on the Hadza, almost all studies have been conducted on populations in the USA, UK, and Japan.

The degree of sexual dimorphism in human facial traits is not as extreme as that which occurs in many non-human primates. Species that are highly sexually dimorphic in body size due to effects of sexual selection, such as gorillas, proboscis monkeys, golden monkeys, hamadryas baboons, orangutans, and mandrills, are often characterized by the possession of striking masculine secondary sexual adornments of the head and face. In none of these monkeys and apes do males invest as heavily in offspring and show the same degree of paternal care as is typical of *H. sapiens*. Evidence acquired from studies of facial attractiveness and mate choice in human beings supports the view that facial traits have undergone sexual selection in both the sexes. There may have been selective forces constraining the development of masculine facial cues, however, so that men's faces are not necessarily rendered more attractive by over-emphasis of traits such as heavy jaws and brow ridges. A striking sexually dimorphic trait which is stimulated by testosterone in men concerns growth of facial hair. The evolutionary significance of the human beard is discussed in the next section. It is also relevant to note that those non-human primates which are monogamous and which form long-term family groups tend to be sexually monomorphic as regards their facial traits. Examples include many marmosets and tamarins, the titi monkeys and owl monkeys of the New World, as well as most of the smaller apes (gibbons) of Southeast Asia. There are exceptions, however, such as the white-faced saki (*Pithecia pithecia*) of South America, in which the striking white facial colouration occurs only in males.

Human beings display the morphological characteristics of a species in which sexual selection has influenced the evolution of facial traits more than would be expected in a primarily monogamous primate. This again hints at the likely occurrence of polygyny as well as monogamy in human precursors. The final section of this chapter examines, in greater detail, the relationships between the degree of development of masculine secondary sexual visual traits and mating systems in monkeys, apes, and human beings.

Masculine secondary sexual adornments: a comparative perspective

Many of the secondary sexual adornments that are visually distinctive and sexually dimorphic in primates are difficult to quantify. It is probably for this reason that most authors have concentrated on sex differences in more readily measurable traits, such as body weight or canine tooth size. Exact measures of masculine adornments, such as the lengths of capes of hair, beards, or crests, or quantitative data on the distribution and brightness of sexual skin colours, are difficult to obtain for comparative studies. When 'fleshy' secondary sexual characters are of interest, such as the large nose of the male proboscis monkey or the bulbous flaps at the mouth corners of the male golden snub-nosed monkey, comparative measurements can be especially challenging (Figure 7.18).

Figure 7.18 Secondary sexual adornments in adult male anthropoids. Upper row: left: Red uakari (*Cacajao calvus*), centre: Proboscis monkey (*Nasalis larvatus*), right: Golden snub-nosed monkey (*Rhinopithecus roxellana*). Lower row: left: Mandrill (*Mandrillus sphinx*), centre: Bornean orangutan (*Pongo pygmaeus*), right: Man (*Homo sapiens*).

Source: From Dixson (1998*a*).

In order to overcome these problems and to quantify sex differences in visual traits across a wide range of primate species, rating scales have been devised (Dixson, Dixson, and Anderson 2005). In our studies, a 6-point scale was used, ranging from zero (no difference occurs between adult males and females, e.g. both sexes have a crest of hair of the same size and colour) up to five (maximum dimorphism, with males possessing a prominent trait that is absent, or virtually so, in females). For example, in Figure 7.18, the massive pendulous nose of the male proboscis monkey receives a score of 5, by comparison with the female's much smaller snub nose. The male orangutan (the Bornean species shown in Figure 7.18) has large cheek flanges, absent in females (score = 5), a beard (score = 4) and a fibrous or fatty hump on the crown of the head (score = 4). Using this approach, it was possible to rate the degrees of sexual dimorphism in visual traits involving the trunk, limbs, and head for a total of 124 species (representing thirty-eight genera) of monkeys and apes, as well as *H. sapiens*.

Figure 17.19 shows scores for sexual dimorphism in visual trait scores for adult male primates analysed according to their principal mating systems (monogamy, polygyny, or multi-male/multi-female). Polygyny is associated with significantly higher ratings for expression of masculine secondary sexual visual traits across the genera and species included in the sample. This finding applies whether the human mating system is classified as either principally polygynous or monogamous. In general, the order of visual trait development in male anthropoids according to their mating systems is (from highest to lowest): polygyny > monogamy > multi-male/multi-female. Multi-male/multi-female species, such as macaques, mangabeys, and chimpanzees, have very low ratings for male-biased sexual dimorphism in visual traits. Sexually dimorphic visual traits are poorly developed in males of most monogamous primate species; however, there are some notable exceptions, as in certain gibbon species and in the white-faced saki (*Pithecia pithecia*). Adult males of this species exhibit a distinctive white facial disc and black pelage on the body and limbs (total score = 10, Figure 7.20). It is important to note, however, that these colour changes develop in male sakis *before they reach puberty* and are not equivalent to androgen-stimulated secondary sexual traits, such as the human beard.

The relatively conspicuous visual traits of adult human males include facial hair, male pattern baldness, and body hair; all these traits vary considerably in their expression both within and between human populations. A score of 10 was allocated for these sexually dimorphic characters in men (Dixson et al. 2005), which is high, given the relatively modest sex differences in human body weights. Among the monkeys and apes which have polygynous mating systems, scores for male-biased sexual dimorphism in visual adornments are significantly correlated with the degree of sexual dimorphism in body size; the largest males have the most striking secondary sexual adornments (Figure 17.19). It is probable that the human male would justify even higher scores using this rating scale, if sex differences in facial morphology were to be taken into consideration.

Human sex differences in body composition are also more pronounced than differences in body weight alone would indicate. Thus, as discussed earlier, the ratio of muscle mass between men and women is 1.53, using data obtained by Clarys et al. (1984). Inclusion of this sex difference in adult muscle mass ratio in Figure 7.19 places *H. sapiens* firmly in line with polygynous non-human primates as regards the degree of expression of masculine visual adornments. These data may be interpreted as strengthening the case for effects of sexual selection operating within polygynous mating systems during evolution of the genus *Homo*.

Very little is known about the origins of the human beard, or of male pattern baldness, and the marked reduction of human body hair. One suggestion is that male pattern baldness might have a function in heat loss and in thermoregulation, especially in bearded men (Cabanac and Brinnel 1988). The argument is not convincing. Body hair is sparse in human beings; its reduction during human evolution, together with increased sweat gland activity, has been linked to specializations for thermoregulation in tropical savannah environments (Wheeler 1984; Jablonski 2006). It has also been suggested that hair reduction may have been adaptive in lessening ectoparasite loads (such as lice) in ancestral humans (Rantala 1999; Pagel and Bodmer 2003). Indeed, of

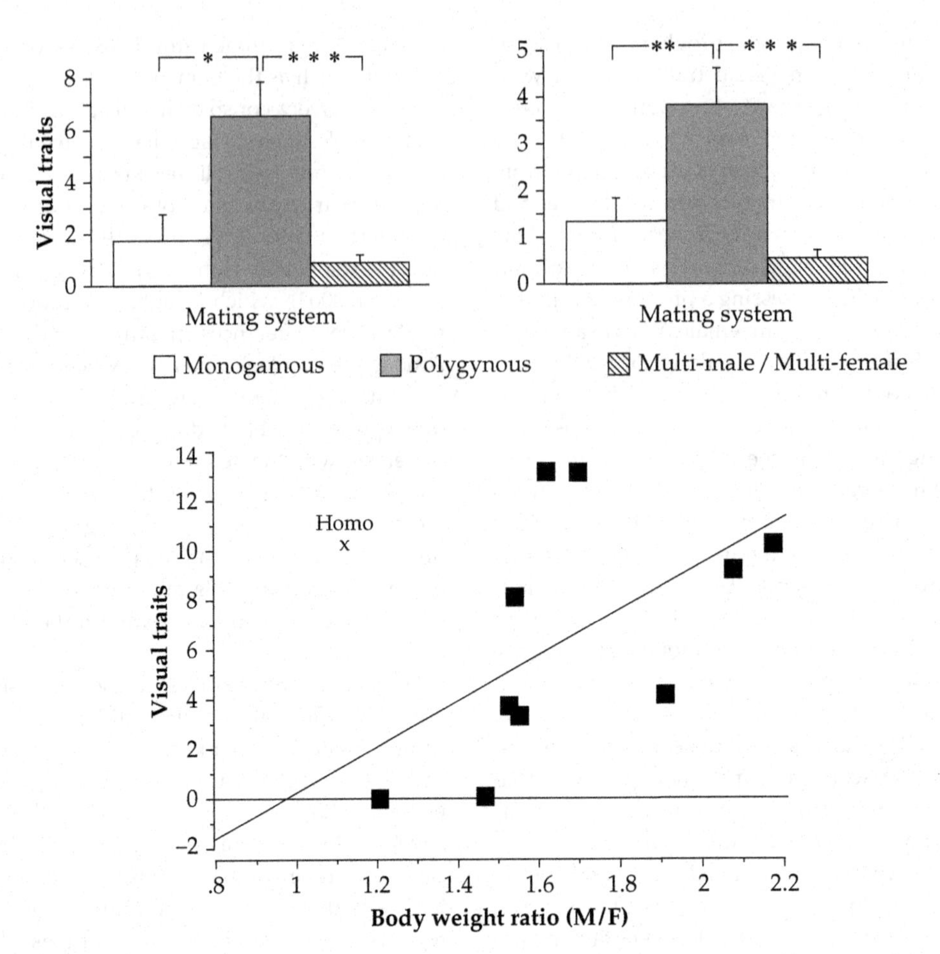

Figure 7.19 Upper: Ratings of the degree of male-biased sexual dimorphism in visual traits in monkeys, apes, and humans, in relation to their principal mating systems. The genus *Homo* is included among the polygynous taxa in this analysis. $^{*}P < .05$; $^{**}P < .01$; $^{***}P < .001$ for analyses at the genus level and (on the right) at the species level after statistical correction for possible phylogenetic biases in the data set. Lower: Correlation between ratings for male biased sexual dimorphism in visual traits and sexual dimorphism in adult body weight in polygynous anthropoid genera. Slope of the regression line $y = 9.738\,x - 9.811$; $r^2 = .78$, $P < .001$.

Source: Redrawn from Dixson, Dixson, and Anderson (2005).

the multiple theories advanced to explain the loss of body hair during human evolution, Marcus Rantala (2007) suggests that the parasite hypothesis is the most compelling. He proposes that once early members of the genus *Homo* began to live in more 'fixed home bases' the risk of infestation by ectoparasites would have increased considerably. Reduction in body hair would have been highly adaptive under these conditions.

The thicker distribution of body hair in men, as compared to women, may represent an epiphenomenon in many modern populations, rather than having significant effects upon attractiveness. In one study conducted in the UK, women found male chest and abdominal hair to be attractive (Dixson et al. 2003). Attractiveness ratings for mesomorphic and endomorphic male images increased if hair was added to the chest and abdomen. Images including

Figure 7.20 The male white-faced saki (*Pithecia pithecia*). This South American monkey lives in small family groups and is thought to be monogamous. It is highly sexually dimorphic in appearance, as only males have the white face and black pelage.

Source: Author's drawing from a photograph by Dr Russ Mittermeier.

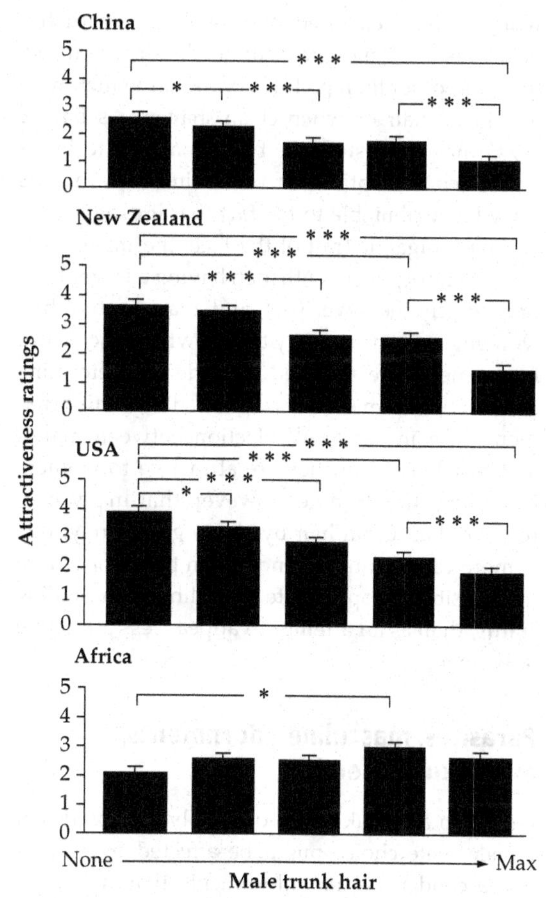

Figure 7.21 Women's ratings for the attractiveness of images of men that vary in amounts of hair on the chest and abdomen. Studies were conducted in China, New Zealand, USA, and Africa. $*P < .05$; $***P < .001$.

Source: Data are from Dixson et al. (2007*a*; 2007*b*; in press).

trunk hair were also rated as being older by women. Given that there is a strong cross-cultural propensity for women to prefer men somewhat older than themselves as marriage partners, masculine body hair might have represented one of the many signals of masculine maturity in ancestral populations of *H. sapiens*. Contrary to this view, it must be stressed that women in China, New Zealand and the USA (California), find the *absence of male trunk hair* to be more attractive; the inclusion of greater amounts of hair in male images renders them correspondingly less attractive to women (Figure 7.21). In the only study conducted in Africa, Cameroonian women did not show any consistent preference for male trunk hair or its absence (Dixson et al. 2007*b*).

Beard growth in men is stimulated by testosterone, and the size and colour of the beard differs with age and also between individuals. Darwin (1871) noted the considerable variations which occur in different human populations as regards the presence or relative absence of facial hair in men. He concluded that 'as far as the extreme intricacy of the subject permits us to judge, it appears that our male ape-like progenitors acquired their beards as an ornament to charm or excite the opposite sex.' However, it is by no means clear that the human beard plays a significant role in female assessments of male attractiveness. It is the case that some studies have produced evidence for positive effects of male facial hair upon judgments of attractiveness (Pelligrini 1973; Hatfield and Sprecher 1986). In Pelligrini's study, both women and men rated photos of men with full beards as being more masculine, dominant, attractive, and older than pictures of the same men with less facial hair or when clean-shaven. As Barber (1995) has suggested, the beard's ability to influence such perceptions of masculine social status 'may be attributable to the fact that it exaggerates another epigamic trait of the face, the male chin'. Sexual dimorphism, with men having a larger face, particularly the lower part, and a larger chin than women, is accentuated by the growth of the beard. Experiments are required, in order to determine how far beards may have originated as status ornaments, via intra-sexual selection between males, and to what extent they are attractive to women. It is interesting to note, however, that many men remove their facial hair by shaving each day; this is more readily understandable in terms of reducing possibly inappropriate signalling of masculine status, than as an attempt to appear less attractive to women.

Parasites, masculine adornments, and sexual selection

Hamilton and Zuk (1982) originally proposed that female mate choice might be affected by masculine secondary sexual adornments that signal the ability to tolerate or resist parasitic infections. The *Hamilton–Zuk hypothesis* was formulated as a result of comparative studies of blood parasites in avian species. Clearly, the secondary sexual adornments of male primates and their greater development in polygynous forms could provide opportunities to further explore the Hamilton–Zuk hypothesis, including its possible relevance during human evolution. The theory predicts that, in comparative terms, those species in which males have the most striking adornments should be subject to higher disease risks: selection having favoured the evolution of masculine adornments precisely because risks of infection are high. Intriguing examples of effects of infectious diseases upon secondary sexual traits in male non-human primates have been reviewed by Nunn and Altizer (2006). In the red uakari (*Cacajao calvus*), for example, the red skin covering the head and face becomes pale when these animals are ill, and such changes may signal susceptibility to malaria and other infections. However, as far as the Order Primates as a whole is concerned, there has been no detailed comparative analysis of disease risk in relation to mating systems or the degree of development of masculine secondary sexual adornments. This is not surprising, as the task of quantifying disease risk in diverse taxa inhabiting tropical environments where infectious organisms abound is truly formidable. Where sexually transmitted diseases are concerned, for example, it is likely that a huge number of these remain to be identified in monkeys and apes (Nunn and Altizer 2006). Cross-cultural studies of humans have produced some evidence for correlations between risk of parasitic infections and polygyny (Low 1987) and the importance that women give to physical traits when judging men's attractiveness (Gangestad and Buss 1993). However, these findings are not yet strong enough to justify conclusions about the importance of polygyny, parasite loads, and the emergence of male secondary sexual adornments during human evolution.

Conclusions

A careful examination of the human body and face reveals a wide variety of sexually dimorphic traits (Figure 7.22), some of which have been examined in this chapter. Sex differences in human body size (weight and height) and composition (especially amounts of muscle and fat) are consistent with an evolutionary inheritance from ancestors who were polygynous. The occurrence of sexual bimaturism in humans and the greater time taken by males to pass through puberty and reach reproductive maturity are also consistent with this conclusion. Comparative studies of male-biased sexual dimorphism in visual traits in monkeys, apes, and humans likewise support the view that polygyny played a role in the emergence of masculine secondary sexual traits such as the human beard.

Both natural selection and sexual selection have played important roles in the evolution of human body shape and composition. Cross-cultural studies have shown that women find mesomorphic (muscular) and average male somatotypes, with

Sexual dimorphism in *Homo sapiens*

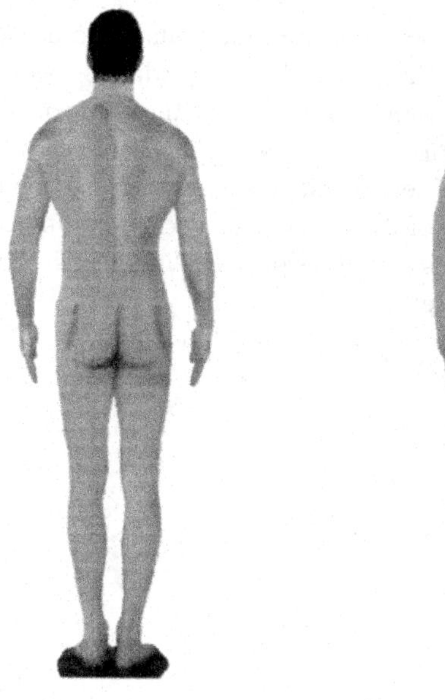

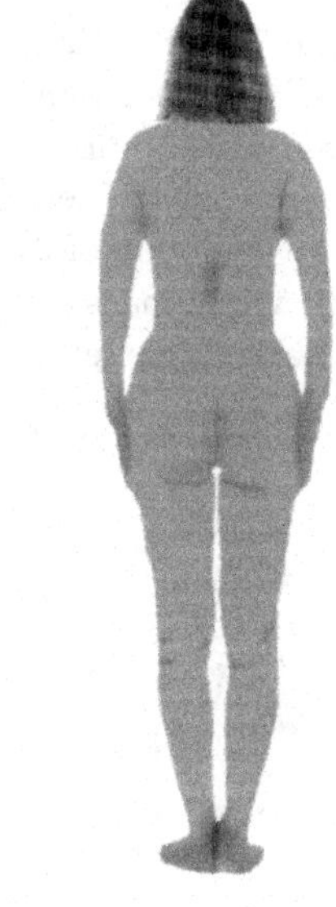

Height	Facial / head hair
Body weight	Body hair
Body composition	Pubic hair
Body shape	Adam's apple
Sexual bimaturism	Breast morphology
Facial morphology	Axillary organ
Hands and feet	2nd / 4th digit ratio
Skin colour	Forearm length

Figure 7.22 Images depicting a mesomorphic masculine somatotype and an hourglass feminine somatotype (WHR = 0.7) together with a list of the morphological sex differences which occur in *Homo sapiens*. The possible origins of some of these sex differences have been considered in this chapter.

broad shoulders and narrow waists, most attractive. Very early during evolution of the genus *Homo* (e.g. in *H. ergaster* and in *H. erectus*) selection probably favoured these physical traits because of their survival value (e.g. during endurance running and persistence hunting) as well as during inter-male competition for status and access to mates. Female preferences for these same masculine traits may thus spring from effects of sexual selection operating in the earliest hunter–gatherers of the genus *Homo*. Body composition in women is greatly affected by the fat reserves which are essential for successful reproduction. This *gynoid* pattern of fat deposition is under oestrogenic control and involves the hips, buttocks, thighs, and breasts. The 'hourglass' female shape, with a low WHR, has been found to be attractive to men in diverse cultures throughout the world. The effects of hypothetical cultural biases (including advertising and the media in various cultures) upon judgements of what is attractive are sometimes invoked as explanations for these findings. However, a more parsimonious interpretation may be that such consistent male preferences, in populations as diverse as those of California and Komodo Island (Indonesia) or university students in China and Hadza hunter–gatherers in Tanzania, reflect fundamental, evolved preferences for the female hourglass figure. Singh's hypothesis posits that a low female WHR acts as an honest signal of female health and reproductive potential. It is likely that sexual selection (as well as natural selection) began to influence these traits early in the evolution of the genus *Homo* in forms such as *H. ergaster* and *H. erectus*.

Sexually dimorphic facial and cutaneous traits are also considered here. There is now compelling evidence that female endocrine condition (including higher levels of oestrogen during menstrual cycles) is reflected in facial cues which influence attractiveness. Sexual selection as well as natural selection may thus have influenced the evolution of human facial morphology, including masculine traits (e.g. size of the jaw and lower face). However, it appears that extreme expression of masculinized facial cues is not necessarily attractive to women, at least in western cultures. Cross-cultural tests of theories concerning human facial sex differences and attractiveness are still at an early stage, however.

Sex differences in skin colour (lighter in women in various populations) were known to Darwin, who considered that they might have been influenced by sexual selection during human evolution. Some cross-cultural evidence supports this view, but again these studies are still at a relatively early stage. The sex differences in body hair that occur in some human populations are not consistently linked to sexual attractiveness. Thus although younger women in the UK prefer male images which include trunk (chest and abdominal) hair; this is not the case in China, New Zealand, USA (California), or in Cameroon. Whether there might be age-related female preferences for this masculine trait remains to be determined.

Among the sexual dimorphic traits listed in Figure 7.22, the size of the larynx ('Adam's Apple') provides a striking example, as there are major associated differences in vocal pitch between men and women. The possible role of sexual selection in the evolution of the larynx and vocal communication in humans and other primates is considered in the next chapter.

CHAPTER 8

Adam's Apple

> 'It is interesting to conjecture as to the appearance and mode of life of the ancestors of present day Man, and in such a study much assistance may be derived from an examination of the larynx.'
>
> Negus (1949)

An obvious anatomical sex difference in adult humans concerns the size and shape of the larynx, with resultant differences in vocal pitch; this is usually deeper in men than in women. These structural and vocal sex differences develop during puberty when, under the action of testosterone, the cartilages and muscles of the male larynx enlarge and the vocal cords increase in length. The thyroid cartilage, the largest of the nine cartilages which make up the larynx, becomes especially prominent in some males, causing a bulge to appear in the midline at the front of the neck; this is the *pomum Adami*, or Adam's apple (Figure 8.1).

Darwin (1871) was well aware of this sex difference in *Homo sapiens*, and noted parallels in the greater size of the larynx and deeper vocalizations of various male mammals. He considered that 'man appears to have inherited this difference from his early progenitors. His vocal cords are about one third longer than in woman, or than in boys.'

In this chapter, I shall examine the question of whether sex differences in laryngeal anatomy and vocal pitch can tell us anything about the likely mating system of our early progenitors. From a comparative standpoint, an important question concerns the extent to which such sex differences correlate with mating systems in other mammals. Darwin knew that in cervids, mature stags exhibit marked enlargement of the larynx and emit powerful roaring vocalizations during the annual mating (rutting) season. Yet, surprisingly, he was unable to establish any firm connection between sexual selection and the evolution of roaring displays in stags, saying that 'As the case stands, the loud voice of the stag during the mating season does not seem to be of any special service to him, either during his courtship or battles, or in any other way.' Modern research has shown, however, that the roar of a red deer stag (*Cervus elaphus*) may transmit information about that individual's dominance status, strength, and reproductive potential. Roaring is valuable as a distance signal to rival males as it may transmit information about body size (Reby and McComb 2003) and probable fighting ability (Clutton-Brock and Albon 1979). The stag's roars can also evoke physiological responses in females, serving to synchronize their oestrus cycles. Putman (1988) states that deer species in which stags compete to hold harems (i.e. polygynous species such as the red deer) are characterized by the occurrence of striking male vocal displays during the rutting season. 'In harem holding species, the call is essentially male competitive display; in stand holding species the call serves the double function of male challenge and female attraction.' The calls made by stags vary a great deal, depending upon the species concerned, and range from high-pitched whistles (sika), roars (red deer), and deep bellows (wapiti and bull moose), to belches and groans (fallow deer). 'Whatever the actual tone, however, the call always has a curious carrying quality' (Putman 1988).

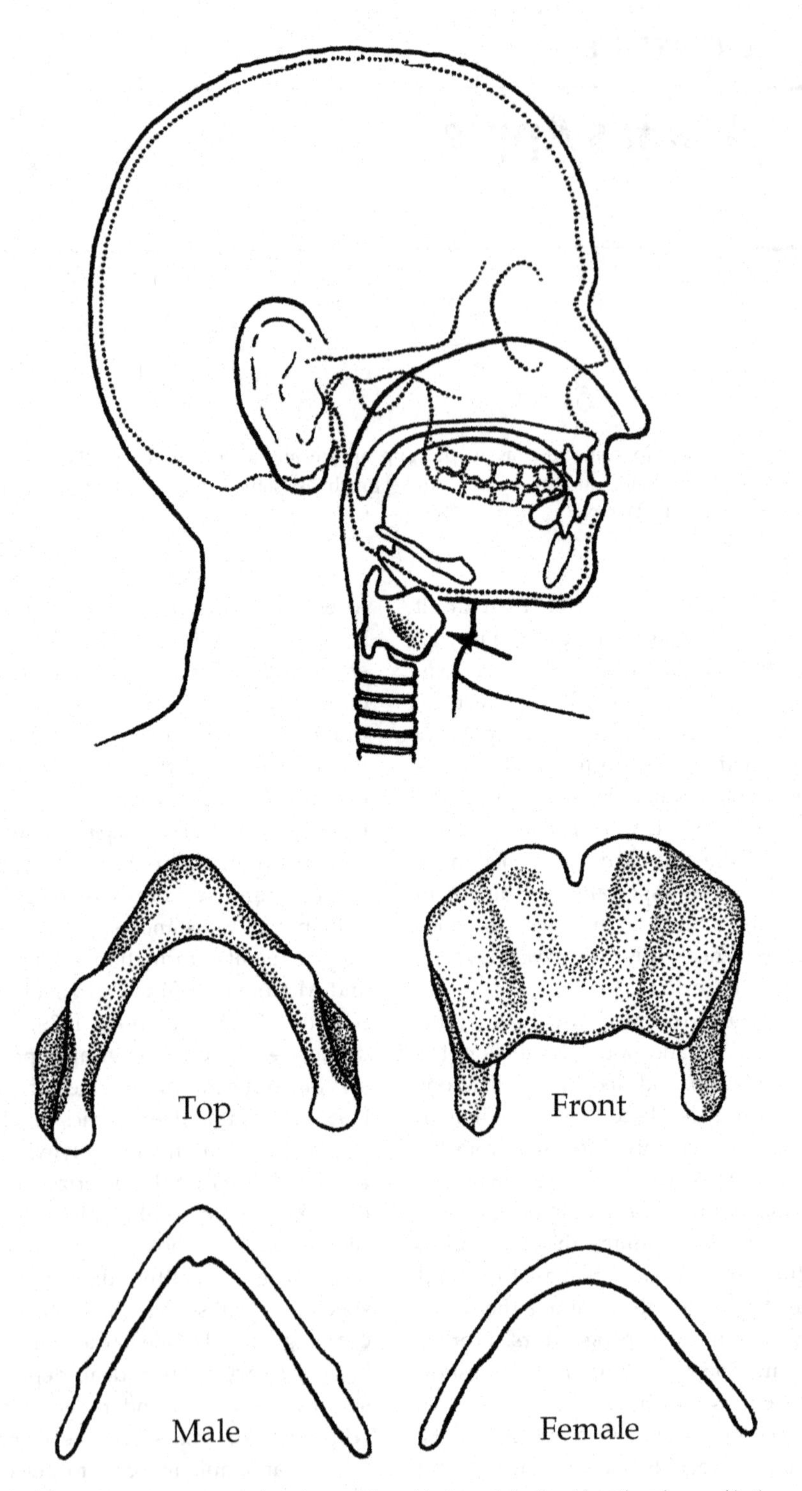

Figure 8.1 The human vocal tract in schematic sagittal section, together with views of the thyroid cartilage (arrowed in the upper diagram) and an illustration of sex differences in the angle of the thyroid cartilage (lower diagram).

Source: From Balaban (1994); after Dickson and Maue-Dickson (1982).

Given that the larynx is markedly sexually dimorphic in adult human beings, and that testosterone stimulates laryngeal growth during puberty (Beckford et al. 1985; Hollien, Green and Massey 1994), it is important to determine whether sexual selection has influenced the evolution of these traits. Firstly, we may ask whether the human voice conveys any information concerning an individual's hormonal or physical status. The answer to these questions appears to be 'Yes'. Indeed, many years ago Eberhard Nieschlag showed that masculine vocal register, body condition, and circulating testosterone are correlated in the human male. Thus, he found that bass singers have higher testosterone/oestradiol ratios and ejaculatory frequencies than tenors, as well as being, on average, taller and more athletic in their physique (Figure 8.2). Subsequently, Dabbs and Mallinger (1999) reported that men who have deeper voices also exhibit significantly higher levels of salivary testosterone. Low-pitched male voices are judged as more pleasant or attractive by women in at least some studies (Collins 2000; Feinberg et al. 2005). Bruckert et al. (2006) found that women tended to find lower-pitched men's voices more pleasant, but also that they based their judgements on changes in intonation and increases in temporal pitch. The men in these studies were recorded as they spoke a series of five vowel sounds. The series was spoken only twice, by twenty-six men ranging in age from 18–32 years. Remarkably, women were able to assess accurately the ages and body weights of these men on the basis of their voices alone. Although they failed to accurately estimate the heights of the men concerned, body weight and height were positively correlated in these same subjects, and taller men also had significantly lower vocal frequencies. Thus it seems reasonable to conclude that masculine voice tone can convey accurate information to women concerning an individual's age, body size, and androgenic status. Women find deeper male voices more pleasant and attractive.

Deeper voices may also be linked to physical and social dominance in men. In a study conducted by Puts, Gaulin, and Verdolini (2006), men rated competitors as more socially and physically dominant if their recorded voices were manipulated so as to be lower in pitch. These investigators also found that men who regarded themselves as physically dominant tended to lower their own voices when responding to recordings of male competitors.

Other studies have shown that women also rate deeper, masculine voices as being more dominant (Feinberg et al. 2005). The pitch of men's voices

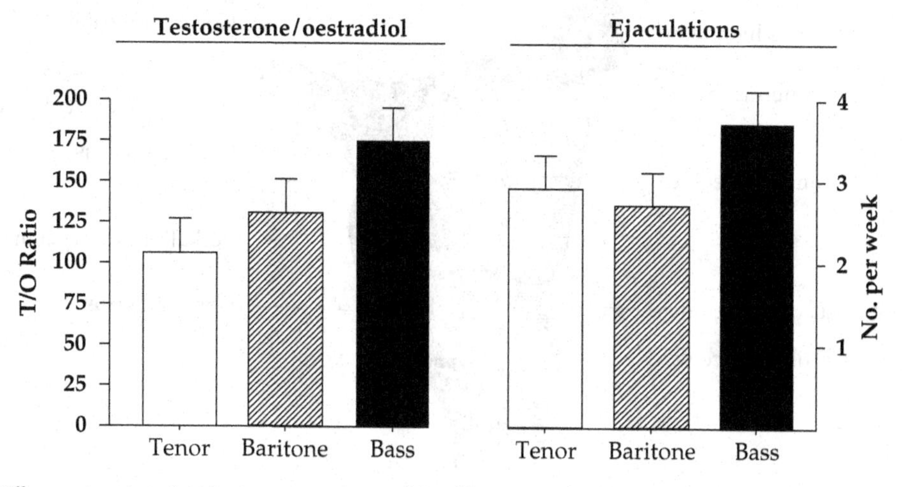

Figure 8.2 Differences in molar ratios of testosterone and oestradiol, and frequencies of sexual activity (ejaculatory frequencies) in male singers (tenor, baritone, and bass) aged 20–40 years.

Source: Based on data from Nieschlag (1979).

tends to be about half as high as that of women (Titze 2000) and vocal pitch correlates with hormonal, physical, and behavioural traits which affect inter-male competition and female mate preferences. Sex differences in the anatomy of the vocal tract and in voice quality in human beings are thus likely to be due to sexual selection, and not simply to sex differences in adult body size. As Darwin suggested, these effects upon the larynx and human voice are likely to be ancient characteristics, deriving from the early progenitors of *Homo sapiens*.

It is important to note that it is the pitch of the voice and sex differences in vocal pitch which are relevant here. Such traits are likely to have arisen before the origin of human language and may have been present in the australopithecines, or in the earliest representatives of the genus *Homo*.

There is evidence that vocal pitch is related to measures of male reproductive success in human beings. Thus, Apicella, Feinberg, and Marlowe (2007) were able to show that among the Hadza of Tanzania, who still live as traditional hunter–gatherers, men with the deepest voices father significantly more children.

In order to further grasp the relevance of sex differences in vocal traits to discussions of human mating systems and their evolution, it is helpful to consider how these traits relate to the mating systems of the extant non-human primates. Are there sex differences in vocal tract anatomy and associated differences in vocal pitch in male and female monkeys and apes? If so, do these differences relate in a consistent fashion to the occurrence of monogamy, polygyny, or multi-male/multi-female mating systems among the anthropoids? The larynx is composed of the same cartilaginous elements in monkeys and apes as it is in *H. sapiens*: as an example, the larynx of a chimpanzee is shown in Figure 8.3. However, the relative sizes of these cartilages and their associated musculature differ between species and, additionally, there are extensions of the larynx called laryngeal air sacs. These can be enormous in some cases (as in the orangutan and gorilla) and they are often inflated during vocal displays. All primates possess a pair of lateral ventricular air sacs and these are especially well developed in the great apes and in the siamang. Human beings are no exception to this rule, but the human lateral laryngeal sacs are very small, being situated in the wall of the larynx between the (upper) false vocal cords and (lower) true vocal cords in both sexes. These pouches are richly supplied with mucous secreting glands; this mucous acts as a lubricant for the vocal cords. The situation is very different in the great

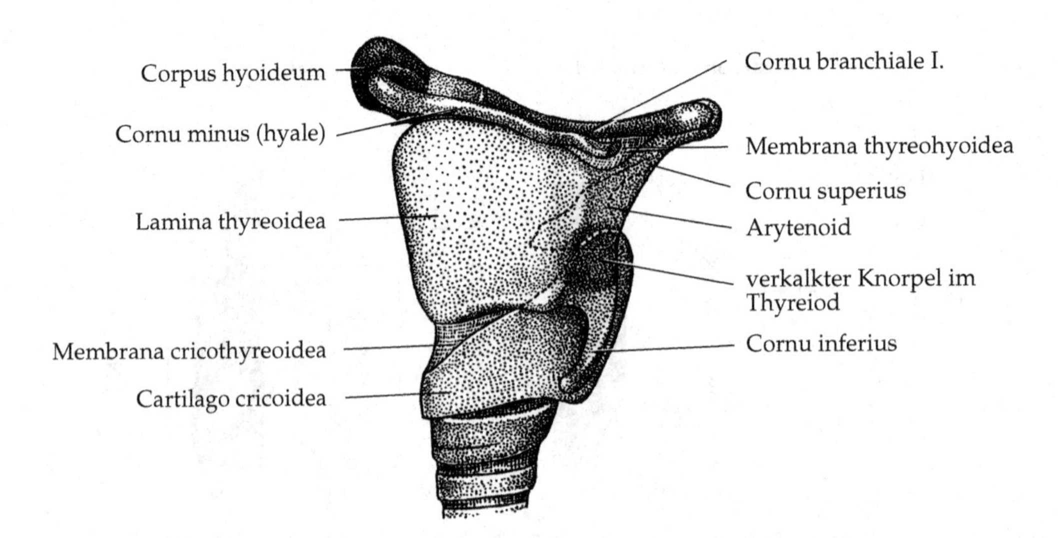

Figure 8.3 Lateral view of the larynx of a chimpanzee (*Pan troglodytes*), dissected to show the arrangement of the laryngeal cartilages.
Source: From Starck and Schneider (1960).

apes, for in all cases the lateral ventricular sacs are much enlarged and extend into the neck, chest, and axillae. This is especially the case in the adult male gorilla and orangutan, in which the laryngeal sacs are also much better developed in adult males than in females (Figures 8.4 and 8.5).

Mature male orangutans emit spectacular, deep-pitched long calls which are audible for over 1 km in the rainforest. Males inflate their vocal sacs during these displays, which may last for 1–2 min, and are thought to serve primarily for communication between individuals, as orangutans are widely dispersed across the forest canopy (Mitra Setia and Van Schaik 2007). The massive size, striking cheek flanges, and other secondary sexual adornments of adult male orangutans are indicative of the effects of evolution by sexual selection, such as occurs among polygynous primates. Because these apes are relatively non-gregarious and do not form permanent social groups, it is especially important for adult males to be able to communicate over long distances. The development of the laryngeal sacs and long call is significant in this respect. In Chapter 3 the occurrence of delayed secondary sexual development in non-flanged adult male orangutans was discussed in relation to their mating strategies and the relevance of sperm competition in orangutan reproduction. Non-flanged males not only lack the striking visual secondary sexual adornments and massive size of territorial males, they also have a smaller larynx, lack the throat sac, and do not give long calls. Development of all these traits is, presumably, influenced by testosterone, or by greater target organ sensitivity to testosterone, in those males which progress to become fully flanged individuals.

Silverback male gorillas also possess an extensive array of air sacs derived from the paired lateral laryngeal sacs (Figure 8.5). Schaller (1963) noted that silverback mountain gorillas make use of the laryngeal sacs during their chest-beating displays. The chest-beats, which resemble hollow 'pock-pocking' sounds, may carry for as much as 2 km. Schaller also observed that 'the prominent air sacs also act as resonators, for their sudden inflation on each side of the throat is sometimes readily apparent before the chest-beat.' Gorillas are polygynous and the adult male's chest-beating display serves a variety of functions, including inter-group communication and intimidation of potential rivals. Although adult

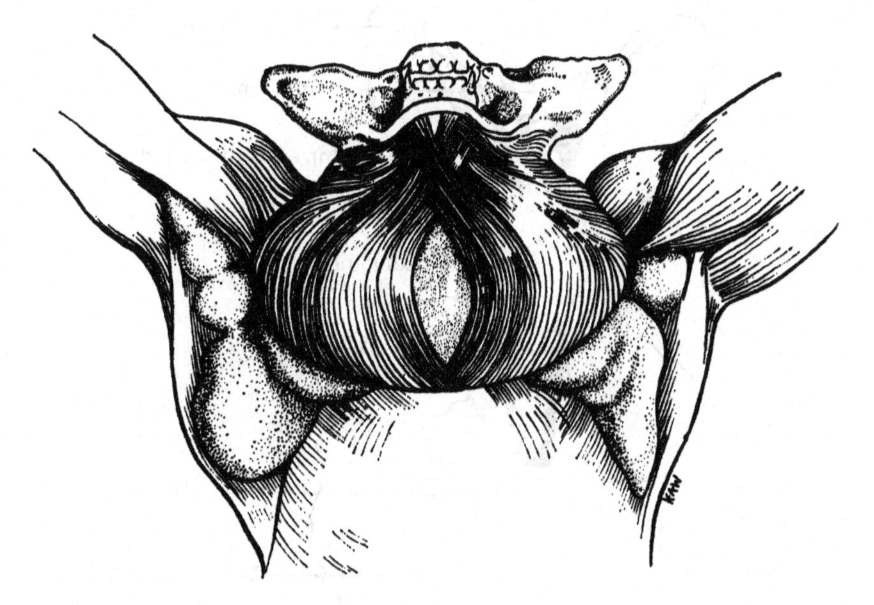

Figure 8.4 Dissection of an adult male orangutan (*Pongo pygmaeus*) to show the extent of the (inflated) laryngeal sacs. These sacs are inflated during the male's 'long-call' displays.

Source: From Dixson (1998a); after Hill (1972).

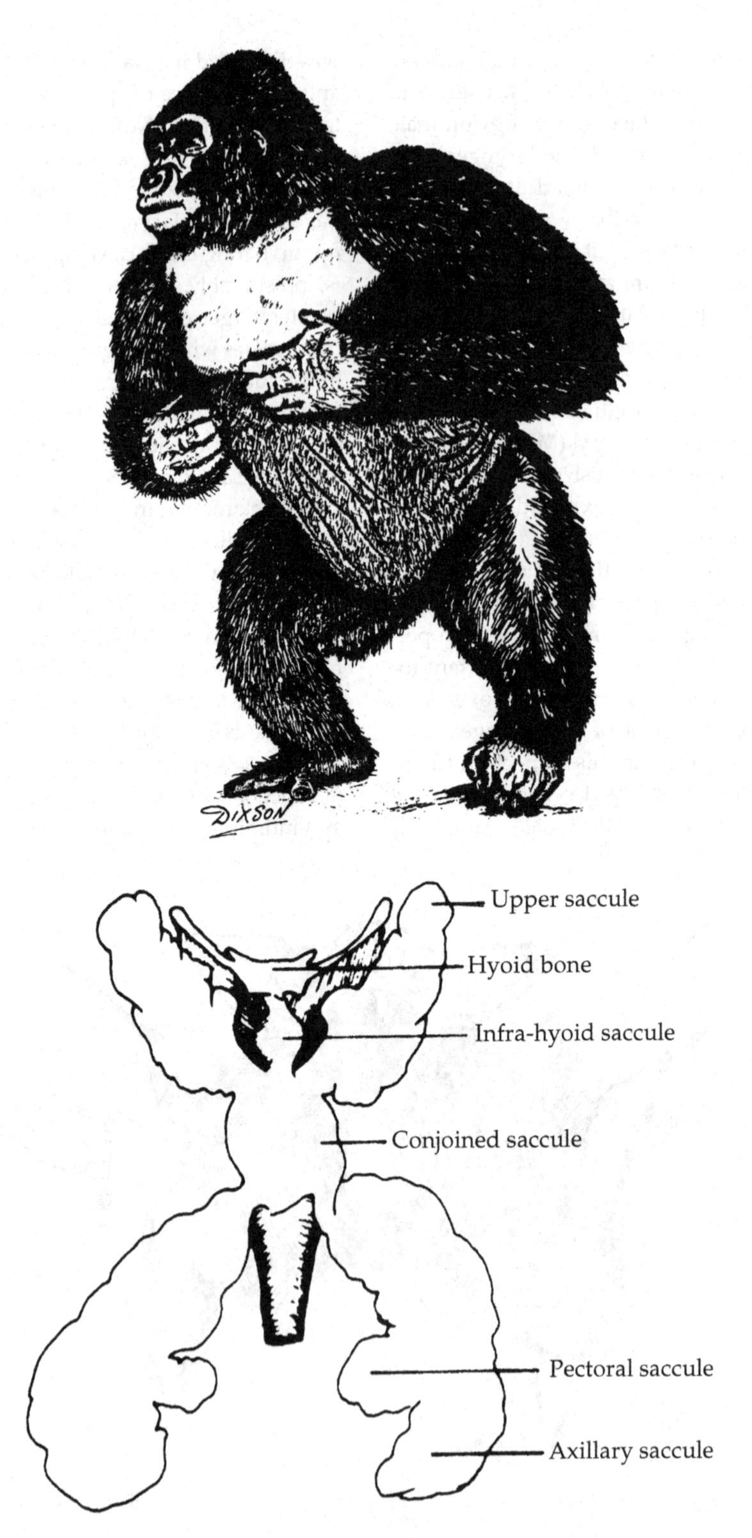

Figure 8.5 Chest-beating display of an adult male western lowland gorilla (*Gorilla g. gorilla*) and a dissection of the laryngeal sacs, which are inflated to act as resonators during the display.

Source: From Dixson (1981).

females and juveniles also chest-beat, they produce much duller sounds which are not audible over long distances. Silverbacks may also vocalize when alarmed or aggressive to produce explosive, deep-throated roars. Again, immature males are unable to emulate the depth of pitch and vocal power of silverbacks, presumably because the larynx and associated musculature is not yet fully developed. Schaller (1963) considered that black-backed (adolescent) males 'were at an awkward age, vocally speaking, in that their voices appeared to be changing. Thus young males screamed when angry, and the older ones produced rather squeaky roars. Only silver backed males were heard to emit the full roar, the clear hooting preceding the chest beat, and the staccato copulation call.' Thus, black-backed males are probably at a stage where the larynx is enlarging, and the vocal cords are lengthening during puberty, a situation which is homologous with the breaking of the voice which occurs in human males at this time.

Chimpanzees and bonobos also exhibit large extensions of the lateral laryngeal sacs. However, the larynx is not markedly sexually dimorphic in these species and nor does the pitch of the voice differ so greatly between the sexes. Both male and female chimpanzees emit loud 'pant-hoots' when communicating over a distance (Marler and Hobbett 1975). These vocalizations are important as contact calls between the sub-groups which make up the multi-male/multi-female, or fusion–fission communities in which chimpanzees live. There is thus no equivalent among chimpanzees of the highly sexually dimorphic displays used by orangutans and gorillas.

In the smaller apes or gibbons (Hylobatidae), the principal mating system is monogamous, and adults of both sexes engage in complex vocal 'duets' which exhibit interesting inter-specific variations (Haimoff 1984). Only the siamang (*Hylobates* (*Symphalangus*) *syndactylus*) possesses prominent lateral laryngeal sacs, but in this case they are equally developed in adults of both sexes. Males and females inflate the sacs during their prolonged displays (Figure 8.6). It is likely that the duets produced by gibbon pairs may transmit complex information between groups; such distance communication has apparently undergone positive selection in the rainforest environments in which the hylobatids have evolved. Although traditionally thought of as territorial displays between rival family groups (Ellefson 1974), these duets probably also convey information about the strength of the pair bond (relationship) between the sexes as well as the health and physical condition of the individual partners (Haimoff 1984). Thus, it is significant that neither the larynx nor the vocal pitch of male and female gibbons is sexually dimorphic, a situation which contrasts markedly with that encountered in the orangutan and gorilla. The only known exception to this generalization concerns the white-cheeked gibbon (*Nomascus concolor*) in which the adult male possesses a small throat sac. Deputte (1982) described a low-pitched sound given by adult males, in association with air entering this vocal sac, and he says that 'this sound is a new acquisition in the mature, adult male vocal repertoire.' However, as Deputte notes, loud calls are given by both sexes in mated pairs of white-cheeked gibbons, so that the functional significance of the throat sac in adult males remains

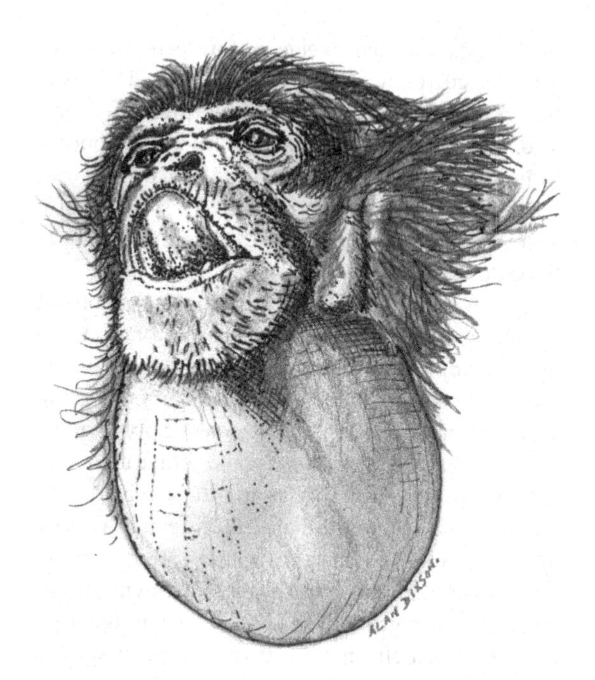

Figure 8.6 A siamang (*Hylobates* (*Symphalangus*) *syndactylus*) with the vocal sac inflated. Both sexes possess large lateral laryngeal sacs, which are inflated during vocal duets.

Source: Author's drawing from a photograph in Rowe (1996).

unclear. Among the twelve to fourteen species of gibbons currently recognized (Groves 2001; Bartlett 2007), absence of sexual dimorphism in body size, laryngeal structure, and vocal pitch appears to be the rule. However, detailed studies of vocal anatomy remain to be carried out for some gibbon species, so it is possible that sex differences remain to be described.

Can the principles derived from observations of the vocal displays and laryngeal anatomy of the apes be applied to the primates as a whole? Is the larynx or its extensions (vocal sacs) larger, and are loud calls more specialized, in males of polygynous monkey species? Are there consistent differences between polygynous species and those which have monogamous or multi-male/multi-female mating systems? Are the effects of sexual selection upon vocal tract anatomy greatest in forest-living monkeys, in which groups are spatially separated and, if so, are such effects most pronounced in species which form polygynous one-male units? In this regard it is instructive to consider the guenons (Genus *Cercopithecus*) and the colobine monkeys of Africa and Asia, because species with polygynous and multi-male/multi-female mating systems are well represented in both these groups. Many of them are highly arboreal and inhabit rainforests. Gautier (1971; 1988) studied the loud calls of forest guenons in relation to the functions of the adult male's ventral (sub-hyoid) laryngeal sac. Adult males of species such as *Cercopithecus mona*, *C. pogonius*, *C. neglectus*, and *C. nictitans* have much larger vocal sacs than females and they emit distinctive resonant 'booms' as part of their loud call displays. These species all have predominantly polygynous mating systems. Gautier has shown that the male's laryngeal sac is filled with air prior to booming (Figure 8.7) and that making a fistula in the sac of *C. neglectus* markedly decreases sound production (Gautier 1971). These findings strengthen the conclusion that vocal sacs act in some way as resonators during loud calls. This is likely to be the case, for example, in the male orangutan, which can retain up to 6 litres of air in its vocal sacs (Starck and Schneider 1960). Some authorities have rejected the idea that laryngeal sacs evolved to enhance vocal

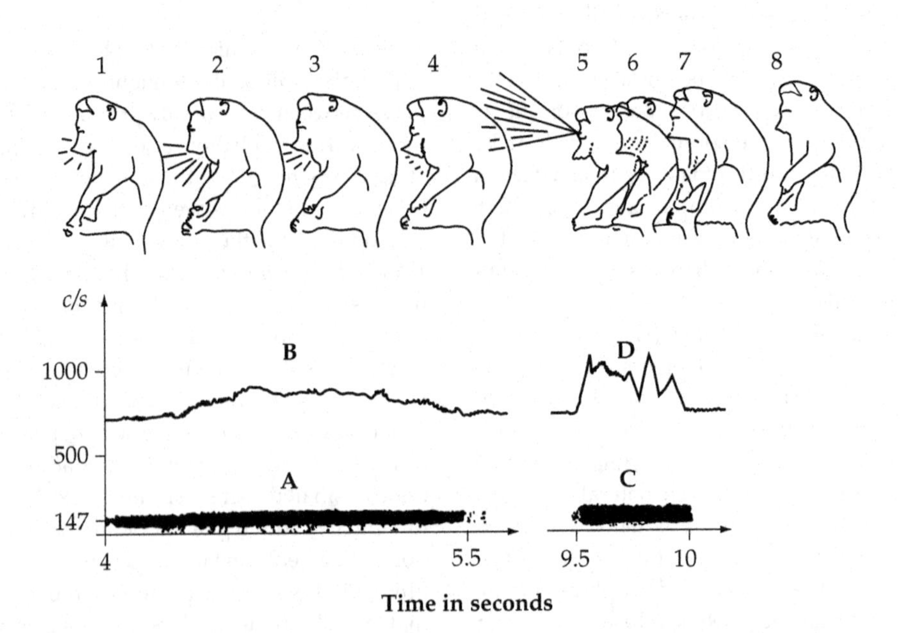

Figure 8.7 Functions of the laryngeal sac during 'booming' vocalizations by an adult male De Brazza's monkey (*Cercopithecus neglectus*). 1–4: inflation of vocal sac accompanied by sound (spectra A, B); 5–7: emission of booming call (spectra C, D); 8: conclusion of call.

Source: From Dixson (1998a); modified after Gautier (1971).

displays (Negus 1949; Harrison 1995). Yet the bulk of the behavioural evidence supports the view that the laryngeal sacs act as resonators in various non-human primates, and that their greater size in adult males of some species is linked to the production of deeper, more resonant calls. In addition, Hewitt, MacLarnon, and Jones (2002) have proposed that the vocal sacs of the great apes and the siamang may also function to recycle exhaled air during fast call sequences, and to reduce the risk of hyperventilation during these displays.

In the leaf-eating monkeys (Subfamily Colobinae) of Africa and Asia, the larynx is sexually dimorphic in some species and once again this is associated with the emission of loud resonant calls by adult males. In the King colobus (*Colobus polykomos*) and the closely related mantled guereza (*C. guereza*), the larynx is especially large in mature males which produce loud roars during displays between polygynous one-male units. Hill and Booth (1957) measured the laryngeal dimensions of the African King colobus, olive colobus, and red colobus, as well as five species of Asiatic leaf-eating monkeys. They noted that the male King colobus has by far the largest larynx and that its size is 'equal to that of an adult human male.' Given that an adult male King colobus weighs only 10 kg, the size of its larynx in relation to body size is truly remarkable. By contrast, the red colobus (*Procolobus badius*) and olive colobus (*P. verus*), both of which have multi-male/multi-female mating systems, lack the vocal power of the King colobus and have much smaller larynges. John Oates and his colleagues have argued that the distinctive roars of black and white colobus monkeys, which are best developed in *C. guereza*, are derived in evolution from an originally multi-male/multi-female call type, such as currently occurs in the black colobus (*C. satanas*). The male's call may have functioned originally as an alarm call, and then become modified for use in inter-male competition within the social group. Sexual selection may have favoured the evolution of a larger larynx in adult males under these conditions to produce calls of deeper pitch. Assumption of polygynous, single-male social groupings in the King colobus and mantled guereza would have been associated with still greater sexual selection for laryngeal enlargement to enhance vocal capacity. All this is derived, however, from the possession of a small larynx in ancestral forms, as this is 'the presumed primitive condition for African colobines' (Oates, Bocian, and Terranova 2000).

Hill and Booth (1957) also noted the occurrence of an enlarged larynx in Asiatic leaf monkeys such as the lutongs (*Trachypithecus*), Gray langurs (*Semnopithecus*), and the proboscis monkey (*Nasalis*). They comment that possession of a sub-hyoid laryngeal sac is associated with greater vocal resonance in these monkeys, many of which have polygynous mating systems. This is the case, for example, in the Northern plains gray langur of India (*Semnopithecus entellus*), in which adult males emit loud 'whooping' calls during inter-group encounters. In a related species, *S. johnii*, only the leader male in a one-male unit calls in this fashion, whereas several males may call together in larger groups of *S. entellus* (Hohmann 1989). Hohmann also made the interesting observation that males inflate their vocal sacs during their whooping displays. However, one male whose sac had been punctured due to a bite wound was unable to emit whooping calls at full volume.

Loud calls are given by adult males in a number of Old World monkeys which live in multi-male/multi-female social groups. Examples include the 'whoopgobble' calls of the forest-dwelling mangabeys (*Lophocebus* and *Cercocebus*) and the 'roar-grunts' of baboons (*Papio*). In his review of the evolution of these calls, Wasser (1982) points out that they tend to be of lowest frequency and most stereotyped in structure among forest-dwelling species. Individual males may be readily distinguishable by their loud calls, as is the case in the arboreal mangabeys. Thus the calls may play some role in communication between males within and between groups, under conditions where visual signaling is limited due to the dense vegetation. However, these calls are neither as loud, nor as resonant, as the loud calls of the polygynous guenons and black and white colobus monkeys described above. Nor is sexual dimorphism in laryngeal size and structure as marked in the mangabeys and baboons.

Among the New world primates, a substantial number of genera contain species which live in small family groups with pair-living (monogamous) mating systems. Examples include the owl monkeys (*Aotus*), titi monkeys (*Callicebus*), marmosets and tamarins (Family Callitrichidae), and the white-faced saki (*Pithecia pithecia*). Many of these monkeys emit long calls and members of a mated pair may synchronize their calls to produce duets. These function for territorial defence and group cohesion (e.g. in cotton-top tamarins, *Saguinus oedipus*: Snowdon and Soini 1988; pygmy marmosets, *Cebuella pygmaea*: Soini 1988, and *Callicebus* species: Kinzey 1981). In the golden lion tamarin (*Leontopithecus rosalia*), Kleiman, Hoage, and Green (1988) state that long calls 'serve to cement and maintain the pair bond but also to advertise presence in a territory.' Golden lion tamarins have a ventral laryngeal air sac, but this is situated lower down in the larynx, rather than being sub-hyoid in position, as in colobines. Hershkovitz (1977) noted that this air sac was better developed in three adult males than in two adult females that he dissected. However, sex differences in vocal tract anatomy are not well represented among the monogamous New World primates as a whole. It appears that, like the gibbons, these pair-living New World monkeys accomplish their duets by means of laryngeal mechanisms which are the same, or at least very similar, in the two sexes. In the nocturnal owl monkeys (*Aotus*), males and females are both capable of giving very low frequency 'hoot calls', but these are emitted by individual monkeys and not as part of territorial duets. Both sexes also give a 'resonant whoop', rising in pitch and volume, as a prelude to aggression between family groups. Males and females inflate their throat sacs before uttering these calls (Moynihan 1964; Wright 1994).

The most spectacular specializations of primate vocal tract anatomy concern the howler monkeys (*Alouatta*), which are widely distributed in the rainforests of Central and South America. Their social groups usually contain 'one or a small number of adult males and several adult females'.

Males are from 20 per cent to 50 per cent heavier than females, depending upon the species considered (Di Fiore and Campbell 2007). Howlers are famed for their loud, resonant roaring vocalizations, and for the remarkable development of the vocal apparatus and laryngeal sacs, especially in adult males. The hyoid bone at the base of the tongue is hollow and egg-shaped in adults of both sexes, but in males the bone is so large that it causes a marked swelling beneath the chin. The red howler (*Alouatta seniculus*) shows the most extreme sex differences in the size of its vocal apparatus (Figure 8.8) and it emits the most powerful calls. The volume of the male's hyoid is five times greater than that of the adult female, whereas in male mantled howlers (*A. palliata*) it is a little more than twice as large (Table 8.1).

The precise functions of the loud vocalizations of howlers have been much debated. Both sexes may vocalize, especially in the early morning, so that group choruses may play some role in intertroop spacing (mantled howler: Carpenter 1965). However, Sekulic (1982) has shown that howlers often call at other times of day and that they roar in order to challenge solitary individuals outside the group, as well as neighbouring groups. She suggested that howlers assess and repel competitors by means of vocal displays, thus reducing the likelihood of energetically costly chases and fights. Competition occurs between males in reproductive contexts, and the sex differences in hyoid volume and vocal pitch are intriguing in this respect. In the red howler, genetic studies have shown that dominant males sire the majority of the offspring (Pope 1990). The pronounced development of the hyoid in male red howlers may relate to the importance of polygyny in its mating system. Thus in the red howler, and in several other species (e.g. *A. caraya*, *A. guariba*, and *A. pigra*), there is usually only one adult male in the social group (Di Fiore and Campbell 2007). However, in the mantled howler, which has the least sexually dimorphic vocal apparatus,

Table 8.1. Hyoid volumes and socionomic sex ratios in howler monkeys (*Alouatta*).

Species	Socionomic sex ratio*	Hyoid volume (ml)		
		Male	Female	Female/Male %
A. palliata	2.63	4.5	2.0	44
A. seniculus	1.55	69.5	12.5	18
A. belzebul	2.0	56.5	12.5	22
A. fusca	–	39.0	8.5	22
A. caraya	1.73	23.0	8.0	35

*Socionomic sex ratio = nos. of adult females/nos. of adult males per group. Data from Crockett and Eisenberg, 1987; Di Fiore and Campbell, 2007, and after Dixson, 1998a.

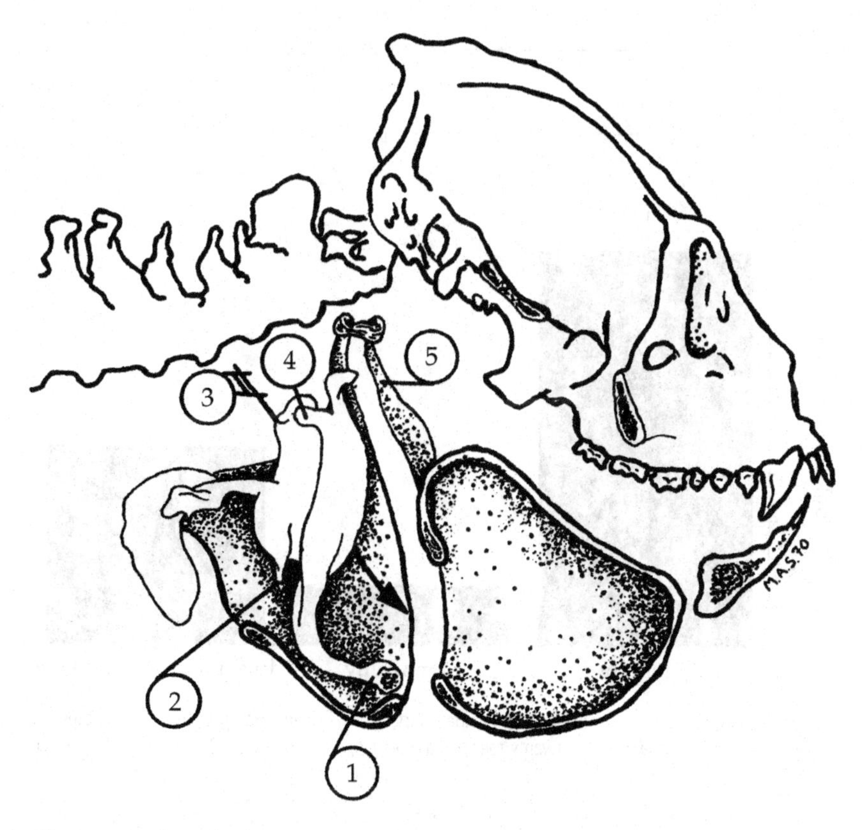

Figure 8.8 The hyoid apparatus and associated anatomical specializations in an adult male howler monkey (*Alouatta seniculus*). 1. *Processus furcatus*; 2. lingula of cuneiform cartilage; 3. Arrow through thyrocuneiform space of left side; 4. lingula of epiglottic cartilage; 5. *Cornu branchiale*.

Source: From Dixson (1998*a*), after Schön (1971).

larger multi-male/multi-female social groups are more common.

One interpretation to be made on the basis of this brief overview of primate vocal anatomy and mating systems is that sexual selection has favoured the greatest development of laryngeal and associated specializations in adult males of species which are polygynous. Sex differences certainly occur in multi-male/multi-female species, but sexual dimorphism tends to be less pronounced under these conditions. By contrast, in most primates that live in small family groups, and which have principally monogamous mating systems, the anatomy of the vocal tract and pitch of calls produced are very similar in males and females. In order to quantify sex differences in the structure of the larynx, laryngeal sacs, and hyoid in relation to primate mating systems, I have employed a rating scale, similar to that which was used in the last chapter to quantify sex differences in visual traits. This scale extends from zero (no sex differences in vocal tract anatomy), up to a maximum score of 5 (marked sexual dimorphism: adult males have a much larger larynx/vocal sacs/hyoid). In the absence of precise anatomical measurements of both sexes across a wide range of primate species, the rating scale provides a useful way of facilitating statistical comparisons of sexual dimorphism in primates that are monogamous, polygynous, or that have multi-male/multi-female mating systems. It was possible to assign ratings to representatives of twenty-four anthropoid genera, including *Homo*. The results, shown in Figure 8.9, indicate that adult males of polygynous forms have consistently larger and more specialized vocal tracts

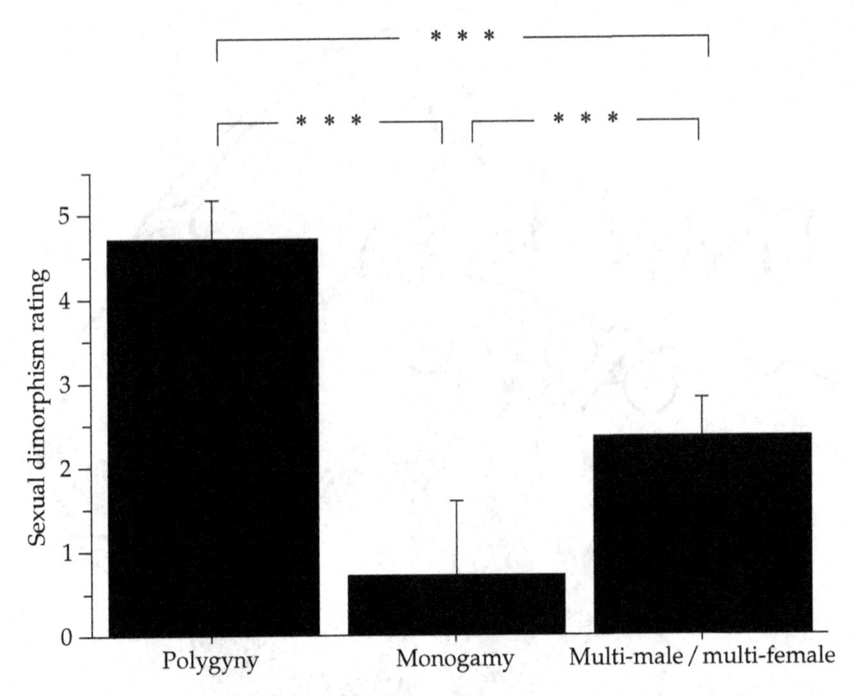

Figure 8.9 Sexual dimorphism in the vocal apparatus (larynx, laryngeal sacs, and hyoid) and mating systems in primates. Data are ratings of the degree of sexual size dimorphism (male>female) in twenty-four primate genera, including *Homo*. ***$P<.001$. Further details are provided in the text.

than adult females, and that their rating scores are significantly greater than either the multi-male/multi-female or monogamous genera. *Homo sapiens*, with a markedly larger larynx in males, scores 3.5 on the rating scale, and thus falls among the polygynous primate genera. Multi-male/multi-female primate genera have average ratings of 2.35 for sexual dimorphism of the vocal tract, whereas in monogamous genera the mean score is 0.55.

These results are preliminary, as detailed studies will be required to obtain exact measurements of the vocal tracts of both sexes for a large enough sample of primates to facilitate a definitive test of the hypothesis. This will be a challenging task, however, due to the scarcity of suitable material for anatomical study. In the interim, the results presented here support the view that the larger larynx and deeper voice of the human male results from evolution by sexual selection, and that the emergence of this trait is likely to have occurred in ancestors of *Homo sapiens* which had polygynous mating systems. The ancestral forms concerned may have been very ancient indeed, perhaps as part of the australopithecine lineage that still lived in partially forested environments. It is probable that sex differences in vocal tract anatomy were well established early in human ancestry and prior to the emergence of language. In this regard, a fascinating discovery was made by Alemseged et al. (2006) during their restoration of the partial skeleton of a 3.3 million year juvenile specimen of *Australopithecus afarensis*. Amazingly, the hyoid bone has survived in this specimen, which is from the Dikika region of Ethiopia. The structure of the hyoid shows similarities to that of the gorilla and chimpanzee and 'almost certainly reflects the presence of laryngeal air sacs characteristic of African apes'.

Sex differences in human vocal pitch relate primarily to anatomical differences in the size of the larynx and vocal cord length in men and women, although there are also sex differences in functioning of the vocal tract above the level of the larynx

(Balaban 1994). The lateral laryngeal sacs, which are much enlarged in the great apes and in the siamang, are represented only by small pouches in human beings. All primates possess lateral laryngeal sacs, so that their occurrence in *Homo sapiens* is a primitive feature. Their small size may be due to reduction from larger structures in ape-like ancestors (e.g. in australopithecines), perhaps in association with changes in respiratory control which accompanied the evolution of language (Hewitt et al. 2002). On the other hand, it is possible to argue, and with equal justification, that the lateral laryngeal sacs may never have been especially large in human ancestors. Enlargement of the cartilages of the male larynx and lengthening of the vocal cords in *Homo* may have occurred independently of changes in the vocal sacs.

Previous chapters have emphasized that many features of the human reproductive system are consistent with an evolutionary ancestry involving polygyny or monogamy rather than a multi-male/multi-female mating system with its associated specializations for sperm competition and cryptic female choice. The sex differences in laryngeal structure and vocal pitch described here take us one step further in the quest to understand whether human ancestors had either monogamous or polygynous mating systems. Human sex differences in the anatomy and functions of the vocal tract are consistent with effects of sexual selection operating within a polygynous mating system, early in the evolutionary line which gave rise to human beings. Even now, in a modern hunter–gatherer population, the deeper voice of the human male correlates with traits which positively impact masculine reproductive success (Apicella et al. 2007). Thus, insights gained from comparative studies of the primate larynx support the conclusion reached in Chapter 7 that polygyny has played an important role in the ancestry of *Homo sapiens*.

Conclusion

Sexual dimorphism in the structure of the human larynx is associated with differences in vocal pitch between the sexes, and these in turn have been shown to influence attractiveness to the opposite sex and perceptions of dominance. The non-human primates present a rich array of specializations of the vocal tract, including the laryngeal sacs and hyoid as well as the cartilages of the larynx. These specializations are related to sex differences in the production of sound for long-distance communication, and to differences in the mating systems of various monkeys and apes. The greatest sex differences in vocal tract anatomy occur in polygynous anthropoids, and the least dimorphism is apparent in monogamous forms. Thus although complex vocal anatomy occurs in some monogamous species (such as the siamang), it is present in both sexes in such cases, rather than being more highly developed in adult males. A preliminary comparative analysis places *Homo sapiens* among the polygynous primates, as regards the degree of sexual dimorphism of the vocal apparatus. These findings strengthen the conclusion that polygyny played a significant role in human ancestry, perhaps among those australopithecines which ultimately gave rise to the genus *Homo*.

CHAPTER 9

The Road to Truth

Much more is now known about human origins and evolution than was the case almost 40 years ago, when I began to study primate reproductive biology. In the final essence, it is the fossil evidence of hominid evolution, reviewed briefly in the first chapter of this book, which provides the backcloth against which the sexual behaviour and reproductive anatomy and physiology of the extant primates may be viewed in true evolutionary perspective. This has been the approach adopted throughout this book, and the goal of this final chapter is to present overall conclusions concerning the origins of human mating systems, patterns of copulatory behaviour, and mate choice.

Darwin's insights concerning the probable African origin of human beings have been confirmed by palaeontological discoveries as well as by the results of research on comparative anatomy and molecular genetics. Anatomically modern human beings originated in Africa and our species has spread far beyond its origins to invade and colonize the rest of the world. The human genus (*Homo*) is most closely related to the chimpanzee and bonobo (*Pan*) among extant primates, and the common ancestors of these two genera existed in Africa approximately 8 million years ago. It is still the case that little is known of the putative stem forms, although on current evidence *Ardipithecus* is thought to be a possible candidate. Earlier still, perhaps 10–8 million years ago, the ancestors of modern gorillas diverged from the lineage which led to the precursors of *Pan* and *Homo*. Recently, fossilized remains of a robust ape (*Chlorapithecus abyssinicus*), which lived in Ethiopia about 10 million years ago, have been described by Suwa et al. (2007), who conclude that this new species 'may be a basal member of the gorilla clade'.

The earliest members of the hominid lineage, the australopithecines, were relatively small-brained and bipedal apes, which are known primarily from fossils discovered in Southern and Eastern Africa. These include more gracile species such as *Australopithecus africanus* and *A. afarensis*, as well as forms with more massive jaws and large molar teeth (e.g. *Paranthropus boisei* and *P. robustus*). It is interesting to cast one's mind back and to recall debates as to whether the australopithecines were efficient bipeds and whether they were ancestral to humans (Day 1973; Zuckerman et al. 1973). The discovery of more complete skeletal material from a specimen of *A. afarensis* ("Lucy": Johanson and White 1980) made it increasingly clear that bipedalism was well developed in these creatures, although the anatomy of their forelimbs is also indicative of the ability to climb well. Most recently, Whitcome, Shapiro, and Lieberman (2007) have reported that, like women, female *A. africanus* exhibited specializations of lumbar vertebral anatomy; serving to strengthen the spine during 'bipedal obstetric load' (i.e. when walking upright during pregnancy).

The cranial capacities of the australopithecines ranged from 400–500 cc, but their brains may have been larger, in relation to body size, than those of extant apes such as the chimpanzee (Martin 1983, 1990). Chimpanzees are able to make and use simple tools (McGrew 1992), so it is likely that the australopithecines also used rudimentary technologies. However, given the crude or ephemeral nature of tools used by the great apes, it should not surprise us that little trace has been found of those used by the earliest human precursors. Raymond Dart interpreted the remains of animals excavated from the same limestone caves as *Australopithecus*, as evidence that it had developed an 'osteodontokeratic' culture; making use of the bones, teeth, and horns of its prey to fashion tools. Although these early ideas were subsequently shown to be incorrect, it still remains a daunting task to reconstruct any credible picture of the lives of the australopithecines or of the earliest representatives of the genus *Homo*. The relevance of possible sexual dimorphism in body size is a case in point. In Chapter 1, evidence was reviewed which indicates that some of these hominids were sexually dimorphic, adult males being much larger than females (see Table 1.1). Among the extant Old World anthropoids, extreme sex differences in body weight occur in species, such as the gorilla, which have polygynous mating systems. However, sex differences in body weight are also typical of monkeys and apes that have multi-male/multi-female mating systems, such as the macaques and chimpanzees. Although body size sexual dimorphism is *on average* less pronounced in such multi-male/multi-female forms, there is considerable inter-specific variability and overlap with the polygynous anthropoids. Where the sexes are similar in size, a monogamous mating system often occurs. Because it is so difficult to identify the sexes and to calculate body sizes from fragmentary fossil evidence, current estimates of sexual dimorphism in australopithecines and other extinct hominids must be treated with the utmost caution. *A. afarensis*, for example, has been characterized as being highly sexually dimorphic. If correct, this finding could be consistent with effects of intra-sexual selection for increased male body size within a polygynous mating system. Yet, detailed studies by Reno et al. (2003) led them to reject this conclusion, and to suggest that the sexes might have been very similar in size, as is more consistent with monogamy. Is it even possible that some larger and smaller specimens, currently assigned to *A. afarensis*, might represent members of two separate species, rather than being males and females of the same species?

Recent evidence of small skull size in an African specimen of *H. erectus* has also fuelled speculation about sexual dimorphism and polygyny in this species (Spoor et al. 2007). However, the cranial capacities of specimens of *H. erectus* vary tremendously and brain size enlarged progressively during the long time span that this taxon existed. Perhaps the small cranial capacity of the recently described African specimen relates to its greater geological age, as well as the degree of its geographical separation from specimens of *H. erectus* discovered in Asia (see Figure 1.10).

Lovejoy (1981) posited that monogamy is an ancient trait among hominids. However, many conflicting views have been expressed as to whether human ancestors had monogamous, polygynous, or multi-partner-mating systems. Cross-cultural anthropological evidence shows that the majority of recent human populations include polygyny, as well as monogamy within their mating system (Ford and Beach 1951; Murdock 1981). The phenomenon of romantic love is a human universal, as is the propensity for men and women to engage in long-term relationships, and to raise their children together. As David Buss has put it

> Some continue to argue that people are fundamentally evolved to be monogamous....Others argue that humans are naturally promiscuous, and that marriage represents an unnatural cultural imposition. Both these simplistic notions are wrong.

Given the extreme difficulties of calculating sexual dimorphism in body size based upon existing fossil evidence, the conclusion reached in Chapter 1 was that it would be unwise to attempt to deduce the mating systems of the australopithecines, or of early members of the genus *Homo,* on this basis alone. In the circumstances, the best route that we may follow involves attempts to reconstruct the likely origins of human sexuality from comparative data on anatomy, physiology, and behaviour obtained by studying the extant non-human primates and other mammals. This approach, coupled with insights derived from anthropology and evolutionary psychology, may help us to understand the likely mating systems and sexual behaviour of extinct hominids.

It is a huge disappointment that none of the australopithecines or members of the genus *Homo* besides *H. sapiens* have survived to the present day; humankind is alone, although it was not always so. In the absence of any more closely related species, the chimpanzee and bonobo have been extensively studied and compared to *H. sapiens*. However, comparisons at the level of their sexuality are of limited value. To cite examples, Tanner and Zihlman (1976) attempted to construct an evolutionary scenario of human sexuality based upon a chimpanzee model.

Smith (1984) likewise injected a strong flavour of chimpanzee-like behaviour, involving multiple-partner matings and specializations for sperm competition, into his reconstruction of the mating system of *Homo habilis*. Such attempts are flawed because despite the much-vaunted close genetic relationships between *Pan* and *Homo* (Diamond 1991), the chimpanzee differs markedly from human beings as regards its mating system, reproductive anatomy, and sexual behaviour. The huge relative testes sizes of chimpanzees and bonobos relate to their multi-male/multi-female mating systems and represent the evolutionary outcome of intense sexual selection at the copulatory and post-copulatory levels. Sperm competition has profoundly influenced many other aspects of chimpanzee reproductive physiology and behaviour, in ways that are very different to the situation which pertains in *H. sapiens*.

Despite the fact that testes sizes are much smaller in man than in the chimpanzee, many authors adhere to the view that sperm competition must have played a significant role in the evolution of human reproductive biology (Smith 1984; Baker and Bellis 1995; Buss 2003; Shackelford et al. 2005; Shackelford and Pound 2006). This problem was not addressed by Darwin (1871), as he was unaware of the important role played by sexual selection in shaping the anatomy and physiology of the primary genitalia. These insights were achieved much later, principally as a result of Parker's (1970) pioneering work on sperm competition and Eberhard's (1985; 1996) seminal contributions to the concept of sexual selection by cryptic female choice. Unfortunately, when considered in isolation, comparative measurements of mammalian relative testes sizes are not sufficient to resolve the question of whether sperm competition has played any significant role in human evolution. This problem was discussed in some detail in Chapter 2. Cross-cultural data on testes size in relation to adult body weight were assembled for more than 7,000 men, representing 14 modern human populations (see Table 2.2). These range from the unusually small testes of Hong Kong Chinese to the much larger testes that occur in men of European and African descent. Although the acquisition of more data would clearly be desirable for comparative purposes, current evidence indicates that the testes are smaller in relation to body weight in some Asiatic populations than is the case for men from Africa and Europe. However, a consistent feature shared by all populations is the statistical tendency for the right testis to be slightly larger than the left testis.

Short's (1984) hypothesis, linking smaller-sized testes in Asiatic males to genetic factors governing gonadal function in both sexes, and to the lower frequencies of dizygotic twinning in these same populations, is supported by the cross-cultural comparisons included in Chapter 2 (Table 2.6). However, it is essential to keep in mind the caveat that diverse methods have been used to measure human testes volumes and that some of them (especially those involving orchidometers) can result in inaccuracies. Nonetheless, the smaller testes sizes of some modern Asiatic populations probably have no connection with the evolution of their mating systems or with sperm competition. It is more probable that selective forces acting upon gonadal function in women, to limit the likelihood of multiple ovulation and dizygotic twinning, may have also impacted genetic mechanisms which affect gonadal size in males. Reduced twinning rates among women in these populations may perhaps be linked to their lighter body build, and to the greater health risks associated with twin pregnancies and births.

Measurements of relative testes size provide only a crude index of the likelihood of sperm competition, albeit a most useful one. My own interpretation of the data on relative testes sizes in mammals (Figures 2.1, 2.2, and 2.3) remains that human testes are quite small, as is consistent with an evolutionary history involving monogamy or polygyny. However, given the continuing controversies which surround the question of human sperm competition, especially in the literature on evolutionary psychology, it is most helpful to explore the effects of sexual selection upon other reproductive traits and to consider sperm morphology, the structure and functions of the accessory reproductive glands, and the penis. Chapters 3 and 4 dealt with the roles played by sperm competition and cryptic female choice in the evolution of mammalian reproductive anatomy and physiology. Sperm competition has favoured the evolution of larger sperm midpiece volumes in mammals (Anderson and Dixson 2002; Anderson et al. 2005). This is not the case in birds (Immler and

Birkhead 2007). However, mammals and birds are widely separated in phylogeny, and avian gamete morphologies differ markedly from those of most mammals (Jamieson 2007).

Human sperm have small midpieces, and this indicates that sperm competition is unlikely to have played a significant role in the evolution of human gametes. This conclusion is strengthened by comparisons of in vitro staining of mitochondria in the sperm midpieces of man and the chimpanzee. The sperm midpiece is not only much larger in chimpanzees than in human males, but it also contains a larger volume of mitochondria, which stain red in response to the dye JC-1 (Anderson et al. 2007). Sexual selection may have favoured the evolution of greater mitochondrial loading in chimpanzees, and in other mammals where sperm competition is significant, in order to enhance the energy production required for sperm motility. Relative testes sizes and sperm midpiece volumes are positively correlated for a large sample of mammalian taxa (Anderson et al. 2005). Thus, sexual selection has acted to increase the volume of seminiferous tissue in the testes and to improve the energetics of individual gametes. Human sperm have even smaller midpieces than those of the gorilla (Figure 3.4 and Table 3.1).

Although human testes are of modest size in relation to body weight, one theoretical possibility is that selection might have increased the efficiency of human spermatogenesis, compared to that of other mammals. Thus, human testes might produce gametes more rapidly and hence make larger numbers than expected, given their size and the amount of seminiferous tissue they contain. This question was also considered in Chapter 3 but, far from being exceptional, human testes produce fewer spermatozoa per gram of parenchyma (seminiferous tissue) than in almost all mammals for which accurate data exist. This finding relates, in turn, to the fact that individual Sertoli cells in the human testis each support the development of relatively few sperm by comparison with other mammals. The total duration of spermatogenesis in men (74–76 days) is also long by comparison with many other mammals and the total daily production of sperm and numbers stored in the cauda epididymis are comparatively low and unexceptional (see Table 3.2).

More unusual is the occurrence of significant numbers of morphologically abnormal sperm in the human ejaculate; sperm pleiomorphism is more pronounced in *H. sapiens* than in most other primates. Human founder populations are thought to have been quite small, and were subject to genetic bottlenecks during their migrations out of Africa and throughout the world (Manica et al. 2007). Abnormalities of sperm morphology may have been one of the consequences of the reductions in genetic variability in human founder populations. One interpretation of this might be that the occurrence of a small percentage of morphologically imperfect sperm in the ejaculate, once established, is more likely to persist under conditions where females copulate with a single principal male (as is the case in polygynous and monogamous mating systems) rather than with multiple partners. There is less possibility of sperm competition occurring and less selection pressure for improvements in sperm quality in such circumstances. Thus, if human founder populations had primarily polygynous or monogamous mating systems, a small degree of sperm abnormality might have persisted in the absence of negative selection pressure. Interestingly, sperm pleiomorphism is also pronounced in the gorilla, which is polygynous, whereas relatively few abnormal sperm occur in the ejaculates of bonobos and chimpanzees which have multi-male/multi-female mating systems. (Figure 3.7).

Contrary to these interpretations, the notion that pleiomorphic human sperm represent distinct morphs, which evolved in order to undertake specialized roles during sperm competition, was advanced by Baker and Bellis (1988; 1995). They posited that Kamikaze sperm exist in human ejaculates, and that these serve to block the progress of the gametes of rival males within the female reproductive tract. Kamikaze sperm thus increase the chances that specialized egg-getter sperm can reach the oviduct. These theories and others concerning the functions of pleiomorphic human sperm are not supported by credible experimental evidence, and have been rebutted by critical authorities (Harcourt 1991; Short 1997; Dixson 1998*a*; Moore et al. 1999; Dixson and Anderson 2001; Birkhead 2000; Lloyd 2005).

Studies of gamete biology provide little support for the idea that sperm competition has played a significant role in the evolution of human reproduction. Indeed, this conclusion is strengthened when one considers the results of comparative studies of the accessory reproductive organs of mammals. For example, the vasa deferentia are relatively shorter and more muscular in those mammals in which females mate with multiple partners during the fertile period. Sperm competition has presumably favoured the evolution of such specializations in order to maximize the efficiency of rapid sperm transport during sexual activity. In particular, it is the thicknesses of the two layers of longitudinal muscles in the wall of the duct which have increased due to sperm competition (Anderson et al. 2004). By contrast, mammals in which females mate primarily with a single male, and which are monogamous or polygynous, have less well-developed longitudinal muscles, and a relatively thicker (central) layer of circular muscles in the wall of the vas deferens. The human vas is structurally of this latter type (see Figures 3.13 and 3.14). This fact provides further evidence that sperm competition is unlikely to have played a significant role in the evolution of the human reproductive system. These findings are contrary to the ideas advanced by Smith (1984), who suggested that specializations for sperm competition might be revealed by studying the structure of the human vas deferens. However, at the time Smith wrote his review no quantitative studies had been carried out to test such ideas.

The sizes and functions of the seminal vesicles vary tremendously in different groups of mammals, and their secretions constitute 60 per cent of the human ejaculate. From an evolutionary standpoint it is interesting that the seminal vesicles of primate species in which sperm competition pressures are greatest are significantly larger than those of monogamous and polygynous species (Dixson 1998*b*). Likewise, seminal coagulation and copulatory plug formation are more pronounced in primates which engage in multi-partner matings and sperm competition (Dixson and Anderson 2002). The seminal vesicles are of medium size in human males, and seminal coagulation is likewise less pronounced than in primates where sperm competition is significant (see Figures 3.16 and 3.17). This example of sexually selected variations of male reproductive physiology is further bolstered by molecular studies of rates of evolution of the genes that encode for production of the semenogelins; these are the proteins which act as substrates for seminal coagulation (Dorus et al. 2004).

Comparisons of mammalian testicular function, sperm morphology, and the structures and functions of the reproductive ducts and accessory glands, all support the view that human sexual behaviour does not spring from an evolutionary background involving specializations for sperm competition. Despite this, numerous authors argue that human penile morphology is exceptional and specialized by comparison with other primates (Fisher 1982; Smith 1984; Baker and Bellis 1995; Diamond 1997; Jolly 1999; Miller 2000; Buss 2003; Rolls 2005). Gallup and Burch (2004) attempted to test hypotheses advanced by Baker and Bellis, proposing that the human penis is morphologically specialized to displace the semen of rival males during copulation. Gallup et al. (2003) have also argued that penile size and the large glans with its pronounced coronal ridge aid sperm displacement during bouts of deep pelvic thrusting. Such patterns of deep thrusting are said to occur if men suspect that their partners have previously engaged in extra-pair copulations. Mechanisms of semen displacement were not assessed in vivo during these experiments. Instead, models of the human genitalia were used (vaginae, dildos, and artificial semen formulated using corn–starch solutions: Gallup et al. 2003). The results of these experiments are likely to be invalid, due to the unrealistic nature of the models employed and the risks of subjective bias in using them to simulate natural events that occur during copulation.

Quantitative comparisons of penile morphologies in forty-eight primate genera, including *Homo*, confirm that human phallic morphology is not exceptional, and is consistent with an evolutionary background of either monogamy or polygyny rather than one that involved any significant degree of sperm competition (see Figure 3.22). Aside its breadth, the human penis is not unusually large, and some monkeys also have large and thick penes (e.g. spider and woolly monkeys: Figure 3.20). Moreover, the possession of a rounded or helmet-shaped glans penis is a phylogenetically ancient trait, present in many Old World monkeys and also in the gorilla. The chimpanzee, in accord with its other specializations for sperm competition, has a highly derived penile morphology, involving loss of a distinct glans to produce an elongated, filiform structure.

Given these facts, it is difficult to understand why so many authors continue to state that human beings have exceptionally large and complex penes, as the morphological evidence does not support this conclusion. The problem may result, in part, from the use of inaccurate or incomplete data on penile sizes in the great apes (as cited by Smith 1984) that have entered the literature and have been repeated by authors, often at several removes from original sources. Indeed, a lack of first-hand experience characterizes many of these comparisons between the human genitalia and those of other primates. There also appears to be a certain coyness where metric studies are involved, as if the measurement of penile morphology is a less acceptable topic than measurements of other external structures. In reality, the human penis is unremarkable when compared to those of many non-human primates and other mammals. There is no substance to the notion that the evolution of human penile morphology has been affected by sperm competition.

Eberhard (1985; 1996) has proposed that the intromittent organs of animals may also function as *internal courtship devices*; especially so in those cases where females mate with multiple partners. Sexual selection may thus have influenced the evolution of phallic morphology via *cryptic female choice*, as well as through sperm competition. In practice, both these processes are likely to entwine during the co-evolution of the male and female genitalia.

Chapter 4 explored the relevance of cryptic female choice to studies of mammalian genitalia and copulatory behaviour. Primate penile morphologies are, indeed, most complex in those species where females mate with multiple partners. This finding represented the first quantitative test of Eberhard's (1985) ideas to be reported for a vertebrate group (Dixson 1987*a*). The inclusion of human phallic morphology in a new analysis reported in Chapter 3 confirms that the human condition is relatively simple, and like those of the other polygynous or monogamous primates. In the chimpanzee, by contrast, the penis is morphologically specialized. It is elongated and filiform in order to deposit ejaculates close to the cervical os, and to negotiate the female's large sexual skin swelling, which increases vaginal length by up to 50 per cent during the fertile period (see Figure 4.2).

A second possible avenue for cryptic female choice concerns the transport or storage of sperm. Because women as well as men have the capacity to exhibit orgasm during sexual activity, it has been suggested that female orgasm may have evolved in relation to patterns of sperm transport and sperm competition. Thus, Baker and Bellis (1993*b*; 1995) proposed that high sperm retention and low sperm retention orgasms occur in women, and that the former may be activated during extra-pair copulations with preferred partners. These ideas have frequently been cited in the literature on evolutionary psychology. However, at the opposite pole of opinion, Symons (1979) interpreted the occurrence of orgasm in women as a non-adaptive homologue of the male phenomenon. This view aroused criticism from some quarters (e.g. Hrdy 1979), fuelled perhaps by a clash between scientific reasoning and feminist or sociological perspectives concerning the significance of orgasm in women.

In fact, there is no robust evidence that female orgasm plays any role in human sperm transport or fertility. Unfortunately, this subject has not received the experimental attention it deserves, and some of the few studies which do exist have been cited selectively and given undue weight in the literature. A recent and most thorough critique of work in this area by Elisabeth Lloyd (2005) shows how limited evidence or flawed experiments have been over-interpreted to promote the view that women's orgasms might influence sperm transport. These matters were discussed in some detail in Chapter 4, so that it is not necessary to repeat the arguments here. Suffice it to say that whilst female orgasm certainly does occur in some primates, it probably represents a non-adaptive homologue of the male capacity to exhibit orgasm during ejaculation. Male primates also display non-adaptive homologues of feminine traits, such as nipples and (in some species) an atrophic *uterus masculinus*. What is worrying, as Lloyd (2005) points out, is the extent to which poorly grounded arguments and limited evidence concerning female orgasm have been incorporated into the literature of evolutionary psychology, as if they represent established facts of human physiology.

The oviduct constitutes the final arena for sperm competition, or cryptic choice, within the female reproductive tract. Indeed, it is likely that differential activation or suppression of motility in subpopulations of sperm within the isthmus of the oviduct may play a key role in mammalian sperm selection (Sakate et al. 2006). Fertilization of ova occurs in the ampullary portion of the human oviduct, as is the case among mammals in general. Studies of insects point the way to effects of sexual selection upon the evolution of female sperm storage organs (spermathecae) and their associated ducts (Eberhard 1996; Simmons 2001). Among mammals, it is of considerable interest that oviductal length varies with respect to the mating system (Anderson, et al. 2006). Thus, the oviducts are longest in those mammals where females commonly mate with multiple partners, and in which males have larger testes in association with increased sperm competition pressures (see Figures 4.5 and 4.6). Perhaps elongation of the oviduct in some way tests the relative fitness of gametes from rival males. In women, the oviducts are relatively short in relation to body size. There is no support from this quarter for possible effects of sexual selection upon the evolution of the human genitalia via cryptic female choice.

The limited data currently available on mammalian cryptic female choice are consistent with conclusions reached by studying the anatomy and physiology of the male reproductive system. *Homo sapiens* exhibits traits consistent with a long history of polygyny or monogamy, and a relative absence of sperm competition. This does not mean that sperm competition never occurs, or that it is impossible for it to occur in modern-day human populations. The anatomical and physiological evidence does indicate, however, that sperm competition is unlikely to have played any significant role in human evolution. Some evolutionary psychologists adopt a contrary view, and propose that human sexual behaviour and mating strategies have been affected by sperm competition pressures (Baker and Bellis 1995; Gallup et al. 2003; Shackelford et al. 2005; Platek and Shackelford 2006). However, the absence of anatomical and physiological specializations for sperm competition in *H. sapiens* argues very strongly against the validity of such conclusions.

Comparative studies also offer useful insights concerning the origins of human copulatory behaviour. Copulatory postures and patterns among the non-human primates present some interesting phylogenetic variations. These have been subject to the effects of both natural and sexual selection, as was discussed in Chapter 5. Among the lorisines, for example, the African potto, and slender and slow lorises of Asia all exhibit distinctive inverted copulatory postures, with the female clinging underneath a branch as the male mounts in a dorso-ventral position. This probably represents a phylogenetically ancient trait in this group of prosimians. Similarly, among the New World primates, most of the ateline monkeys exhibit a peculiar leg-lock variation of the dorso-ventral mounting posture, in which the male places his legs over, and medial to those of the female partner (see Figure 5.3). Such postural specializations are interesting from a phylogenetic perspective, but all of them are variations of the typical dorso-ventral mounting postures which occur throughout the Order Primates, and in mammals in general. However, among the apes as well as in human beings, ventro-ventral and female superior copulatory postures have also been recorded. Face-to-face mating postures, such as the 'missionary position' are thus not unique to human beings, and probably represent phylogenetically primitive traits. In Chapter 5, it was argued that the evolution of brachiation and the postural flexibility conferred by specializations of the shoulder joint and shoulder musculature allowed apes to adopt more variable postures during arboreal copulation. Orangutans, despite their great size, copulate using a variety of suspensory postures, or with the male reclining upon his back whilst the female mounts him. Although the African apes are largely terrestrial, face-to-face matings occur quite commonly among wild bonobos, and have occasionally been reported in captive and free-ranging gorillas. Thus, it is likely that ventro-ventral or female superior copulatory positions in the apes and human beings derive from postures used by their common ancestors, and would have been present also in the earliest hominids. These observations provide no support for the view that ventro-ventral copulatory postures represent uniquely human specializations, serving to increase clitoral stimulation and enhance orgasmic responses in women (Ford and Beach 1951). Rather, the three basic human copulatory positions (ventro-ventral, female superior, and dorso-ventral) all derive from patterns homologous with those which occur in extant apes.

Copulatory durations also vary considerably among mammals, lasting only for a few seconds, or for hours depending upon the species considered. Among the primates, examples of these two extremes are provided by the common marmoset, which reaches ejaculation in 5 s, and the greater galago which may intromit for 2 h or more, and probably exhibits multiple ejaculations during that time. The evolution of intromission durations, patterns of pelvic thrusting, and occurrences of a genital lock between the sexes and other features were also addressed in some detail in Chapter 5. There, the conclusion was reached that the evolution of complex patterns, involving single prolonged, or multiple brief, intromissions with pelvic thrusting has occurred due to sexual selection. Such patterns are found almost exclusively in those primates where the females engage in multiple-partner matings. The behavioural specializations exhibited by such males during mating thus correlate with their genital specializations. These are due to sexual selection by sperm competition, cryptic female choice, or a combination of both processes.

Human beings sometimes prolong periods of copulation. However, it is noteworthy that large-scale surveys of human sexual behaviour such as those conducted by Kinsey and his colleagues (1948) have reported that the average time taken by men to reach ejaculation is relatively brief. Likewise, consideration of the anthropological literature leads to the conclusion that techniques for extending the duration of intercourse and delaying ejaculation have to be practised and learned by men and women. Although the acquisition of such techniques may be encouraged in some cultures, they appear not to represent fundamental patterns of human behaviour. On the contrary, men are most likely to reach ejaculation relatively quickly once intromission is attained and pelvic thrusting has been initiated. The phenomenon of premature ejaculation, whereby the male is over-responsive and reaches orgasm quickly, is relatively common, and may have its origin in the evolutionary history of *H. sapiens*.

Copulatory frequencies vary in primates depending upon their mating systems and the likely occurrence of sperm competition. Thus, ejaculatory frequencies are greatest in males of those species which have multi-male/multi-female mating systems, such as macaques, baboons, and chimpanzees. Polygynous or monogamous species such as gorillas or marmosets copulate much less often. Again, the comparative data indicate that frequencies of human copulation are similar to those of polygynous or monogamous non-human primates (see Table 5.4). Moreover, men's sperm counts decline rapidly as a result of repeated ejaculations, a further indication of the limited human capacity to produce and store sperm.

It has been proposed that despite these limited capacities, men might engage in prudent allocation of sperm numbers in the ejaculate, and that this might be adaptive in terms of human sperm competition (Pound et al. 2006). Prudent allocation of sperm has been demonstrated in domestic fowl, for example, in which cockerels ejaculate greater numbers of sperm if paired with a new hen (the 'Coolidge effect') or one with a larger comb on the head (females with larger combs have greater reproductive success: Pizzari et al. 2003). Some experiments have shown that men also ejaculate larger numbers of sperm under particular circumstances. This occurs, for example, if men are separated from their partners for long periods (Baker and Bellis 1989), or if stimulation is heightened during prolonged bouts of masturbation (Pound et al. 2002). It is not the experimental results which I find problematic in these studies, so much as the interpretation that these effects must necessarily reflect adaptations for sperm competition in human beings. As an alternative, I suggest that male vertebrates may share a fundamental physiological capacity to increase sperm numbers in the ejaculate in response to situations which heighten sexual arousal (e.g. the presence of a highly attractive female, or the sight of other conspecifics mating). It may be these very capacities which sexual selection acts upon in species, such as the domestic fowl, where males commonly mate with multiple partners and prudent sperm allocation becomes adaptive in order to maximize fertility.

Taken as a whole, the results of research on comparative reproductive anatomy, physiology, and copulatory patterns indicate that *H. sapiens* was originally polygynous or monogamous (or both), and did not arise from forms which had a multi-male/multi-female mating system. Current knowledge of reproductive anatomy and physiology does not allow a distinction to be drawn between the relative contributions made by polygyny or monogamy to the remote origins of human mating systems. It is possible that further fine-grained analyses of sperm biology, the functions of the accessory reproductive organs, and penile morphology might reveal differences between polygynous and monogamous primate species. However, it is much more productive, when seeking a solution to this problem, to consider comparative studies of sexually dimorphic traits in physique and secondary sexual adornments. These topics were covered in Chapters 7 and 8, which revisit some of the ground covered by Darwin (1871), in order to reassess matters in the light of modern advances in the fields of anthropology and evolutionary psychology.

In a wide range of human populations, men are, on average, heavier and taller than women. Although sexual dimorphism is not as pronounced in *Homo sapiens* as in polygynous non-human primate species, its significance becomes clearer when sex differences in *body composition*, as well as *body size* are taken into account. Women have much larger reserves of body fat than men (Clarys et al. 1984; Pond 1998).These play essential roles in fuelling reproductive processes, and especially the enormous energetic demands posed by pregnancy and lactation. Men, by contrast, are considerably more muscular and physically stronger than women, traits which were probably favoured by both natural and sexual selection from the earliest phases of the evolution of the genus *Homo* in Africa. The ratio of adult male/female muscle mass is 1.5 in modern human beings, and this is similar to the sexual dimorphism in overall body mass exhibited among non-human primates which have polygynous mating systems. The occurrence of sexual bimaturism in human beings, with earlier growth and reproductive development taking place at puberty in the female sex, also resembles the situation found in those anthropoid primates which display significant sexual (body size) dimorphism in adulthood.

Success in hunting, and especially in endurance running and persistence hunting, as well as intermale competition may have favoured the evolution of masculine physical traits required for survival (Bramble and Lieberman 2004). In all probability, these developments took place relatively early in the evolution of the genus *Homo*, for instance, in *H. ergaster* and *H. erectus* which lacked the degree of cerebral development and hunting technology achieved by *H. sapiens*. Then, as now, it is surmised that males provided resources as well as protection for their female partners and offspring. It is interesting, therefore, that women in a variety of cultures rate images of men with muscular (mesomorphic) or average somatotypes as most attractive. The ideal form may lie somewhere between the average and the mesomorphic male physique. A muscular male torso, with broad shoulders and a narrow waist and hips, is often preferred. Images of very slim men (ectomorphs) tend to be assessed as somewhat less attractive, whilst heavily built (endomorphic) males receive consistently low ratings in cross-cultural studies (see Figure 7.7).

Fat deposition during puberty and adolescence is markedly sexually dimorphic in human beings. It is during this time that young women develop larger fat deposits than young men in their buttocks, thighs, and breasts. Fat deposition and distribution is under oestrogenic control, and it is indicative of female health and reproductive potential. Evidence leading to these conclusions was reviewed in Chapter 7, in relation to Singh's theory concerning evolution of the hourglass figure and female sexual attractiveness. Cross-cultural evidence indicates that a low female waist-to-hip ratio (WHR) is perceived as attractive by men, who prefer female images in which the WHR falls somewhere in the range between 0.6 and 0.8, depending upon the population considered (see Figures 7.9 and 7.10). Women with narrow waists and large breasts have significantly higher levels of oestrogen during their menstrual cycles (Jasieńska et al. 2004). By contrast, women with high WHRs and more obese physiques are at greater risk of diabetic and other health problems (Singh 2002; 2006). Although female body mass index (BMI) has been suggested to play a more important role than WHR in men's judgements of women's attractiveness, it appears that female shape is the crucial cue (see Figure 7.12). Female WHR may provide males with a first pass filter when assessing a prospective partner.

A word of caution is required about defining any single trait as being decisive in judgements of sexual attractiveness, as they impact human mate choice. Meston and Buss (2007) record that men and women express no less than 237 possible reasons for having sex. Factors which influence mating decisions are clearly complex, and attractiveness is likely to involve an amalgam of traits, some of them physical and others linked to behaviour and personality. Buss (1989) has made this point comprehensively in his reports concerning cross-cultural norms of mate choice. He stresses both the similarities and differences in mate preference displayed by men and women around the world:

> Both sexes wanted mates who were kind, understanding, intelligent, healthy and dependable.

However

> Women more than men in all 37 cultures valued potential mates with good financial prospects. Men more than women across the globe placed a premium on youth and physical attractiveness, two hypothesized correlates of fertility and reproductive value.

When we consider the remote origins of human mating preferences, including the effects of selection prior to the evolution of language, and of sophisticated cognitive assessments concerning personality, then judgements of physical traits indicative of female reproductive potential and of masculine status and health were probably of paramount importance. This is the case in the non-human primates, as well as in other animals. Human beings are the descendents of non-human primate ancestors. Deeply implanted in the human psyche, therefore, is the tendency to respond to physical traits displayed by the opposite sex when making initial decisions about attractiveness. Although such *initial filtering* may not always dictate *ultimate decisions* about mate choice, it is probable that even in modern societies it continues to exert important effects. Further, such filtering of *first impressions* concerning members of the opposite sex may be very subtle indeed, and involve subliminal processing, which is not necessarily verbalized in discussions (or when answering questionnaire surveys). Hence the reference to the significance of *love at first sight, not love at first discussion* at the beginning of Chapter 7.

A wide range of cues involving the body, face, skin and hair, as well patterns of movement, vocal and olfactory stimuli, combine in ways which are still not fully understood, to produce individual variations in attractiveness. Many of these traits relate to underlying qualities of physiology, and may provide men and women with clues about the likely health, endocrine status, and developmental history of prospective partners. It is remarkable, for example, that women who have higher levels of oestrogen during the follicular phases of their menstrual cycles are consistently rated as being facially more attractive (Law-Smith et al. 2006). The quality of the skin and the female complexion may be relevant here, as well as differences in other facial traits which are affected by hormones (see Figure 7.17). The reduction of hair on the body and face, which occurred during human evolution, for reasons that are still debated (Jablonski 2006; Rantala 2007), provided further avenues for sexual selection to operate, as the exposed areas of skin vary with age, sex, health, and reproductive status. The significance of sex differences in skin colour, noted originally by Darwin (1871), still remains enigmatic, although there is a little evidence that lighter feminine skin tones are rated as more attractive by men in some cultures (see Figure 7.15). This question is still unresolved, however, as is the significance of sex differences in other cutaneous features. Men, for example, have more body hair than women, although the expression of this trait varies tremendously between populations, as does the development of facial hair. It is interesting that masculine trunk hair is rated as being *unattractive* by women in some countries (Figure 7.21) although most of the available information concerns the preferences of younger subjects, and it would be important to obtain data from older age groups.

Darwin considered that the origin of the human beard might reside in its attractiveness to women.

However, there has been relatively little detailed experimental or cross-cultural work conducted on the significance of masculine facial hair. The origins of the human beard, whether via inter-sexual selection (via attractiveness) or intra-sexual selection in relation to status and dominance, are still unclear. It may be significant that beard growth accentuates the appearance of another masculine trait: the larger size and length of the lower face and jaw (Barber 1995). Because beard growth is stimulated by testosterone, it could provide yet another potential cue concerning physical maturity and reproductive condition. It may be a very ancient trait indeed, representing part of a suite of specializations inherited from australopithecines, or the earliest members of the genus *Homo*. Although Darwin drew attention to the occurrence of beards in some non-human primates, the degree of homology is questionable. This is because growth of human facial hair, like pubic hair, is stimulated by testosterone, and the hair is morphologically distinct from that which occurs elsewhere on the head or body. The beards of non-human primates such as orangutans or mandrills are not morphologically differentiated to the same degree as the human beard and, to the best of my knowledge, there is no experimental evidence that their growth is stimulated by testosterone. The peculiar nature of the human beard and pubic hair may relate to their evolution as conspicuous visual signals in ancestral forms where hair was lost on adjacent areas of the face and body. Thus, these traits may have arisen very early in the genus *Homo*, and may have accompanied the overall reduction of body hair in *H. ergaster* and *H. erectus*.

Whatever the functions of these cutaneous specializations in human beings may be, it is clear that the degree of expression of such secondary sexual traits in human males (beard, male pattern baldness, and body hair) is commensurate with that which occurs in polygynous non-human primates. Comparative studies of visually striking, secondary sexual traits in male monkeys, apes, and humans (see Figure 7.19) strengthen the view that polygyny has played a significant role during the origins of human sexuality.

This conclusion is also reinforced when sex differences in the structure of the larynx and vocalizations are compared among the anthropoid primates. This topic was addressed in Chapter 8. Men have larger larnyges, longer vocal cords, and deeper voices than women. Further, there is evidence that women rate the lower pitch of the male voice as pleasant and attractive, and that men may lower the pitch of the voice in situations involving social dominance (Bruckert et al. 2006; Puts et al. 2006). Testosterone stimulates laryngeal growth in boys when they reach puberty, and there is a positive relationship between masculine testosterone levels and vocal pitch in adulthood (Dabbs and Malinger 1999). Comparisons of sexual dimorphism of the larynx and its associated structures in the non-human primates show that such differences are greatest in polygynous species, less marked in forms with multi-male/multi-female mating systems, and minimal or absent in monogamous species (see Figure 8.9). It is among the polygynous non-human primates that the most sexually dimorphic and low-pitched calls are given by adult males, especially in those species that inhabit rainforest. The enlargement of the larynx and deep voice of the human male is also likely to represent another ancient trait, derived from early ancestral forms among australopithecines which were polygynous in their mating behaviour.

All these observations lead to the prediction that, as further fossil evidence accumulates regarding the question of body size in early hominids, it may confirm that the direct ancestors of the genus *Homo* were indeed markedly sexually dimorphic, as is consistent with sexual selection via inter-male competition within polygynous mating systems.

A major question concerns the nature of social organization and group composition in ancestral hominids. Recently Chapais (2008) has produced a most scholarly review and analysis of the origins of kinship and pair bonding in human evolution. He points out that kinship is important in many non-human primates, and that homologues of these processes would have existed also in early hominids. Thus, monkeys such as the macaques and baboons exhibit strong matrilineal social bonds. Mating between mothers and offspring, or between brothers and sisters is very rare, as in Barbary macaques (Kuester, Paul, and Arnemann 1994), or chimpanzees (Goodall 1986). This is the non-human primate equivalent of what is called an *incest taboo* and

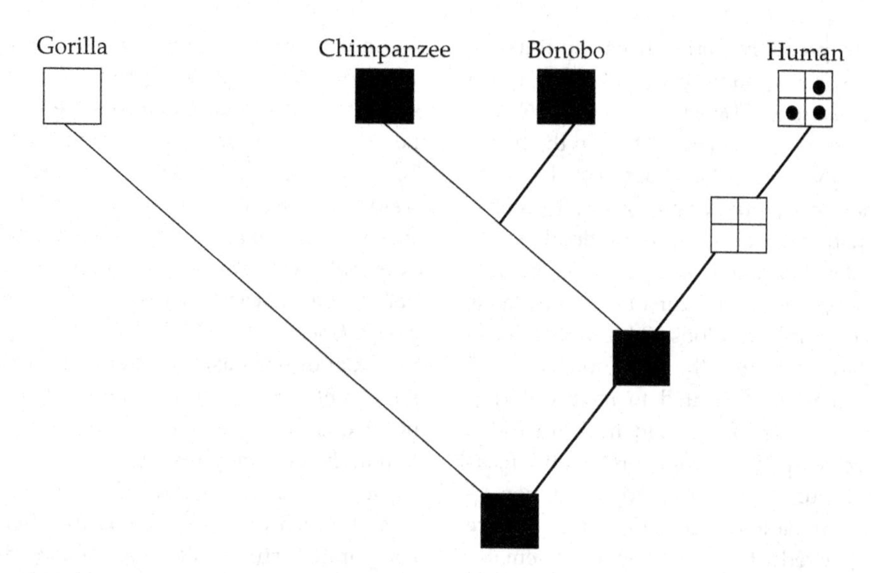

Figure 9.1 Hypothetical stages in the evolution of the mating systems of humans and the African apes. ■ = multi-male/multi-female mating system (males are philopatric). □ = Polygynous, one-male unit. In the gorilla, individual units are spatially separate. In australopithecines ancestral to *Homo*, groups consisted of a number of polygynous units. ▣ = monogamy. Ancestral human groups were made up of monogamous pairs, together with smaller numbers of polygynous units.

probably preceded its cultural elaboration in human beings. Long-term familiarity between female apes or monkeys and their offspring ('uterine kin' as defined by Chapais) reduces sexual attractiveness and discourages the likelihood of copulation. These observations of non-human primates provide parallels with Westermarck's hypothesis, by which he sought to explain lack of sexual attraction due to familiarity between close relatives in humans (as reviewed in Dixson 1998*a*). Among most non-human primates, however, there is no evidence that fathers recognize their offspring, or vice versa, so that the likelihood of incestuous matings cannot be inhibited by social familiarity in these cases. Chapais (2008) stresses that the chimpanzee and the bonobo are unusual among extant anthropoids in that their multi-male/multi-female social groups exhibit *male philopatry*. Thus, unlike most Old World monkeys, male chimpanzees and bonobos remain in their natal groups, whilst females emigrate as they mature and join neighbouring communities in order to reproduce. Wrangham (1987) has proposed that this form of male philopatry is a primitive trait among African apes, and would likely have occurred also in the earliest hominids. Indeed, even in the gorilla, females transfer out of their natal groups at maturity, whilst males may either leave and become lone silverbacks, or remain in their natal groups until they achieve alpha breeding status (Harcourt and Stewart 2007).

These arguments lead Chapais to propose that the common ancestor of *Pan* and *Homo* probably formed large multi-male/multi-female groups, in which males were philopatric, and females emigrated at sexual maturity. Polygyny emerged later, in the hominid line, and in forms which were ancestral to human beings. Chimpanzees, by contrast, retained a multi-male/multi-female social organization and mating system, in which sperm competition has been a major driving force (Figure 9.1). It is important to consider the mechanism by which polgyny might have developed in those australopithecines which gave rise to the genus *Homo*. Here Chapais posits (and I concur) that spatially separate one male units (as in gorillas) were much less likely to have been involved. It is more parsimonious to suggest that a group structure containing numbers of adults of both sexes would have been retained. In various African and Asian monkeys, polygynous one-male units are nested within much larger social groups, which probably derive from multi-male/multi-female social groups in evolutionary terms (e.g. *Papio hamadryas, Theropithecus gelada, Nasalis larvatus, Rhinopithecus roxellana*). Thus, in the hominid case, polygyny may have developed due to enhanced male mate guarding of a small number of partners within multi-male/multi-female social groups. This development would have been present in the australopithecine precursors of the genus *Homo*. Under these conditions, intra-sexual selection would have favoured the evolution of sexual size dimorphism, as is posited to have occurred (on the basis of limited fossil evidence) in various australopithecine species. From this condition, as Chapais points out, the transition to mixed monogamy/polygyny in bands of early *Homo* would have required only a reduction in numbers of females associated with each male. Monogamous pair bonds gradually replaced polygyny as the prevalent mating system (Figure 9.1).

Conclusions

Comparative studies of mammalian reproduction and insights gained from anthropology and evolutionary psychology support a number of conclusions concerning the origins of human sexuality. These are listed in Table 9.1, and summarized below.

1. It is likely that *Homo sapiens* evolved from a primarily polygynous non-human primate precursor, and that the earliest members of the genus *Homo* were to some degree polygynous. In this context, the absence of marked sexual dimorphism in canine tooth size in the australopithecines and members of the genus *Homo* is perplexing, as larger male canines would be expected to occur in association with polygyny. It is possible that modest canine size represents part of the overall changes in the shape and size of the dental arcade which occurred in these hominids, for selective reasons which have not been fully explained. Whatever factors affected the evolution of hominid canine tooth size, the occurrence of multiple sexually dimorphic traits in *H. sapiens* (in body size, shape and composition, facial traits, secondary sexual adornments, the larynx and vocal pitch, and age at reproductive maturity) all indicate that human ancestors were polygynous. Fossil evidence for the occurrence of pronounced body size sexual dimorphism in some of the australopithecines is far from conclusive. Thus, further research is required to clarify whether these hominids were sexually dimorphic, as is posited here to be the case for the direct ancestors of the genus *Homo*.

Sexual dimorphism in extant anthropoids which have multi-male/multi-female mating systems is not strong evidence for the occurrence of such a system in the australopithecine precursor of the genus *Homo*, or in those forms (e.g. *H. ergaster*; *H. heidelbergensis*) which preceded *H. sapiens*. This is because comparative studies of reproductive anatomy and patterns of copulatory behaviour consistently indicate that *H. sapiens* did not evolve directly from a multi-male/multi-female precursor (see below: points 3 and 4).

The propensity for men and women to form long-lasting relationships for reproductive purposes is an ancient trait, probably present in the earliest members of the genus *Homo*. The occurrence of such relations between the sexes is universal across extant human populations, and its existence in ancestral forms provided the biological foundation for the later emergence of marriage. Lengthy periods of human infancy, childhood, and adolescence require considerable parental investment by both sexes if offspring are to survive. These selective pressures may have contributed to the formation of long-term relationships between men and women early in human evolution. Nowadays, as throughout recorded history, marriages might be either polygynous or monogamous, depending upon the culture considered. Most human cultures allow polygynous marriage, but for the vast majority of people, monogamy is the accepted norm. During evolution of the genus *Homo*, a shift towards paired relationships and monogamy has occurred. In those modern cultures which allow polygynous marriage, it tends to be prevalent among men of higher status, and those who control most resources. Nonetheless, a contributory factor to occurrences of extra-pair copulations by men in primarily

Table 9.1 Status of selected traits and mating systems, during three phases of human evolution.

	Common ancestor Of *Pan* and *Homo*	**Australopithecine precursor of *Homo***	***H. ergaster; H. erectus***
Mating system	Multi-male/multi-female	Polygyny	Monogamy > polygyny
Male philopatry?	Yes	Yes	Yes
Sperm competition pressure	High	Low	Very low
Incest avoidance?	Yes	Yes	Yes
Male genitalic complexity	Pronounced	Reduced	Minor
Copulatory posture	DV;VV;FS	DV;VV;FS	VV;FS;DV
Copulatory pattern*	Nos 10, 11, 12	No. 12?	No.12?
Female sex skin?	Swelling likely	Unlikely	Absent
Oestrus occurs?	No	No	No
Female orgasm?	Yes	Yes	Yes
Copulatory frequency	High	Low	Low
Sexual dimorphism			
Body size ♂>♀?	Moderately so	Yes: major	Yes
Larynx ♂>♀?	Moderately so	Yes: major	Yes: major
Secondary sexual adornments			
Males	Not pronounced	Pronounced e.g. capes and crests of hair	Pronounced e.g. facial hair
Females	Sexual skin?	?	Breasts, low WHR
Hirsuteness	Pronounced	Reduced?	Body hair lost: sweat glands well developed

*Definitions of these copulatory patterns are provided in Table 5.1.
Abbreviations: DV = Dorso-ventral; VV = ventro-ventral; FS = female superior mating position.

monogamous relationships may be an underlying tendency towards polygyny. Chapais (2008) rejects the notion that joint parental care acted as the selective force to promote pair formation and monogamy during human evolution. He proposes instead that the development of weapons during human evolution made inter-male competition for access to mates more equal, and potentially more lethal. Under these conditions, most males may have opted for monogamous rather than polygynous mating strategies. Contrary to this view, there is some evidence from molecular biological studies that polygyny may have been more widespread until relatively recently in *H. sapiens*. Dupanloup et al. (2003) suggest (on the basis of studies of Y-chromosome diversity) that a recent shift from polygyny to monogamy has occurred in humans, coincident with the rise of agriculture and expansion of settled populations. However, the evidence presented and reviewed throughout this book and arguments advanced by Chapais (2008) support the conclusion that the transition from polygyny to monogamy began much earlier, in taxa that were ancestral to *H. sapiens*.

The occurrence of a polygynous mating system in human ancestors does not mean that the social system necessarily comprised spatially separate 'one-male units', such as are found in modern day gorillas. Larger social groups with a multilevel social system seem more probable, especially in early hominids such as the australopithecines, which inhabited more open savannah/woodland habitats. Some extant monkeys, such as the hamadryas baboon, live in large groups composed of many one-male units. It is possible that the australopithecine ancestors of the genus *Homo* formed groups in which individual males associated with several females for mating purposes. Based upon such a polygynous system, a transition to monogamy as well as polygyny would have characterized the evolution of mating systems within the genus *Homo*.

2. Because the human species is intellectually and culturally sophisticated, it is easy to overlook

the contribution made by ancient traits and adaptations to our sexual lives. The anthropoid primates are par excellence visually orientated creatures, and we have inherited a rich repertoire of nonverbal communication from non-human primate ancestors. There may be a very good reason why cues serving for visual communication and facial recognition are stored in the human memory and retrieved more effectively than linguistic labels. How often, for example, have we heard someone say 'I never forget a new face, but I can't always remember a new name?' All human cultures share the propensity to exhibit particular facial expressions in the same emotional contexts (Ekman 1973) and homologues of these have been identified in the great apes. Fundamental aspects of human sexual communication such as eye contact and other nuances of facial communication derive from periods of evolution before the emergence of language. Likewise, our assessments of sexual attractiveness relate, initially at least, to physical cues involving the body and face of members of the opposite sex. These physical cues provide information about age, health, reproductive condition, and a host of variables which were (and still are) important for mate choice. Sexual selection, as well as natural selection, has affected the evolution of traits such as masculine somatotype, female waist-to-hip ratios, and facial cues in both sexes. Moreover, these outcomes of sexual selection predated the more complex decision making facilitated by language, e.g., concerning matters such as a potential partner's kindness or sense of humour. Such complex abstractions represent more recent developments in human evolution, and they may be less important in people's initial rankings of mate preferences than is currently recognized.

3. Darwin was apparently unaware that the primary genitalia and patterns of copulatory behaviour have also been influenced by sexual selection during the evolution of many animals, including the primates. The more recent concepts of sperm competition and cryptic female choice provide rich sources of comparative information concerning the evolution of genitalic complexity in the anthropoids, including *H. sapiens*. Genitalic complexity is, in turn, strongly linked to patterns of sexual behaviour. The information considered throughout this book does not support the conclusion that sperm competition has played a significant role in the evolution of anatomical or physiological specializations of the human reproductive system. By implication, therefore, sperm competition is unlikely to have exerted a significant effect on the evolution of human sexual behaviour. Whatever the role of sexual selection in evolution of the human brain (Miller 2000), it is unlikely that sperm competition has impacted the evolution of neurological mechanisms which facilitate human mate choice and copulatory patterns. Nor, unfortunately, can the evolution of such neurological mechanisms be addressed by analyses of overall brain size. As an example of this approach, Pitnik, Jones, and Wilkinson (2006) reported that mating systems and brain sizes are correlated in bats; those species which engage in sperm competition have smaller brains than those in which females mate primarily with a single male. Applying this line of reasoning to human evolution and the importance of human sperm competition posited by evolutionary psychologists, one might expect human brain size to be reduced, compared to that of more exclusively monogamous or polygynous primates. Plainly this is not the case. Nor is it useful to engage in this type of speculation, because so many correlations between brain size and behavioural traits have been reported in mammals and birds. In their review of fifty such studies, Healy and Rowe (2007) concluded that overall measures of brain size are too crude to facilitate meaningful correlations with complex behavioural traits. Even the cautious statements I have made in this book concerning brain sizes in fossil hominids, with respect to likely tool use or language acquisition, should be viewed with more caution. Of brain size and fossil hominid mating systems nothing of value may be said.

It has become commonplace to use comparisons of human and great ape testes sizes as evidence that sperm competition has influenced human evolution (the argument often presented is that human testes are larger than those of the polygynous gorilla, but smaller than those of the chimpanzee; thus sperm competition played a significant role in human evolution). However, the occurrence of relatively small testes sizes in human males nests within an extensive array of traits shared by those mammals which exhibit little evidence of sperm competition. These traits include facets of sperm morphology, the structures and functions of the reproductive ducts and accessory glands, phallic morphology, and patterns and frequencies of copulatory behaviour. It is regrettable that traits such as penile morphology are often misrepresented in the scientific and popular literature as being exceptional in human beings. Quantitative comparisons of primate genital morphology and mating systems place *H. sapiens* consistently among those species which have low relative testes sizes, small accessory sexual glands, a less muscular vas deferens, and monogamous or polygynous mating systems.

4. Evidence pertaining to sexual selection by cryptic female choice in mammals, although limited, is consistent with the conclusions reached from studies of the male reproductive system. The oviduct is elongated in mammals in which males have larger relative testes sizes and large sperm midpieces. The human oviduct, however, is relatively short, as would be expected, given the small testes and sperm midpiece volumes of men. The notion that female orgasm represents an adaptation to selectively transport or reject sperm of different male partners is based upon misinterpretations of false and limited evidence (Lloyd 2005).

5. Human copulatory postures, far from being unique, are derivable in evolutionary terms from those exhibited by other anthropoids. Ventro-ventral or female superior copulatory postures have been described in a number of ape species. These traits are probably primitive for the ape–human lineage, and would have been present in early hominids, including the australopithecines. Face-to-face mating postures are thus not a uniquely human specialization. Nor did their evolution have any connection with a requirement to enhance clitoral stimulation and to facilitate female orgasm. The dorso-ventral copulatory postures sometimes employed by humans have a still more ancient origin, for these are found throughout the primates, as well as in the majority of mammals.

6. The occurrence of extended periods of intromission during human copulation is likely to represent an effect of cultural evolution in various populations. The primitive human pattern involves attainment of ejaculation during an intromission of a few minutes in duration, accompanied by pelvic thrusting. The human copulatory pattern, like those of other primates, and mammals in general, may be classified using a modified version of the schema originally developed by Dewsbury (1972). This modified schema includes measures of the presence or absence of extended intromission as a component of the copulatory pattern. This is important because intromission duration has been subject to both natural and sexual selection during mammalian evolution. The human copulatory pattern, when viewed in comparative perspective, is unremarkable and there is no evidence that sperm competition or cryptic female choice would have impacted its evolution.

7. Oestrus was reduced or lost in the anthropoid lineage long before human beings appeared, and would already have been lacking in early hominids such as the australopithecines. There has been no loss of oestrus during human evolution. Female Old World monkeys and apes share with *H. sapiens*, to varying degrees, the ability to dissociate their sexual behaviour from a rigid control by ovarian hormones secreted during the menstrual cycle. Very few New World monkeys exhibit menstruation. However, like the Old World anthropoids, female New World primates allow and invite copulation when conception is unlikely to occur. Thus, the anthropoids as a whole do not exhibit oestrus, whereas it is present in the prosimian primates (Keverne 1981; Dixson 1983; 1998*a*; Loy 1987; Goy 1992; Martin 2003). Measurements of female proceptivity, receptivity, and sexual attractiveness, as defined by Frank Beach (1976*b*), are far more informative when analyzing how hormones control female sexuality. Cyclical changes in female attractiveness and proceptivity (sexual invitational behaviour) are more pronounced in monkeys and apes than are changes in receptivity (willingness to accept the male). Ovarian hormones, and oestrogens in particular, facilitate increases in attractiveness and/or proceptivity,

during the follicular and peri-ovulatory phases of the cycle.

Effects of the menstrual cycle upon human sexual behaviour are relatively subtle, subject to individual variability, and are more difficult to quantify than is the case for monkeys and apes. Nonetheless, mid-cycle or follicular phase increases in women's sexual interest and proceptivity have been identified in some studies, as have increases in female attractiveness (e.g. in olfactory and facial cues). Likewise, some studies report that women in the fertile phase of the menstrual cycle may be more responsive to masculinized faces, and other sexually dimorphic traits indicative of male fitness.

Two points should be emphasized with respect to these findings. Firstly, although the effects of the menstrual cycle on women's behaviour are subtle, in some cases they follow the same pattern as the mid-cycle, oestrogen-dependent increases in proceptivity which occur in many non-human primates. Although considerable weight has been given in the clinical literature to the importance of androgens in regulating women's sexual interest, it is possible that oestrogen is more significant in this context than currently recognized. Secondly, the modest shifts in behaviour measured in some women during the fertile phase of the menstrual cycle are insufficient to justify hypotheses concerning the evolution of *dual mating strategies* in women. It has been argued, for example, that women in the fertile phase of the cycle may switch their mating preferences from long-term partners towards extra-pair copulations with men who possess more attractive traits, indicative of superior genes (Gangestad and Simpson 2000). This hypothesis is rejected here for a number of reasons. Investment by women in long-term relationships probably outweighs genetic benefits they might theoretically derive from extra-pair matings. More probable is the possibility that hormonal (putatively oestrogenic) facilitation of female sexual interest is part of the mechanism by which women are attracted to men, seek to establish enduring relationships, and reproduce successfully. From puberty onwards, young and un-paired women undergo repeated menstrual cycles, and initially many of these cycles are anovulatory. Women are potentially able to mate at any time, but periodic increases in their interest in men, though slight, may reinforce the development of their sexual relationships with particular partners. Whilst acknowledging the fact that women may engage in extra-pair copulations in a number of cultures, I think it unwise to ascribe an evolutionary basis to what is an artificially created dichotomy between long-term versus short-term mating strategies. What appear to be short-term mating choices may have their origins in the requirement for women to cultivate potential partners in case of problems involving a long-term mate, such as desertion, illness, death, and inability to provide or to procreate.

More important than the relatively small, cyclical changes in female attractiveness or behaviour that occur during menstrual cycles are the long-term effects of female reproductive condition and health upon attractiveness. Thus, women whose cycles are fertile, regular, and involve higher follicular phase levels of oestrogen tend also to be facially more attractive, and to have a more healthy distribution of body fat. It is these longer-term hormonal effects which have probably been of greatest importance in shaping the evolution of human mate choice, rather than the transitory shifts which may take place during the fertile phase of an individual menstrual cycle. Perhaps this is why women do not exhibit graded signals of attractiveness and fertility status during their menstrual cycles. This contrasts with the occurrence of prominent, oestrogen-dependent sexual skin swellings which act as graded signals in female chimpanzees and some other Old World anthropoids. It is therefore most unlikely that in the direct ancestors of *H. sapiens*, females would have had sexual skin swellings.

Darwin: the Parthian shot

During the preparation of this book it was essential to read widely and to explore a number of fields that impinge directly or indirectly upon the subject of human sexuality and its evolution. These included palaeontology, anthropology, primatology, reproductive biology, genetics, comparative anatomy, sexology, behavioural ecology, and evolutionary psychology. I believe that a better understanding of human sexuality would result from the removal of the barriers which currently tend to divide many of these areas of scientific specialization. Unfortunately, human beings have a strong tendency to form small groups, and to protect their group's interests against those of perceived 'outsiders'. Xenophobia can afflict scientific disciplines, as well as other facets of human culture. In particular, a better dialogue is needed between the comparatively recent and burgeoning field of evolutionary psychology and more traditional disciplines such as anthropology, reproductive physiology, behavioural biology, and sexological research.

Publications within the sphere of evolutionary psychology that deal with sexuality have too often become disconnected from the knowledge base of human behaviour and the physiology of reproduction. This has been detrimental at all levels, including the teaching of university courses, as well as the quality of published research. As examples, here are ten highly questionable findings and views which have been reported and cited during the last two decades:

1. Masturbation, in the human male, serves to optimize sperm numbers, and sperm quality in the ejaculate; this trait evolved because it is advantageous in human sperm competition, and to reduce risks of polyzoospermy (Baker and Bellis 1993*a*).
2. Female partners of men who are morphologically more symmetrical have more orgasms (Thornhill, Gangestad, and Comer 1996). This is adaptive, because their orgasms influence sperm transport.
3. Women's orgasms affect sperm transport and sperm choice in different ways, depending upon preferences for particular males in sperm competition contexts (Baker and Bellis 1993*b*, 1995).
4. Women exhibit oestrus (Thornhill 2006; Gangestad and Thornhill 2008), and nightclub 'lapdancers', who are in oestrus, are given larger cash rewards by men (Miller et al. 2007).
5. Human semen acts as an anti-anxiolytic in women (Gallup et al. 2002) whether they receive it during vaginal intercourse or oral sex. An additional evolutionary benefit of oral sex may be to reduce risks pre-eclampsia in pregnant women, via chemicals contained in the male's semen (Davis and Gallup 2006).
6. The human penis is morphologically specialized to displace sperm of rival males from the vagina; *ergo* it is possible that such sperm might then be transferred on the penis, from one woman to another, resulting in fertilization (Gallup and Burch 2006).
7. Men prefer to view pornography which involves interactions between a woman and multiple males; this is because men are more aroused by images indicative of high sperm competition risk (Pound 2002).
8. Partner rape, by men, is more likely to occur if the woman has been unfaithful (or is suspected of being so). This tactic has evolved to reduce possible sperm competition risk (Goetz and Shackelford 2006*a*; 2006*b*).
9. Men with larger testes are more likely to engage in extra-pair copulations and sperm competition (Baker 1997).
10. Male genital mutilation (such as circumcision) evolved as a 'signal of sexual obedience'. It reduces the male capacity for extra-pair copulations, by impairing sperm competition. Genital mutilation is 'a hard-to-fake signal of a man's reduced ability to challenge the paternity of older men who are already married' (Wilson 2008).

Such reports, and many others contained in the literature on evolutionary psychology, are deeply troubling in a scientific sense. However, just as it is wise to assess the terrain carefully before embarking upon a course of action, it is also prudent to take care when quitting the field of battle. In ancient times, the Parthian cavalry deceived their enemies by feigning flight, only to turn in their saddles and unleash a salvo of arrows at their pursuers. This tactic was employed with deadly effect by the Parthians during their battles with the Romans. It thus became known as the *Parthian shot*. The epithet later acquired a literary connotation, denoting the use of a decisive phrase to end a debate. In the current context, it is Darwin who fires a Parthian shot across the centuries, as he cautions against the use of faulty data and flawed scientific reasoning. His views on these matters remain as powerful today as when he wrote them in *The Descent of Man and Selection in Relation to Sex*:

False facts are highly injurious to the progress of science, for they often endure long; but false views, if supported by some evidence, do little harm, for every one takes a salutary pleasure in proving their falseness: and when this is done, one path towards error is closed and the road to truth is often at the same time opened.

Bibliography

Abernethy, K.A., White, L.J.T., and Wickings, E.J. (2002). Hordes of Mandrills (*Mandrillus sphinx*): Extreme group size and seasonal male presence. *J. Zool. Lond.*, 258: 131–37.

Adams, D.B., Gold, A.B., and Burt, A.D. (1978). Rise in female sexual activity at ovulation blocked by oral contraceptives. *New Engl. J. Med.*, 299: 1145–50.

Adler, N.T. (1978). Social and environmental control of environmental processes in animals. In *Sex and Behavior: Status and Prospectus*. (eds. T.E. McGill, D.A. Dewsbury, and B.D. Sachs), pp. 115–60, Plenum: New York.

Ajmani, M.L., Jain, S.P., and Saxena, S.K. (1985). Anthropometric study of male external genitalia of 320 healthy Nigerian adults. *Anthrop. Anz.*, 43: 179–86.

Alemseged, Z., Spoor, F., Kimbel, W.H., et al. (2006). A juvenile early hominin skeleton from Dikika, Ethiopia. *Nature, Lond.*, 443: 296–301.

Alexander, R.D. and Noonan, K.M. (1979). Concealment of ovulation, parental care and human evolution. In *Evolutionary Biology and Human Behavior: An Anthropological Perspective*. (eds. N.A. Chagnon and W. Irons), pp. 436–53, Duxbury Press: North Scituate, MA.

Alexander, R.D., Hoogland, J.L., Howard, R.D., et al. (1979). Sexual dimorphism and breeding systems in pinnipeds, ungulates, primates and humans. In *Evolutionary Biology and Human Social Behavior: An Anthropological perspective*. (eds. N.A. Chagnon, and W. Irons), pp. 402–35, Duxbury Press: North Scituate, MA.

Alexeev, V.P. (1986). *Origin of the Human Race*, Progress Publishers: Moscow.

Allen, M.L. and Lemmon, W.B. (1981). Orgasm in female primates. *Am J. Primatol.*, 1: 15–34.

Amann, R.P. and Howards, S.S. (1980). Daily spermatozoal production and epididymal spermatozoal reserves of the human male. *J. Urol.*, 124: 211–5.

Andelman, S.J. (1987). Evolution of concealed ovulation in vervet monkeys (*Cercopithecus aethiops*). *Am. Nat.*, 129: 785–99.

Andersen, A.G., Jensen, T.K., Carlsen, E., et al. (2000). High frequency of sub-optimal semen quality in an unselected population of young men. *Hum. Reprod.*, 15: 366–72.

Anderson, M.J. and Dixson, A.F. (2002). Sperm competition: Mobility and the midpiece in primates. *Nature, Lond.*, 416: 496.

Anderson, M.J. and Dixson, A.F. (in press). Sexual selection affects the sizes of the mammalian prostate gland and seminal vesicles. *Current Zool.*

Anderson, M.J., Chapman, S.J., Videan, E.N., et al. (2007). Functional evidence for differences in sperm competition in humans and chimpanzees. *Am. J. Phys. Anthropol.*, 134: 274–80.

Anderson, M.J., Dixson, A.S., and Dixson, A.F. (2006). Mammalian sperm and oviducts are sexually selected: Evidence for co-evolution. *J. Zool. Lond.*, 270: 682–6.

Anderson, M.J., Nyholt, J., and Dixson, A.F. (2004). Sperm competition affects the structure of the mammalian vas deferens. *J. Zool. Lond.*, 264: 97–103.

Anderson, M.J., Nyholt, J., and Dixson, A.F. (2005). Sperm competition and the evolution of sperm midpiece volume in mammals. *J. Zool. Lond.*, 267: 135–42.

Apicella, C.L., Feinberg, D.R., and Marlowe, F.W. (2007). Voice pitch predicts reproductive success in male hunter-gatherers. *Biol. Lett.*, doi: 10.1098/rsbl.2007.0410

Arechiga, J., Prado, C., Canto, M., et al. (2001). Women in transition—menopause and body composition in different populations. *Coll. Anthropol.*, 25: 443–8.

Asdell, S.A. (1964). *Patterns of Mammalian Reproduction*, 2nd edn. Cornell University Press: Ithaca, NY.

Ashton, G.C. (1980). Mismatches in genetic markers in a large family study. *Am. J. Hum. Genet.*, 32: 601–13.

Baker, R.R. (1997). Copulation, masturbation and infidelity. In *New Aspects of Human Ethology*. (eds. A. Schmitt, K. Atzwanger, K. Grammer, and K. Schäfer), pp. 163–88, Plenum: New York.

Baker, R.R. and Bellis, M.A. (1988). "Kamikaze" sperm in mammals. *Anim. Behav.*, 36: 936–9.

Baker, R.R. and Bellis, M.A. (1989). Number of sperm in human ejaculates varies in accordance with sperm competition theory. *Anim. Behav.*, 37: 867–9.

Baker, R.R. and Bellis, M.A. (1993*a*). Human sperm competition: Ejaculate adjustment by males and the function of masturbation. *Anim. Behav.*, 46: 861–85.

Baker, R.R. and Bellis, M.A. (1993*b*). Human sperm competition: Ejaculate adjustment by females and a function for the female orgasm. *Anim. Behav.*, 46: 887–909.

Baker, R.R. and Bellis, M.A. (1995). *Human Sperm Competition*, Chapman and Hall: London.

Balaban, E. (1994). Sex differences in sounds and their causes. In *The Differences Between the Sexes.* (eds. R.V. Short and E. Balaban), pp. 243–72, Cambridge University Press: Cambridge, UK.

Balter, M. (1995). Did *Homo erectus* tame fire first? *Science*, 268: 1570.

Bancroft, J. (1989). *Human Sexuality and its Problems*, Churchill Livingstone: Edinburgh.

Barber, N. (1995). The evolutionary psychology of physical attractiveness: Sexual selection and human morphology. *Ethol. Sociobiol.*, 16: 395–424.

Bardin, C.W., Cheng, C.Y., Mustow, N.A., et al. (1994). The Sertoli cell. In *The Physiology of Reproduction*, 2nd edn., Vol. 2. (eds. E. Knobil and J.D. Neill), pp. 1291–333, Raven Press: New York.

Barrett, L., Dunbar, R. and Lycett, J. (2002). *Human Evolutionary Psychology*, Princeton University Press: Princeton, NJ.

Bartlett, T.Q. (2007). The Hylobatidae: Small apes of Asia. In *Primates in Perspective.* (eds. C.J. Campbell, A. Fuentes, K.C. MacKinnon, M. Panger, and S.K. Bearder), pp. 274–89, Oxford University Press: New York.

Barton, R.A. (2000). Sociobiology of baboons: The Interaction of male and female strategies. In *Primate Males: Causes and Consequences of Variation in Group Composition.* (ed. P.M. Kappeler), pp. 97–107, Cambridge University Press: Cambridge, UK.

Beach, F.A. (1970). Coital behavior in dogs. VI. Long term effects of castration upon mating in the male. *J. Comp. Physiol. Psychol. Monogr.*, 70: 1–32.

Beach, F.A. (1976*a*). Cross species comparisons and the human heritage. *Arch. Sex. Behav.*, 5: 469–85.

Beach, F.A. (1976*b*). Sexual attractivity, proceptivity and receptivity in female mammals. *Horm. Behav.*, 7: 105–38.

Beckford, N.S., Schain, D., Roor, S.R., et al. (1985). Androgen stimulation and laryngeal development. *Ann. Otol. Rhinol. Laryngol.*, 94: 634–40.

Bedford, J.M. (1974). The biology of primate spermatozoa. In *Contributions to Primatology. Reproductive Biology of the Primates.* (ed. W.P. Luckett), pp. 97–119, Karger, Basel.

Bedford, J.M. and Hoskins, D.D. (1990). The mammalian spermatozoon: Morphology, biochemistry and physiology. In *Marshall's Physiology of Reproduction. Vol. 2. Reproduction in the Male.* (ed. G.E. Lamming), pp. 379–568, Churchill Livingstone: Edinburgh.

Behar, D.M., Villerns, R., Soodyall, H., et.al. (2008). The dawn of human matrilineal diversity. *Am. J. Hum. Gen.*, 82: 1–11.

Benoit, J.C. (1922). Sur les rapports quantitatifs entre le tissu interstitiel, testiculaire, le tissue séminal et la masse du corps chez les oiseaux et quelques mammifères. *R. Séanc. Biol.*, 87: 1387–90.

Bercovitch, F.B. (1988). Coalitions, cooperations and reproductive tactics among adult male baboons. *Anim. Behav.*, 36: 1198–209.

Bercovitch, F.B. (2000). Behavioral ecology and socioendocrinology of reproductive maturation in cercopithecine monkeys. In *Old World Monkeys.* (eds. P.F. Whitehead and C.J. Jolly), pp. 298–320, Cambridge University Press: Cambridge, UK.

Bermúdez de Castro, J.M. Arsuega, J.L., Carbonell, E., et al. (1997). A hominid from the lower Pleistocene of Atapuerca, Spain: Possible ancestor to Neanderthals and modern humans. *Science*, 276: 1392–5.

Betzig, L. (ed.) (1997). *Human Nature.* Oxford University Press: New York.

Betzig, L. (ed.) Borgerhoff Mulder, M., and Turke, P. (eds.) (1988). *Human Reproductive Behavior: A Darwinian Perspective*, Cambridge University Press: Cambridge, UK.

Bielert, C. (1986). Sexual interactions between captive adult male and female chacma baboons (*Papio ursinus*) as related to the female's menstrual cycle. *J. Zool. Lond.*, 209: 521–36.

Bingham, H.C. (1928). Sex development in apes. *Comp. Psychol. Monogr.*, 5: 1–165.

Birkhead, T. (2000). *Promiscuity: An Evolutionary History of Sperm Competition and Sexual Conflict*, Faber and Faber: London.

Birkhead, T.R. and Fletcher, F. (1995). Male phenotype and ejaculate quality in the zebra finch *Taeniopygia guttata. Proc. Roy. Soc. Lond. B.*, 262: 329–34.

Birkhead, T.R. and Pizzari, T. (2002). Postcopulatory sexual selection. *Nat. Rev. Genet.*, 3: 262–73.

Blandau, R.J. (1945). On factors involved in sperm transport through the *cervix uteri* of the albino rat. *Am. J. Anat.*, 73: 253–72.

Bloom, W. and Fawcett, D.W. (1962). *A Textbook of Histology*, 8th edn, W.B. Saunders: Philadelphia, PA.

Boesch, C. and Boesch-Achermann, H. (2000). *The Chimpanzees of the Taï Forest: Behavioural Ecology and Evolution*, Oxford University Press: Oxford.

Bolunchuk, W.W., Siders, W.A., Lykken, G.I., et al. (2000). Association of dominant somatotype of men with body structure, function during exercise, and nutritional assessment. *Am. J. Hum. Biol.*, 12: 167–80.

Borgerhoff Mulder, M. and Caro, T.M. (1983). Polygyny: Definition and application to human data. *Anim. Behav.*, 31: 609–10.

Bouzouggar, A., Barton, N., Vanhaeren, M. et al. (2007). 82,000-year-old shell beads from North Africa and implications for the origins of modern human behavior. *Proc. Natl. Acad. Sci., USA*, 104: 9964–9.

Bowman, E.A. (2008). Why the human penis is larger than in the great apes. *Arch. Sex. Behav.*, 37: 361.

Boyd, R. and Silk, J.B. (2000). *How Humans Evolved*, 2nd edn., Norton and Co: New York.

Bradley, B.J., Robbins, M.M., Williamson, E.A., et al. (2005). Mountain gorilla tug-of-war: Silverbacks have limited control over reproduction in multimale groups. *Proc. Nat. Acad. Sci. USA*, 102: 9418–23.

Bramble, D.M. and Lieberman, D.E. (2004). Endurance running and the evolution of Homo. *Nature, Lond.*, 432: 345–52.

Brock, D.J.H. and Shrimpton, A.E. (1991). Non-paternity and prenatal genetic screening. *Lancet*, 338: 1151.

Brown, P., Sutikna, T., Morwood, M.J., et al. (2004). A new small-bodied hominin from the late Pleistocene of Flores, Indonesia. *Nature, Lond.*, 431: 1055–61.

Brownell, R.W. Jr. and Ralls, K. (1986). Potential for sperm competition in baleen whales. *Rep. Int. Whal. Comm.* (special issue 8), 97–112.

Bruckert, L., Liénard, J.S., Lacroix, A., et al. (2006). Women use voice parameters to assess men's characteristics. *Proc. Roy. Soc. Lond. B.*, 273: 83–9.

Brunet, M. et al. (2002). A new hominid from the Upper Miocene of Chad, Central Africa. *Nature, Lond.*, 418: 145–52.

Burley, N. (1979). The evolution of concealed ovulation. *Am. Nat.*, 114: 835–58.

Burton, F.D. (1971). Sexual climax in the female *Macaca Mulatta*. In *Proceedings of 3rd International Congress of Primatology*, pp. 180–91, Karger: Basel.

Buss, D.M. (1989). Sex differences in human mate preferences: evolutionary hypotheses tested in 37 cultures. *Behav. Brain. Sci.*, 12: 1–14.

Buss, D.M. (1994). *The Evolution of Desire*, Basic Books: New York.

Buss, D.M. (2000). *The Dangerous Passion: Why Jealousy is as Necessary as Love and Sex*, Free Press: New York.

Buss, D.M. (2003). *The Evolution of Desire*, 2nd edn., Basic Books: New York.

Buss, D.M. (ed.) (2005). *The Handbook of Evolutionary Psychology*. John Wiley and Sons: Hoboken, NJ.

Buss, D.M. and Schmidt, D.P. (1993). Sexual strategies theory: An evolutionary perspective on human mating. *Psychol. Rev.*, 100: 204–32.

Cabanac, M. and Brinnel, H. (1988). Beards, baldness and sweat secretion. *Eur. J. Appl. Physiol.*, 58: 39–46.

Campbell, C.J. (2007). Primate sexuality and reproduction. In *Primates in Perspective*. (eds. C.J. Campbell, A. Fuentes, K.C. Mackinnon, M. Panger, and S.K. Bearder), pp. 423–37, Oxford University Press: New York.

Campbell, C.J., Fuentes, A., Mackinnon, K.C., et al. (eds.) (2007) *Primates in Perspective*, Oxford University Press: New York.

Carbonell, E., Bermúdez de Castro, J.M., Parés, J.M., et al. (2008). The first hominin of Europe. *Nature, Lond.*, 452: 465–9.

Carlsen, E., Giwercman, A., Keiding, N., et al. (1992). Evidence for decreasing quality of semen during the past 50 years. *Br. Med. J.*, 305: 609–13.

Carmichael, M.S., Humbert, R., Dixen, J., et al. (1987). Plasma oxytocin increases in human sexual response. *J. Clin. Endocrin. Metab.*, 64: 27–31.

Carpenter, C.R. (1965). The howlers of Barro Colorado Island. In *Primate Behavior: Field Studies of Monkeys and Apes*. (ed. I. DeVore), pp. 250–91, Holt, Rinehart and Winston: New York.

Carrier, D.R. (2007). The short legs of great apes: evidence for aggressive behavior in australopiths. *Evolution*, 61: 596–605.

Carter, J.E.L. and Heath, B.H. (1990). *Somatotyping—Development and Applications*, Cambridge University Press: Cambridge, UK.

Caulfield, M.D. (1985). Sexuality in human evolution: What is natural in sex? *Feminist Studies*, 11: 343–63.

Cavalli-Sforza, L., Piazza, A., Menozzi, P., et al. (1988). Reconstruction of human evolution: Bringing together genetic, archaeological and genetic data. *Proc. Natl. Acad. Sci. USA*, 85: 6002–6.

Cerda-Flores, R.M., Barton, A.A., Marky-Gonzalez, L.F., et al. (1999). Estimation of nonpaternity in the Mexican population of Nuevo Léon: A validation study with blood group markers. *Am. J. Phys. Athropol.*, 109: 281–93.

Chagnon, N.A. (1979). Mate competition, favouring close kin, and village fissioning among the Yanomamo Indians. In *Evolutionary Biology of Human Social Behavior*. (eds. N. Chagnon and W. Irons), pp. 86–132, Duxbury Press: North Scituate, MA.

Chang, K.S.F., Hsu, F.K., and Chan, Y.B. (1960). Scrotal asymmetry and handedness. *J. Anat.*, 94: 543–8.

Chapais, B. (2008). *Primeval kinship: How Pair-Bonding Gave Birth to Human Society*, Harvard University Press: Cambridge, MA.

Civetta, A. (1999). Direct visualization of sperm competition and sperm storage in *Drosophila*. *Curr. Biol.*, 9: 841–4.

Clarke, R.J. (2002). Newly revealed information on the Sterkfontein Member 2 *Australopithecus* skeleton. *S. Afr. J. Sci.*, 98: 523–36.

Clarys, J.P., Martin, A.D., and Drinkwater, D.T. (1984). Gross tissue weights in the human body by cadaver dissection. *Hum. Biol.*, 56: 459–73.

Clutton-Brock, T.H. (1991).The evolution of sex differences and consequences of polygyny in mammals. In *The Development and Integration of Behaviour: Essays in Honour of Robert Hinde.* (ed. P. Bateson), pp. 229–53. Cambridge University Press, Cambridge, UK.

Clutton-Brock, T.H. and Albon, S.D. (1979). The roaring of red deer and the evolution of honest advertisement. *Behaviour*, 69: 145–70.

Clutton-Brock, T.H. and Isvaran, K. (2007). Sex differences in ageing in natural populations of vertebrates. *Proc. Roy. Soc. Lond. B.*, 274: 3097–104.

Clutton-Brock, T.H., Guinness, F.E., and Albon, S.D. (1982). *Red Deer: Behavior and Ecology of Two Sexes*, University of Chicago Press: Chicago, IL.

Clutton-Brock, T.H., Harvey, P., and Rudder, B. (1977). Sexual dimorphism, socionomic sex ratio, and body weight in primates. *Nature, Lond.*, 269: 797–9.

Cohen, J. (1967). Correlation between chiasma frequency and sperm redundancy. *Nature, Lond.*, 215: 862–3.

Collard, M. (2002). Grades and transitions in human evolution. In *The Speciation of Modern Homo Sapiens, Vol. 106, Proceedings of the British Academy.* (ed. T.J. Crow), pp. 61–100, Oxford University Press: Oxford.

Collins, S.A. (2000). Men's voices and women's choices. *Anim Behav.*, 60: 773–80.

Conaway, C.H. (1971). Ecological adaptation and mammalian reproduction. *Biol. Reprod.*, 4: 239–47.

Conaway, C.H. and Koford, C.B. (1965). Estrous cycles and mating behavior in a free-ranging band of rhesus monkeys. *J. Mammal.*, 45: 577–88.

Connor, R.C., Read, A.J., and Wrangham, R. (2000). Male reproductive strategies and social bonds. In *Cetacean Societies: Field Studies of Dolphins and Whales.* (eds. J. Mann, R.C. Connor, P.L. Tyack, and H. Whitehead), pp. 247–69. University of Chicago Press: Chicago, IL.

Conway, C.A., Jones, B.C., DeBruine, L.M., et al. (2008). Evidence for adaptive design in human gaze preference. *Proc. Roy. Soc. Lond. B.*, 275: 63–9.

Coon, C.S. (1962). *The Origin of Races*, Alfred A. Knopf: New York.

Cordoba-Aguilar, A. (1999). Male copulatory sensory stimulation induces female ejection of rival sperm in a damselfly. *Proc. Roy. Soc. Lond. B.*, 266: 779–84.

Cornwell, R.E., Law Smith, M.J., Boothroyd, L.G., et al. (2006). Reproductive strategy, sexual development and attraction to facial characteristics. *Phil. Trans. Roy. Soc. Lond. B.*, 361: 2143–54.

Courtenay, J., Groves, C., and Andrews, P. (1988). Inter-or intra- island variation? An assessment of the differences between Bornean and Sumatran orang-utans. In *Orang-Utan Biology.* (ed. J.H. Schwartz), pp. 19–29, Oxford University Press: Oxford.

Crockett, C.M. and Eisenberg, J.F. (1987). Howlers: Variations in group size and demography. In *Primate Societies* (eds. B. Smuts, D. Cheney, R. Seyfarth, R. Wrangham, and T. Struhsaker), pp. 54–68, University of Chicago Press: Chicago, IL.

Dabbs, J.M. and Mallinger, A. (1999). High testosterone levels predict low voice pitch among men. *Pers. Indiv. Differ.*, 27: 801–4.

Dahl, J.F. (1988). External genitalia. In *Orang-Utan Biology.* (ed. J.H. Schwartz), pp. 133–44, Oxford University Press: New York.

Dahl, J.F. (unpublished manuscript). Testicle size variation in gorillas.

Dahl, J.F. Gould, K.G., and Nadler, R.D. (1993). Testicle size in orang-utans in relation to body size. *Amer. J. Phys. Anthropol.*, 90: 229–36.

Daly, M. and Wilson, M. (1983). *Sex, Evolution and Behavior*, Wadsworth: Belmont, CA.

Dart, R.A. (1925). *Australopithecus africanus*: The man-ape of South Africa. *Nature, Lond.*, 115: 195–9.

Dart, R.A. (1957). The osteodontokeratic culture of *Australopithecus prometheus. Transvaal. Mus. Mem.*, No. 10.

Darwin, C. (1859). *On the Origin of Species by means of Natural Selection or the Preservation of Favoured Races in the Struggle for Life*, John Murray: London.

Darwin, C. (1871). *The Descent of Man and Selection in Relation to Sex*, John Murray: London.

Darwin, C. (1876). Sexual selection in relation to monkeys. *Nature, Lond.*, Nov.2, p. 18.

Davies, A.G. and Oates, J.F. (eds.) (1994). *Colobine Monkeys: Their Ecology, Behaviour and Evolution*, Cambridge University Press: Cambridge, UK.

Davis, J.A. and Gallup, G.G. Jr. (2006). Pre-eclampsia and other pregnancy complications as an adaptive response to unfamiliar semen. In *Female Infidelity and Paternity Uncertainty.* (eds. S.M. Platek and T.K. Shackelford), pp. 191–204, Cambridge University Press: Cambridge, UK.

Daw, S.F. (1970) Age of boys' puberty in Leipzig, 1729–49, as indicated by voice breaking in J.S. Bach's choir members. *Hum. Biol.*, 42: 87–9.

Day, M.H. (1973). Locomotor features of the lower limb in hominids. *Symp. Zool. Soc. Lond.*, 33: 29–51.

DeBruine, L.M., Jones, B.C., Little, A.C., et al. (2006). Correlated preferences for facial masculinity and ideal or actual partner's masculinity. *Proc. Roy. Soc. Lond. B.*, 273: 1355–60.

De Kretser, D.M. and Kerr, J.B. (1994). The cytology of the testis. In *The Physiology of Reproduction*, Vol. 1. 2nd edn. (eds E. Knobil and J.D. Neill), pp. 1177–290, Raven Press: New York.

Dennerstein, L., Gotts, G., Brown, J.B., et al. (1994). The relationship between the menstrual cycle and female sexual interest in women with PMS complaints and volunteers. *Psychoneuroendocrinol.* 19: 293–304.

Deputte, B.L. (1982). Duetting in male and female songs of the white cheeked gibbon (*Hylobates concolor leucogenys*), In *Primate Communication*. (eds. C.T. Snowdon, C.H. Brown, and M.R. Peterson), pp. 67–93, Cambridge University Press: Cambridge, UK.

D'Errico, F., Henshilwood, C., Vanhaeren, M. et al. (2005). Nassarius kraussianus shell beads from Blombos cave: Evidence for symbolic behavior in the middle Stone Age. *J. Hum. Evol.*, 48: 3–24.

De Waal, F.B.M. (1989). *Peacemaking Among Primates*, Harvard University Press: Cambridge, MA.

De Waal, F.B.M. (2003). *My Family Album: Thirty Years of Primate Photography*, University of California Press: Berkeley, CA.

Dewsbury, D.A. (1972). Patterns of copulatory behavior in male mammals. *Q. Rev. Biol.*, 47: 1–33.

Dewsbury, D.A. and Hodges, A.W. (1987). Copulatory behavior and related phenomena in spiny mice (*Acomys cahirinus*) and hopping mice (*Notomys alexis*). *J. Mammal.*, 69: 49–57.

Dewsbury, D.A. and Pierce, J.D. (1989). Copulatory patterns of primates as viewed in broad mammalian perspective. *Am. J. Primatol.*, 17: 51–72.

Diamond, J. (1991). *The Rise and Fall of the Third Chimpanzee*, Radius: London.

Diamond, J. (1997). *Why is Sex Fun? The Evoluton of Human Sexuality*, Weidenfield and Nicolson: London.

Diamond, J.M. (1986). Variation in human testis size. *Nature Lond.*, 320: 488–489.

Dickinson, R.L. (1949). *Human Sex Anatomy*, Robert E. Krieger: Huntington, NY.

Dickinson, R.L. and Beam, L. (1931). *A Thousand Marriages*, Williams and Wilkins Co: Baltimore, MD.

Dickson, D. and Maue-Dickson, W. (1982). *Anatomical and Physiological Bases of Speech*, Little Brown & Co: Boston, MA.

Di Fiore, A. and Campbell, C.J. (2007). The atelines: Variations in ecology, behavior, and social organization. In *Primates in Perspective*. (eds. C.J. Campbell, A. Fuentes, K.C. MacKinnon, M. Panger, and S.K. Bearder), pp. 155–85, Oxford University Press: New York.

Dixson, A.F. (1981). *The Natural History of the Gorilla*, Weidenfeld and Nicolson: London.

Dixson, A.F. (1983*a*). The hormonal control of sexual behavior in primates. *Oxford Revs. Repro. Biol.*, 5: 131–218.

Dixson, A.F. (1983*b*). Observations on the evolution and behavioural significance of 'sexual skin' in female primates. *Adv. Study. Behav.*, 13: 63–106.

Dixson, A.F. (1987*a*). Observations on the evolution of the genitalia and copulatory behaviour in male primates. *J. Zool. Lond.*, 213: 423–43.

Dixson, A.F. (1987*b*). Baculum length and copulatory behavior in primates. *Am. J. Primatol.*, 13: 51–60.

Dixson, A.F. (1987*c*). Effects of adrenalectomy upon proceptivity, receptivity and sexual attractiveness in ovariectomized marmosets. *Physiol. Behav.*, 39: 495–9.

Dixson, A.F. (1989). Sexual selection, genital morphology and copulatory behaviour in male galagos. *Int. J. Primatol.*, 10: 47–55.

Dixson, A.F. (1991). Sexual selection, natural selection and copulatory patterns in male primates. *Folia. Primatol.*, 57: 96–101.

Dixson, A.F. (1992). Observations on postpartum changes in hormones and sexual behavior in callitrichid primates: Do females exhibit postpartum 'estrus'? In *Topics in Primatology*, Vol. 2. (eds. N. Itoigawa, Y. Sugiyama, G.P. Sackett, and R.K.R. Thompson), pp. 141–9, University of Tokyo Press: Tokyo.

Dixson, A.F. (1993). Sexual selection, sperm competition and the evolution of sperm length. *Folia Primatol.*, 61: 221–7.

Dixson, A.F. (1995*a*). Sexual selection and the evolution of copulatory behaviour in nocturnal prosimians. In *Creatures of the Dark: The Nocturnal Prosimians*. (eds. L. Alterman, G.A. Doyle, and M.K. Izard), pp. 93–118, Plenum: New York.

Dixson, A.F. (1995*b*). Sexual selection and ejaculatory frequencies in primates. *Folia Primatol.*, 64: 146–52.

Dixson, A.F. (1997). Evolutionary perspectives on primate mating systems and behavior. *Ann. NY Acad. Sci.*, 807: 42–61.

Dixson, A.F. (1998*a*). *Primate Sexuality: Comparative Studies of the Prosimians, Monkeys, Apes, and Human Beings*, Oxford University Press: Oxford.

Dixson, A.F. (1998*b*). Sexual selection and the evolution of the seminal vesicles in primates. *Folia Primatol.*, 69: 300–6.

Dixson, A.F. (2001). The evolution of neuroendocrine mechanisms regulating sexual behaviour in female primates. *Reprod. Fertil. Dev.*, 13: 599–607.

Dixson, A.F. and Anderson, M.J. (2001). Sexual selection and the comparative anatomy of reproduction in monkeys, apes, and human beings. *Ann. Revs. Sex Res.*, 12: 121–44.

Dixson, A.F. and Anderson, M.J. (2002). Sexual selection, seminal coagulation and copulatory plug formation in primates. *Folia Primatol.*, 73: 63–9.

Dixson, A.F. and Anderson, M.J. (2004). Sexual behavior, reproductive physiology and sperm competition in male mammals. *Physiol. Behav.*, 83: 361–71.

Dixson, A.F. and Hastings, M.H. (1992). Effects of ibotenic acid-induced neuronal degeneration in the hypothalamus upon proceptivity and sexual receptivity in the female marmoset. *J. Neuroendocr.*, 4: 719–26.

Dixson, A.F. and Lunn, S.F. (1987). Post-partum changes in hormones and sexual behavior in captive marmoset groups. *Physiol. Behav.*, 41: 577–83.

Dixson, A.F. and Mundy, N.I. (1994). Sexual behaviour, sexual swelling and penile evolution in chimpanzees (*Pan troglodytes*). *Arch. Sex Behav.*, 23: 267–80.

Dixson, A.F., Bossi, T., and Wickings, E.J. (1993). Male dominance and genetically determined reproductive success in the mandrill (*Mandrillus sphinx*). *Primates*, 34: 525–32.

Dixson, A.F., Dixson, B., and Anderson, M. (2005). Sexual selection and the evolution of visually conspicuous sexually dimorphic traits in male monkeys, apes, and human beings. *Ann. Rev. Sex. Res.*, 16: 1–19.

Dixson, A.F., Everitt, B.J., Herbert, J., et al. (1972). Hormonal and other determinants of sexual attractiveness and receptivity in rhesus and talapoin monkeys. In *Symposium on 4th International Congress on Primatology, Vol. 2., Primate Reproductive Behavior.* (eds. W. Montagna and C.H. Phoenix), pp. 36–63. Karger: Basel.

Dixson, A.F., Halliwell, G. East., R., et al. (2003). Masculine somatotype and hirsuteness as determinants of sexual attractiveness to women. *Arch. Sex Behav.*, 32: 29–39.

Dixson, A.F., Knight, J., Moore, H.D.M., et al. (1982). Observations on sexual development in male orangutans, *Pongo pygmaeus. Int. Zoo. Yrbk.*, 22: 222–7.

Dixson, A.F., Moore, H.D.M. and Holt, W.V. (1980). Testicular atrophy in captive gorillas. *J. Zool. Lond.*, 191: 315–22.

Dixson, A.F., Nyholt, J., and Anderson, M.J. (2004). A positive relationship between baculum length and prolonged intromission patterns in mammals. *Acta. Zool. Sinica*, 50: 490–503.

Dixson, B.J., Dixson, A.F., Bishop, P.J., et al. (in press). Human physique and sexual attractiveness in men and women: A US–New Zealand comparative study. *Arch. Sex Behav.*

Dixson, B.J., Dixson, A.F., Li, B., et al. (2007*a*). Studies of human physique and sexual attractiveness: Sexual preferences of men and women in China. *Am. J. Hum. Biol.*, 19: 88–95.

Dixson, B.J., Dixson, A.F., Morgan, B., et al. (2007*b*). Human physique and sexual attractiveness: Sexual preferences of men and women in Bakossiland, Cameroon. *Arch. Sex Behav.*, 36: 369–75.

Döernberger, V. and Döernberger, G. (1987). Vergleichende volumetrie des menschlichen hodens unter besonderer Berucksichtigung der hodensonographie, Praderorchidometer, Schirrenzirkel und Schublehre. *Andrologia*, 19: 487–96.

Dorus, S., Evans, P.D., Wyckoff, G.J., et al. (2004). Rate of molecular evolution of the seminal gene SEMG2 correlates with levels of female promiscuity. *Nat. Genet.*, 36: 1326–9.

Dunbar, R.I.M. and Barrett, L. (eds.) (2007). *Oxford Handbook of Evolutionary Psychology.* Oxford University Press: Oxford.

Dupanloup, I., Pereira, L., Bertorelle, G., et al. (2003). A recent shift from polygyny to monogamy in humans is suggested by the analysis of worldwide Y-chromosome diversity. *J. Mol. Evol.*, 57: 85–97.

Eberhard, W.G. (1985). *Sexual Selection and Animal Genitalia*, Harvard University Press: Cambridge, MA and London.

Eberhard, W.G. (1996). *Female Control: Sexual Selection by Cryptic Female Choice*, Princeton University Press: Princeton, NJ.

Eckstein, P. and Zuckerman, S. (1956). Morphology of the reproductive tract. In *Marshall's Physiology of Reproduction*, 3rd edn. (ed. A.S. Parkes) Vol.1, pp. 43–155, Longmans, Green and Co.: London.

Edwards, J.H. (1957). A critical examination of the reputed primary influence of ABO phenotype on fertility and sex ratio. *Brit. J. Prevent. Soc. Med.*, 11: 79–89.

Eisenberg, J.F. (1977). The evolution of the reproductive unit in the class Mammalia. In *Reproductive Behavior and Evolution.* (eds. J.S. Rosenblatt and B. Komisaruk), pp. 39–71, Plenum: New York.

Eisenberg, J.F. (1981). *The Mammalian Radiations: An Analysis of Trends in Evolution, Adaptation and Behavior*, Athlone Press: London.

Ekman, P. (ed.) (1973) *Darwin and Facial Expression: A Century of Research in Review.* Academic Press: New York.

Ekman, P. and Friesen, W.V. (ed.) (1971). Constants across cultures in the face and emotion. *J. Pers. Soc. Psychol.*, 17: 124–9.

Ellefson, J.O. (1974). A natural history of white-handed gibbons in the Malay peninsula. In *Gibbon and Siamang*, Vol. 3 (ed. D.M. Rumbaugh), pp. 1–136, Karger: Basel.

Ellis, B.J. (1992). The evolution of sexual attraction: Evaluative mechanisms in women. In *The Adapted Mind: Evolutionary Psychology and the Generation of Culture* (eds. J.H. Barlow, L. Cosmides, and J. Tooby), pp. 267–88, Oxford University Press: Oxford.

Ellison, P. (2001). *On Fertile Ground: A Natural History of Human Reproduction*, Harvard University Press: Cambridge, MA.

Enomoto, T., Matsubayashi, K., Nakano, M., et al. (2004). Testicular histological examination of spermatogenetic activity in captive gorillas (*Gorilla gorilla*). *Am. J. Primatol.*, 63: 183–99.

Everitt, B.J. and Herbert, J. (1972). Hormonal correlates of sexual behaviour in sub-human primates. *Dan. Med. Bull.*, 19: 246–55.

Ewer, R.F. (1968). *Ethology of Mammals*, Logos Press: London.

Falk, D., Hildebolt, C., Smith, K., et al. (2005*a*). The brain of LB1, *Homo floresiensis. Science*, 308: 242–5.

Falk, D., Hildebolt, C., Smith, K., et al. (2005*b*). Response to comment on 'The brain of LB1 *Homo floresiensis*'. *Science*, 310: 236.

Falk, D., Hildebolt, C., Smith, K., et al. (2006). Response to comment on 'The brain of LB1 *Homo floresiensis*'. *Science*, 312:999.

Farkas, L.G. (1971). Basic morphological data of external genitals in 177 healthy central European men. *Am. J. Phys. Anthropol.*, 34: 325–8.

Farley, J. (1982). *Gametes and Spores: Ideas about Sexual Reproduction*, Johns Hopkins University Press: Baltimore, MD.

Fawcett, D.W. (1975). The mammalian spermatozoon. *Devel. Biol.*, 44: 394–436.

Feinberg, D.R., Jones, B.C., Little, A.C., et al. (2005). Manipulations of fundamental and formant frequencies influence the attractiveness of human male voices. *Anim. Behav.*, 69: 561–8.

Finch, C.E. (2007). *The Biology of Human Longevity*, Elsevier, Academic Press: Burlington, MA.

Fisch, H. and Goluboff, E.T. (1996). Geographic variations in sperm counts: A potential cause of bias in studies of semen quality. *Fertil. Steril.*, 65: 1044–46.

Fisher, H. (1992). *Anatomy of Love: A Natural History of Mating, Marriage and Why We Stray*, Fawcett Books: New York.

Fisher, H.E. (1982). *The Sex Contract: The Evolution of Human Behavior*, Paladin: London.

Fisher, S. (1973). *The Female Orgasm*, Basic Books: New York.

Fleagle, J.G. (1999). *Primate Adaptation and Evolution*, 2nd edn., Academic Press: San Diego, CA.

Foley, R. (1995). *Humans Before Humanity*, Blackwell: Oxford.

Folstad, I. and Karter, A.J., (1992). Parasites, bright males, and the immunocompetence handicap. *Am. Nat.*, 139: 603–22.

Ford, C.S. and Beach, F.A. (1951). *Patterns of Sexual Behavior*, Harper and Rowe: New York.

Fox, C.A., Meldrum, S.J., and Watson, B.W. (1973). Continuous measurement by radiotelemetry of vaginal pH during human coitus. *J. Reprod. Fertil.*, 33: 69–75.

Fox, C.A., Wolff, H.S., and Baker, J.A. (1970). Measurement of intra-vaginal and intra-uterine pressures during human coitus by radio-telemetry. *J. Reprod. Fertil.*, 22: 243–51.

Freeman, M.E. (1994). The neuroendocrine control of the ovarian cycle of the rat. In *The Physiology of Reproduction*, 2nd edn., Vol. 2. (eds. E. Knobil and J.D. Neill), pp. 613–58, Raven Press: New York.

Freund, M. (1962). Interrelationships among the characteristics of human semen and factors affecting semen-specimen quality. *J. Reprod. Fertil.*, 4: 143–59.

Freund, M. (1963). Effect of frequency of emission on semen output and an estimate of daily sperm production in man. *J. Reprod. Fertil.*, 6: 269–86.

Frost, P. (1988). Human skin color: A possible relationship between its sexual dimorphism and its social perception. *Perspect. Biol. Med.*, 32: 38–58.

Frost, P. (1994). Geographic distribution of human skin colour: A selective compromise between natural and sexual selection? *Hum. Evol.*, 9: 141–53.

Fujii, M., Kawai, S., Shimuzu, Y., et al. (1982). Recent topics in the evaluation of male infertility. *J. Urol.*, 44: 532–9.

Furuichi, T. (1987). Sexual swelling, receptivity and grouping of wild pigmy chimpanzee females at Wamba, Zaire. *Primates*, 28: 309–18.

Gabunia, L., Vekua, A., Lordkipanidze, D., et al. (2000). Earliest pleistocene homind cranial remains from Dmanisi, republic of Georgia: Taxonomy, geological settings, and age. *Science*, 288: 1019–25.

Gage, M.J.G. (1998). Mammalian sperm morphometry. *Proc. Roy. Soc. Lond. B.*, 265: 97–103.

Gage, M.J.G. and Freckleton, R. (2003). Relative testis size and sperm morphometry across mammals: no evidence for an association between sperm competition and sperm length. *Proc. Roy. Soc. Lond., B.*, 270: 625–32.

Galdikas, B.M.F. (1983). The orang-utan long call and snag crashing at Tanjung Puting Reserve. *Primates*, 24: 371–84.

Galdikas, B.M.F. (1985*a*). Subadult male sociality and reproductive behavior at Tanjung Puting. *Int. J. Primatol.*, 8: 87–99.

Galdikas, B.M.F. (1985*b*). Adult male sociality and reproductive tactics among orang-utans at Tanjung Puting Reserve. *Folia Primatol.*, 45: 9–24.

Gallup, G.G. Jr. and Burch, R.L. (2004). Semen displacement as a sperm competition strategy in humans. *Evol. Psychol.*, 2: 12–23.

Gallup, G.G. Jr. and Burch, R.L. (2006). The semen-displacement hypothesis: Semen hydrolics and the intra-pair copulation proclivity model of female infidelity. In *Female Infidelity and Paternity Uncertainty*. (eds. S.M. Platek and T.K. Shackelford), pp. 129–40, Cambridge University Press: Cambridge, UK.

Gallup, G.G. Jr. and Burch, R.L., and Platek, S.M. (2002). Does semen have anti-depressant properties? *Arch Sex. Behav.*, 31: 289–93.

Gallup, G.G. Jr. and Burch, R.L., Zappieri, M.L., et al. (2003). The human penis as a semen displacement device. *Evol. Hum. Behav.*, 24: 277–89.

Gangestad, S.W. (2007). Reproductive strategies and tactics. In *The Oxford Handbook of Evolutionary Psychology*. (eds. Dunbar, R.I.M. and Barrett, L.), pp. 321–32, Oxford University Press: Oxford.

Gangestad, S.W. and Buss, D.M. (1993). Pathogen prevalence and human mate preferences. *Ethol. Sociobiol.*, 14: 89–96.

Gangestad, S.W. and Cousins, A.J. (2001). Adaptive design, female mate preferences and shifts across the menstrual cycle. *Ann. Revs. Sex. Res.*, 12: 145–85.

Gangestad, S.W. and Simpson, J.A. (2000). The evolution of human mating: Trade-offs and strategic pluralism. *Behav. Brain Sci.*, 23: 573–644.

Gangestad, S.W. and Thornhill, R. (1998). Menstrual cycle variation in women's preferences for the scent of symmetrical men. *Proc. Roy. Soc. Lond. B.*, 265: 927–33.

Gangestad, S.W. and Thornhill, R. (2008). Human oestrus. *Proc. Roy. Soc. Lond. B.*, 275: 991–1000.

Gangestad, S.W., Simpson, J.A., Cousins, A.J., et al. (2004). Women's preferences for male behavioral displays change across the menstrual cycle. *Psychol. Sci.*, 15: 203–7.

Gangestad, S.W., Thornhill, R., and Garver-Apgar, C.E. (2005*a*). Female sexual interests across the ovulatory cycle depend on primary partner developmental instability. *Proc. Roy. Soc Lond. B.*, 272: 2023–7.

Gangestad, S.W., Thornhill, R., and Garver-Apgar, C.E. (2005*b*). Adaptations to ovulation. In *The Handbook of Evolutionary Psychology*. (ed. D.M. Buss), pp. 344–71, John Wiley & Sons: Hoboken, NJ.

Gautier, J.P. (1971). Etude morphologique et fonctionelle des annexes extra-laryngées des Cercopithecinae; liaison avec les cris d'espacement. *Biol. Gabon.*, 7: 229–67.

Gautier, J.P. (1988). Interspecific affinities among guenons as deduced from vocalizations. In *A Primate Radiation: Evolutionary Biology of the African Guenons*. (eds. A. Gautier-Hion, F. Boulière, J.P. Gautier, and J. Kingdon) pp. 194–226, Cambridge University Press: Cambridge, UK.

Gautier-Hion, A. (1968). Etude du cycle annuel de reproduction du talapoin (*Miopithecus talapoin*), vivant dans son milieu naturel. *Biol. Gabon*, 4: 163–73.

Gebhard, P.H. and Johnson, A.B. (1979). *The Kinsey Data: Marginal Tabulations of the 1938–1963 Interviews conducted by the Institute for Sex Research*, Saunders: Philadelphia, PA.

Geissmann, T. (2002). Duet-splitting and the evolution of gibbon songs. *Biol. Revs.*, 77: 57–76.

Gerloff, U., Hartung, B., Fruth, B., et al. (1999). Intracommunity relationships, dispersal pattern and paternity success in a wild living community of bonobos (*Pan paniscus*) determined from DNA analysis of faecal samples. *Proc. Roy. Soc. Lond. B.*, 266: 1189–95.

Glick, B.B. (1980). Ontogenetic and psychobiological aspects of the mating activities of male *Macaca radiata*. In *The Macaques: Studies in Ecology, Behavior and Evolution*. (ed. D.G. Lindburg), pp. 345–69, Van Nostrand Reinhold: New York.

Goetz, A.T. and Shackelford, T.K. (2006*a*). Sperm competition and its evolutionary consequences in humans. In *Female Infidelity and Paternity Uncertainty*. (eds. S.M. Platek and T.K. Shackelford), pp. 103–28, Cambridge University Press: Cambridge, UK.

Goetz, A.T. and Shackelford, T.K. (2006*b*). Sexual coercion and forced in-pair copulation as anti-cuckoldry tactics in humans. In *Female Infidelity and Paternity Uncertainty*. (eds. S.M. Platek and T.K. Shackelford), pp. 82–99, Cambridge University Press: Cambridge, UK.

Goldfoot, D.A., Westerborg-Van Loon, H., Groeneveld, W., et al. (1980). Behavioral and physiological evidence of sexual climax in the female stump-tailed macaque (*Macaca arctoides*), *Science*, 208: 1477–9.

Gomendio, M. and Roldan, E.R.S. (1991). Sperm size and sperm competition in mammals. *Proc. Roy. Soc. Lond. B.*, 243: 181–5.

Gomendio, M. and Roldan, E.R.S. (1993). Co-evolution between male ejaculates and female reproductive biology in eutherian mammals. *Proc. Roy. Soc. B.*, 252: 7–12.

Gomendio, M. Harcourt, A.H., and Roldan, E.R.S. (1998). Sperm competition in Mammals. In *Sperm Competition and Sexual Selection*. (eds. T.R. Birkhead and A.P. Møller), pp. 667–756, Academic Press: San Diego, CA.

González-José, R., Escapa, I., Neves, W.A., et al. (2008). Cladistic analysis of continuous modularized traits provides phylogenetic signals in *Homo* evolution. *Nature, Lond.*, 453: 775–8.

Goodall, J. (1965). Chimpanzees of the Gombe Stream Reserve. In *Primate Behavior: Field Studies of Monkeys and Apes*. (ed. I. DeVore), pp. 425–73, Holt, Rinehart and Winston: New York.

Goodall, J. (1968). The behaviour of free-living chimpanzees in the Gombe Stream area. *Anim. Behav. Monogr.*, 1: 161–311.

Goodall, J. (1986). *The Chimpanzees of Gombe: Patterns of Behavior*, Belknap Press of Harvard University: Cambridge, MA.

Gottelli, D., Wang, J., Bashir, S., et al. (2007). Genetic analysis reveals promiscuity among female cheetahs. *Proc. Roy. Soc. Lond. B.*, 274: 1993–2001.

Goy, R.W. (1979). Sexual compatability in rhesus monkeys: Predicting sexual performance of oppositely sexed pairs of adults. In *Sex, Hormones and Behaviour. CIBA Foundation Symposium*. Vol. 62, pp. 227–55, Excerpta Medica: Amsterdam.

Goy, R.W. (1992). The mating behavior of primates: estrus or sexuality? In *Topics in Primatology, Vol. 2*. (eds. N. Itoigawa, Y. Sugiyama, G.P. Sackett, and R.K.R. Thompson), pp. 151–161, University of Tokyo Press: Tokyo.

Grafenberg, E. (1950). The role of the urethra in female orgasm. *Int. J. Sexol.*, 3: 145–8.

Grammer, K. and Thornhill, R. (1994). Human (*Homo sapiens*) facial attractiveness and sexual selection: The role of symmetry and averageness. *J. Comp. Psychol.*, 108: 233–42.

Gray, H. (1977). *Anatomy, Descriptive and Surgical*, 15th edn. (eds. T.P. Pick and R. Howden), Bounty Books: New York.

Groves, C. P. (2001). *Primate Taxonomy*. Smithsonian Institution Press: Washington, DC.

Gvozdover, M.D. (1989). The typology of female figurines of the Kostenki palaeolithic culture. *Soviet Anthropol. Archaeol.*, 27: 32–94.

Haimoff, E.H. (1984). Acoustic and organizational features of gibbon songs. In *The Lesser Apes: Evolutionary and Behavioural Biology*. (eds. H. Preuschoft, D. Chivers, W. Brockelman, and N. Creel), pp. 333–53, Edinburgh University Press: Edinburgh.

Hall, K.R.L. (1962). The sexual, agonistic, and derived social behaviour patterns of the wild chacma baboon, *Papio ursinus*. *Proc. Zool. Soc. Lond.*, 13: 283–27.

Hall-Craggs, E.C.B. (1962). The testis of *Gorilla gorilla beringei*. *Proc. Zool. Soc. Lond.*, 139: 511–14.

Hamilton, D.W. (1990). Anatomy of mammalian male accessory reproductive organs. In *Marshall's Physiology of Reproduction*, Vol. 2, 4th edn. (ed. G.E. Lamming), pp. 691–746, Churchill Livingstone: Edinburgh.

Hamilton, W.D., and Zuk, M. (1982): Heritable true fitness and bright birds: A role for parasites? Science, 218: 384–387.

Hanby, J.P., Robertson, L.T., and Phoenix, C.H. (1971). The sexual behavior of a confined troop of Japanese macaques. *Folia Primatol.*, 16: 123–43.

Hanken, J. and Sherman, P.W. (1981). Multiple paternity in Belding's ground squirrels. *Science*, 212: 351–3.

Harcourt, A.H. (1978). Strategies of emigration and transfer by primates with particular reference to gorillas. *Z. Tierpsychol.*, 48: 401–20.

Harcourt, A.H. (1989). Deformed sperm are probably not adaptive. *Anim. Behav.*, 37: 863–5.

Harcourt, A.H. (1991). Sperm competition and the evolution of non-fertilising sperm in mammals. *Evolution*, 45: 314–28.

Harcourt, A.H. and Gardiner, J. (1994). Sexual selection and genital anatomy of male primates. *Proc. Roy. Soc. Lond. B.*, 255: 47–53.

Harcourt, A.H. and Stewart, K.J. (2007). *Gorilla Society: Conflict, Compromise, and Cooperation Between the Sexes*, University of Chicago Press: Chicago, IL.

Harcourt, A.H., Fossey, D., Stewart, K., et al. (1980). Reproduction in wild gorillas and some comparison with chimpanzees. *J. Reprod. Fertil. Suppl.*, 28: 59–70.

Harcourt, A.H., Harvey, P.H., Larson, S.G., et al. (1981). Testis weight, body weight and breeding system in primates. *Nature, Lond.*, 293: 55–7.

Harcourt, A.H., Purvis, A., and Liles, L. (1995). Sperm competition: Mating system, not breeding season, affects testes size of primates. *Funct. Ecol.*, 9: 468—76.

Harper, M.J.K. (1982). Sperm and egg transport. In *Reproduction in Mammals, Vol. 1. Germ cells and Fertilization*. (eds. C.R. Austin and R.V. Short), pp. 101–27, Cambridge University Press: Cambridge, UK.

Harper, M.J.K. (1994). Gamete and zygote transport. In *The Physiology of Reproduction Vol. 1*, 2nd edn. (eds. E. Knobil and J.D. Neill), pp. 123–87, Raven Press: New York.

Harris, H. (1995). Human nature and the nature of romantic love, Unpublished doctoral dissertation, University of California, Santa Barbara, CA.

Harrison, D.F.N. (1995). *The Anatomy and Physiology of the Mammalian Larynx*, Cambridge University Press: Cambridge, UK.

Harrison, G.A., Tanner, J.M., Pilbeam, D.R., et al. (1988). *Human Biology: An Introduction to Human Evolution, Variation, Growth, and Adaptability*, 3rd edn. Oxford University Press: Oxford.

Hartung, T.G. and Dewsbury, D.A. (1978). A comparative analysis of copulatory plugs in muroid rodents and their relationship to copulatory behavior. *J. Mammal.*, 59: 717–23.

Hasegawa, T. and Hiraiwa-Hasegawa, M. (1990). Sperm competition and mating behavior. In *The Chimpanzees of the Mahale Mountains: Sexual and Life History Strategies*. (ed. T. Nishida), pp. 115–32, University of Tokyo Press: Tokyo.

Hatfield, E. and Sprecher, S. (1986). *Mirror, Mirror: The Importance of Looks in Everyday Life*, State University of New York Press: Albany, NY.

Haubruge, E., Arnaud, L., Mignon, J. et al. (1999). Fertilization by proxy: Rival sperm removal and translocation in a beetle. *Proc. Roy. Soc. Lond. B.*, 266: 1183–7.

Hausfater, G. (1975). *Dominance and Reproduction in Baboons (Papio cynocephalus): A Quantitative Analysis*, Basel: Karger.

Healy, S.D. and Rowe, C. (2007). A critique of comparative studies of brain size. *Proc. Roy. Soc. Lond. B.*, 274: 453–64.

Heape, W. (1900). The 'sexual season' of mammals and the relation of the 'pro-oestrus' to menstruation. *Quart. J. Microscop. Sci.*, 44: 1–70.

Henns, R. (2000). WHR and female attractiveness. Evidence from photographic stimuli and methodological considerations. *Perspect. Indiv. Diff.*, 28: 501–13.

Herbert, J. (1970). Hormones and reproductive behaviour in rhesus and talapoin monkeys. *J. Reprod. Fertil.*, 11: 119–40.

Hershkovitz, I., Kornreich, L., and Laron, Z. (2007). Comparative skeletal features between *Homo floresiensis* and patients with primary growth hormone insensitivity (Laron Syndrome). *Am. J. Phys. Anthropol.*, 134: 198–208.

Hershkovitz, P. (1977). *Living New World Monkeys*, Vol. 1. University of Chicago Press: Chicago, IL.

Hess, J.P. (1973). Some observations on the sexual behaviour of captive lowland gorillas, *Gorilla g. gorilla* (Savage and Wyman). In *Comparative Ecology and Behaviour of Primates*. (eds. R.P. Michael and J.H. Crook), pp. 507–81, Academic Press: London.

Hewitt, G., MacLarnon, A., and Jones, K.E. (2002). The functions of laryngeal air sacs in primates: A new hypothesis. *Folia Primatol.*, 73: 70–94.

Hill, W.C.O. (1939). Observations on a giant Sumatran orang. *Am. J. Phys. Anthropol.*, 24: 449–510.

Hill, W.C.O. (1958). External genitalia. In *Primatologia: Handbook of Primatology*, Vol. 3. (eds. H. Hofer, A.H. Schultz, and D. Starck), pp. 630–704, Karger: Basel.

Hill, W.C.O. (1966). *Primates, Comparative Anatomy and Taxonomy, Vol. 6. Catarrhini, Cercopithecoidea, Cercopithecinae*, Edinburgh University Press: Edinburgh.

Hill, W.C.O. (1970). *Primates, Comparative Anatomy and Taxonomy, Vol. 8. Cynopithecinae, Papio, Mandrillus, Theropithecus*, Edinburgh University Press: Edinburgh.

Hill, W.C.O. (1972). *Evolutionary Biology of the Primates*, Academic Press: London.

Hill, W.C.O. (1974). *Primates, Comparative Anatomy and Taxonomy, Vol. 7. Cynopithecinae, Cercocebus, Macaca, Cynopithecus*, Edinburgh University Press: Edinburgh.

Hill, W.C.O. and Booth, A.H. (1957). Voice and larynx in African and Asiatic Colobidae. *J. Bombay Nat. Hist. Soc.*, 54: 309–21.

Hill, W.C.O. and Matthews, L.H. (1949). The male external genitalia of the gorilla, with remarks on the *os penis* of other Hominoidea. *Proc. Zool. Soc. Lond.*, 119: 363–78.

Hite, S. (1977). *The Hite Report*, Paul Hamlyn Pty, Summit Books: Sydney.

Hochereau de Riviers, M.-T., Courtens, J.-L., Courot, M., et al. (1990). Spermatogenesis in mammals and birds. In *Marshall's Physiology of Reproduction*, Vol. 2. (ed. G.E. Lamming), pp. 106–82, Churchill Livingstone: Edinburgh.

Hohmann, G. (1989). Comparative study of vocal communication in two Asian leaf monkeys, *Presbytis johnii* and *Presbytis entellus*. *Folia Primatol.*, 52: 27–57.

Hollien, H., Green, R., and Massey, K. (1994). Longitudinal research on voice change in males. *J. Acoust. Soc. Am.*, 96: 2646–54.

Hosken, D.J. (1997). Sperm competition in bats. *Proc. Roy. Soc. Lond. B.*, 264: 385–92.

Houghton, P. (1996). *People of the Great Ocean: Aspects of the Biology of the Early Pacific*, Cambridge University Press: Cambridge, UK.

Hrdlička, A. (1925). Weight of the brain and of the internal organs in American monkeys. *Am. J. Phys. Anthrop.*, 8: 201–11.

Hrdy, S.B. (1979). The evolution of human sexuality: The latest word and the last. *Quart. Rev. Biol.*, 54: 309–14.

Hrdy, S.B. (1981). *The Woman that Never Evolved*, Harvard University Press: Cambridge, MA.

Hunter, R.H.F. (1988). *The Fallopian Tubes: Their Role in Fertility and Infertility*, Springer-Verlag: Berlin.

Hurtado, A.M. and Hill, K.R. (1992). Paternal effect upon offspring survivorship among Ache and Hiwi hunter-gatherers: Implications for modelling pair-bond stability. In *Father-Child Relations: Cultural and Biosocial Contexts*. (ed. B.S. Hewlett), pp. 31–55. Aldine de Gruyter: New York.

Huxley, T.H. (1863). *Evidence as to Man's Place in Nature*, Williams and Norgate: London.

Immler, S. and Birkhead, T.R. (2007). Sperm competition and sperm midpiece size: no consistent pattern in passerine birds. *Proc. Roy. Soc. Lond. B.*, 274: 561–8.

Immler, S., Moore, H.D.M., Breed, W.G., et al. (2007). By hook or by crook? Morphometry, competition and cooperation in rodent sperm. *PLOS*, January 2007, Issue 1: 1–5.

Izor, R.J., Walchuk, S.I., and Wilkins, I. (1981). Anatomy and systematic significance of the penis of the pigmy chimapanzee, *Pan paniscus*. *Folia Primatol.*, 35: 218–24.

Jablonski, N.G. (ed.) (1998). *The Natural History of the Doucs and Snub-Nosed Monkeys*. World Scientific Publishing Co.: Singapore.

Jablonski, N.G. (2006). *Skin: A Natural History*, University of California Press: California.

Jablonski, N.G. and Chaplin, G. (2000). The evolution of skin coloration. *J. Hum. Evol.*, 39: 57–106.

Jamieson, B.G.M. (2007). Avian spermatozoa: Structure and phylogeny. In *Reproductive Biology and Phylogeny of Birds*. (ed. B.G.M. Jamieson), pp. 349–511, Science Publishers: Enfield, NH.

Jankowiak, W.R. and Fischer, E.F. (1992). A cross-cultural perspective on romantic love. *Ethnology*, 31: 149.

Janson, C. (1984). Female choice and the mating system of the brown capuchin monkey, *Cebus apella* (Primates: Cebidae). *Z. Tierpsychol.*, 65: 177–200.

Jasieńska, G., Zomkiewicz, A., Ellison, P., et al. (2004). Large breasts and narrow waists indicate high reproductive potential in women. *Proc. Roy. Soc. Lond. B.*, 271: 1213–7.

Jensen-Seaman, M.I. and Li, W.H. (2003). Evolution of the hominid semenogelin genes, the major proteins of ejaculated semen. *J. Mol. Evol.*, 57: 261–70.

Jit, I. and Sanjeev. (1991). Weight of the testes in northwest Indian adults. *Am. J. Hum. Biol.*, 3: 671–676.

Johanson, D. and Edey, M. (1981). *Lucy: The Beginnings of Humankind*, Simon and Shuster: New York.

Johanson, D. and White, T.D., (1980). On the status of *Australopithecus afarensis*. *Science*, 207: 1104–5.

Johnson, L., Petty, C.S., and Neaves, W.B. (1984). Influence of age on sperm production and testicular weights in men. *J. Reprod. Fertil.*, 70: 211–18.

Johnson, L., Zane, R.S., Petty, C.S., et al. (1984). Quantification of the human Sertoli cell population: Its distribution, relation to germ cell numbers, and age-related decline. *Bio. Reprod.*, 31: 785–95.

Johnson, M. and Everitt, B.J. (1988). *Essential Reproduction*, 3rd edn. Blackwell Scientific Publications: Oxford.

Johnston, V.S. and Franklin, M. (1993). Is beauty in the eye of the beholder? *Ethol. Sociobiol.*, 14: 183–99.

Jolly, A. (1999). *Lucy's Legacy: Sex and Intelligence in Human Evolution*, Harvard University Press: Cambridge, MA.

Jones, C.B. (1985). Reproductive patterns in mantled howler monkeys: Estrus, mate choice and competition. *Primates*, 26: 130–42.

Jones, C.B. (ed.) (2003). *Sexual Selection and Reproductive Competition in Primates: New perspectives and Directions*. American Society of Primatologists: Oklahoma.

Jørgensen, N., Andersen, A-G., Eustache, F., et al. (2001). Regional differences in semen quality in Europe. *Hum. Reprod.*, 16: 1012–9.

Jurmain, R. (1997). Skeletal evidence of trauma in African apes with special reference to the Gombe chimpanzees. *Primates*, 38: 1–14.

Kano, T. (1992). *The Last Ape: Pygmy Chimpanzee Behavior and Ecology*, Stanford University Press: Stanford, CA.

Kappeler, P.M. (1995). Life history variation among nocturnal prosimians. In *Creatures of the Dark: The Nocturnal Prosmians*. (eds. L. Alterman, G.A. Doyle, and K. Izard), pp. 75–92, Plenum Press: New York.

Kappeler, P.M. (ed.) (2000). *Primate Males: Causes and Consequences of Variation in Group Composition*, Cambridge University Press: Cambridge, UK.

Kappeler, P.M. and van Schaik, C. (eds.) (2004). *Sexual Selection in Primates: New and Comparative Perspectives*, Cambridge University Press: Cambridge, UK.

Katzmarzyk, P.T., Malina, R.M., Song, T.M.K., et al. (1998). Somatotype and indicators of metabolic fitness in youth. *Am. J. Hum. Biol.*, 10: 341–50.

Kauffman, A.S. and Rissman, E.F. (2006). Neuroendocrine control of mating-induced ovulation. In *Knobil and Neill's Physiology of Reproduction*, 2nd edn., Vol. 2. (ed. J.D. Neill), pp. 2283–326, Elsevier, Academic Press: New York.

Kauth, M.R. (ed.) (2006). *Handbook of the Evolution of Human Sexuality*, Haworth Press: New York.

Ke, Y., Su, B., Song, X., et al., (2001). African origin of modern humans in East Asia: A tale of 12,000 Y chromosomes. *Science*, 292: 1151–3.

Kenagy, G.J. and Trombulak, S.C. (1986). Size and function of mammalian testes in relation to body size. *J. Mammal.*, 67: 1–22.

Kendrick, K.M. and Dixson, A.F. (1983). The effect of the ovarian cycle on the sexual behaviour of the common marmoset (*Callithrix jacchus*). *Physiol. Behav.*, 30: 735–42.

Kendrick, K.M. and Dixson, A.F. (1986). Anteromedial hypothalamic lesions block proceptivity but not receptivity in the female common marmoset (*Callithrix jacchus*). *Brain Res.*, 375: 221–9.

Keverne, E.B. (1981). Do Old World primates have oestrus? *Malay. Appl. Biol.*, 10: 119–26.

Kim, D.H. and Lee, H.Y. (1982). Clinical investigation of testicular size. 1. Testis size of normal Korean males. *J. Korean. Med. Ass.*, 25: 135–44.

Kingsley, S.K. (1982). Causes of non-breeding and the development of the secondary sexual characteristics in the male orang-utan: A hormonal study. In *The Orang-Utan: Its Biology and Conservation*. (ed. L.E.M. de Boer), pp. 215–29, Dr. W. Junk Publishers: The Hague.

Kingsley, S.R. (1988). Physiological development of male orang-utans and gorillas. In *Orang-Utan Biology*. (ed. J.H. Schwartz), pp. 123–31, Oxford University Press: New York.

Kinsey, A.C., Pomeroy, W.B., and Martin, C.E. (1948). *Sexual Behavior in the Human Male*, W.B. Saunders Co: Philadelphia, PA.

Kinsey, A.C., Pomeroy, W.B., and Martin, C.E., et al. (1953). *Sexual Behavior in the Human Female*. W.B. Saunders Co.: Philadelphia, PA.

Kinzey, W.G. (1981). The titi monkeys, genus *Callicebus*. In *Ecology and Behavior of Neotropical Primates*, Vol. 1. (eds. A.F. Coimbra-Filho and R.A. Mittermeier), pp. 241–76, Academia Brasiliera de Ciências: Rio de Janiero.

Kleiman, D.G., Hoage, R.T., and Green K.M. (1988). The lion tamarins, genus, *Leontopithecus*. In *Ecology and Behavior of Neotropical Primates*, Vol. 2. (eds. R.A. Mittermeier, A.B. Rylands, A. Coimbra-Filho, and G.A.B. Fonseca), pp. 299–347, World Wildlife Fund: Washington, D.C.

Klein, L.L. (1971). Observations on copulation and seasonal reproduction of two species of spider monkeys, *Ateles belzebuth* and *A. geoffroyi*. *Folia Primatol.*, 15: 233–48.

Klein, R. (1999). *The Human Career*, 2nd edn., University of Chicago Press: Chicago, IL.

Klein, R.G. (2000). Archaeology and the evolution of human behavior. *Evol. Anthropol.*, 9: 7–36.

Knobil, E. and Neill, J.D. (eds.) (1994). *The Physiology of Reproduction Vols. 1 & 2*. Raven Press: New York.

Kobayashi, H. and Kohshima, S. (2001). Unique morphology of the human eye and its adaptive meaning: Comparative studies on external morphology of the primate eye. *J. Hum. Evol.*, 40: 419–35.

Komisaruk, B.R., Beyer-Flores, C., and Whipple, B. (2006). *The Science of Orgasm*. Johns Hopkins University Press: Baltimore, MD.

Koyama, N. (1971). Observations on the mating behavior of wild siamang gibbons at Fraser's Hill, Malaysia. *Primates*, 12: 183–9.

Koyama, N. (1988). Mating behavior of ring-tailed lemurs (*Lemur catta*) at Berenty, Madagascar. *Primates*, 29: 163–75.

Krings, M., Stune, A., Schmitz, R.W., et al. (1997). Neanderthal DNA sequences and the origin of modern humans. *Cell*, 90: 14–30.

Ku, J.H., Kim, M.E., Jeon, Y.S., et al. (2002). Factors influencing testicular volume in young men: Results of a community-based survey. *BJU Int.*, 96: 446–50.

Kuester, J., Paul, A., and Arnemann, J. (1994). Kinship, familiarity and mating avoidance in Barbary macaques, *Macaca sylvanus*. *Anim. Behav.*, 48: 1183–94.

Lambert, B. (1951). The frequency of mumps and of mumps orchitis and the consequences for sexuality and fertility. *Acta. Genet. Stat. Med.*, 2 Suppl. 1: 71–2.

Lancaster, J.B. and Lee, R.B. (1965). The annual reproductive cycle in monkeys and apes. In *Primate Behavior: Field Studies of Monkeys and Apes*, (ed. I. DeVore), pp. 486–513, Holt, Rinehart and Winston: New York.

Langlois, J.H. and Roggman, L.A. (1990). Attractive faces are only average. *Psychol. Sci.*, 1: 115–21.

Larsson, K. (1956). *Conditioning and Sexual Behaviour in the Male Albino Rat*, Almquist and Wiksell: Stockholm.

Lassek, W.D. and Gaulin, S.J.C. (2008). Waist-hip ratio and cognitive ability: Is gluteofemoral fat a privileged store of neurodevelopmental resources? *Evol. Hum. Behav.*, 29: 26–34.

Law-Smith, M.J.L., Perrett, D.I., Jones, B.C., et al. (2006). Facial appearance is a cue to oestrogen levels in women. *Proc. Roy. Soc. Lond. B.*, 273: 135–40.

Leakey, L.S.B., Tobias, P.V., and Napier, J.R. (1964). A new species of the genus *Homo* from Olduvai Gorge. *Nature, Lond.*, 202: 7–9.

Leakey, M.D. and Hay, R.I. (1979). Pliocene footprints in the Laetoli beds at Laetoli, Northern Tanzania. *Nature, Lond.*, 278: 317.

Leakey, R.E.F. (1973). Australopithecines and hominines: A summary of the evidence from the early Pleistocene of Eastern Africa. *Symp. Zool. Soc. Lond.*, 33: 53–69.

LeBatard, A.-E., Bourlès, D.L., Duringer, P., et al. (2008). Cosmogenic nuclide dating of *Sahelanthropus tchadensis* and *Australopithecus bahrelghazali*: Mio-Pliocene hominids from Chad. *Proc. Natl. Acad. Sci. USA*, 105: 3226–31.

Lemmon, W.B. and Oakes, E. (1967). Tieing between stump-tail macaques during mating. *Lab. Primate Newsletter*, 6: 14–5.

Le Roux, M.-G., Pascal, O., Andre, M.-T., et al. (1992). Nonpaternity and genetic counseling. *Lancet*, 340: 607.

Le Vay, S. and Valente, S.M. (2002). *Human Sexuality*, Sinauer: Sunderland, MA.

Liebenberg, L. (2006). Persistence hunting by modern hunter-gatherers. *Curr. Anthropol.*, 47: 1017–25.

Linz, B., Balloux, F., Moodley, Y. et al., (2007). An African origin for the intimate association between humans and *Helicobacter pylori*. *Nature, Lond.*, 445: 915–18.

Lisa, J.R., Gioia, J.D., and Rubin, I.C. (1954). Observations on the interstitial portion of the Fallopian tube. *Surgic. Gynecol. Obstet.*, 99: 159–69.

Little, A.C., Apicella, C.L., and Marlowe, F.W. (2007). Preferences for symmetry in human faces in two cultures: Data from the UK and the Hadza, an isolated group of hunter-gatherers. *Proc. Roy. Soc. Lond.* B., 274: 3113–7.

Liu, D., Ng., M.L., Zhou. L.P. et al. (1997). *Sexual Behavior in Modern China*. Continuum: New York.

Liu, H., Prugnolle, F., Manica, A., et al. (2006). A geographically explicit genetic model of worldwide human settlement history. *Am. J. Hum. Genet.*, 79: 230–7.

Lloyd, E.A. (2005). *The Case of the Female Orgasm: Bias in the Science of Evolution*, Harvard University Press: Cambridge, MA.

Lovejoy, O. (1981). The origins of man. *Science*, 211: 241–50.

Lovell, N.C. (1990). *Patterns of Illness and Injury in Great Apes: A Skeletal Analysis*, Smithsonian Institutional Press: Washington, DC.

Low, B.S. (1987). Pathogen stress and polygyny in humans. In *Human Reproductive Behavior: A Darwinian Perspective*. (eds. L. Betzig, M. Borgerhoff Mulder, and P.W. Turke), pp. 115–27, Cambridge University Press: Cambridge, UK.

Low, B.S. (2007). Ecological and socio-cutural impacts on mating and marriage systems. In *The Oxford Handbook of Evolutonary Psychology*. (eds. R.I.M. Dunbar and L. Barrett), pp. 449–62, Oxford University Press: Oxford.

Loy, J. (1970). Peri-menstrual sexual behavior among rhesus monkeys. *Folia Primatol.*, 13: 286–97.

Loy, J. (1971). Estrous behavior of free-ranging rhesus monkeys. *Primates*, 12: 1–31.

Loy, J. (1987). The sexual behavior of African monkeys and the question of estrus. In *Comparative Behavior of African Monkeys*. (ed. E.L. Zucker), pp. 175–95, Alan R. Liss: New York.

Lynch, S.M. and Zellner, D.A. (1999). Figure preference in two generations of men: The use of figure drawings illustrating differences in muscle mass. *Sex Roles*, 40: 833–43.

Mackinnon, J. (1974). The ecology and behaviour of wild orang-utans (*Pongo pygmaeus*). *Anim. Behav.*, 22: 3–74.

Maggioncalda, A.N., Czekala, N.M., and Sapolsky, R.M. (2000). Growth hormone and thyroid stimulating hormone concentrations in captive male oragn-utans: Implications for understanding developmental arrest. *Am. J. Primatol.*, 50: 67–76.

Maggioncalda, A.N., Sapolsky, R.M., and Czekala, N.M. (1999). Reproductive hormone profiles in captive male orang-utans: Implications for understanding developmental arrest. *Am. J. Phys. Anthropol.*, 109: 19–32.

Malo, A.F., Gomendio, M., Garde, J., et al. (2006). Sperm design and sperm function. *Biol. Lett.* doi: 10.1098/rsbl.2006.0449.

Malo, A.F., Roldan, E.R.S., Garde, J., et al. (2005). Antlers honestly advertise sperm production and quality. *Proc. Roy. Soc. Lond. B.*, 272: 149–57.

Manica, A., Amos, W., Balloux, F., et al. (2007). The effect of ancient population bottlenecks on human phenotypic variation. *Nature, Lond.*, 448: 346–8.

Mann, T. (1964). *The Biochemistry of Semen and of the Male Reproductive Tract*, Methuen: London.

Mann, T. and Lutwak-Mann, C. (1981). *Male Reproductive Function and Semen*, Springer: Berlin.

Manning, J.T., Trivers, R.L., Singh, D., et al. (1999). The mystery of female beauty. *Nature, Lond.*, 399: 214–5.

Manson, J.H. (1992). Measuring female choice in Cayo Santiago rhesus macaques. *Anim. Behav.*, 44: 405–16.

Marchetti, C., Jouy, N., Leroy-Martin, B., et al. (2004). Comparison of four fluorochromes for the detection of the inner mitochondrial membrane potential in human spermatozoa and their correlation with sperm motility. *Hum. Reprod.*, 19: 2267–76.

Marler, P. and Hobbett, L. (1975). Individuality in a long-range vocalization of chimpanzees. *Z. Tierpsychol.*, 38: 97–109.

Marlowe, F. (2000). Paternal investment and the human mating system. *Behav. Proc.*, 51: 45–61.

Marlowe, F., Apicella, C.L., and Reed, D. (2005). Men's preferences for women's profile waist-to-hip ratio in two societies. *Evol. Hum. Behav.*, 25: 371–8.

Marlowe, F.W. (1999*a*). Male care and mating effort among Hadza foragers. *Behav. Ecol. Sociobiol.*, 46: 57–64.

Marlowe, F.W. (1999*b*). Showoffs or providers? The parenting effort of Hadza men. *Evol. Hum. Behav.*, 20: 391–404.

Marlowe, F.W. (2003). A critical period for provisioning by Hadza men: Implications for pair bonding. *Evol. Hum. Behav.*, 1: 1–13.

Marshall, D.S. and Suggs, R.C. (1971). *Human Sexual Behavior: Variations in the Ethnographic Spectrum*, Prentice-Hall: Englewood Cliffs, NJ.

Marshall, W.A. (1970). Sex differences at puberty. *J. Biosoc. Sci.*, May Suppl. 2: 31–41.

Marson, J., Gervais, D., Cooper, R.W., et al. (1989). Influence of ejaculation frequency on semen characteristics in chimpanzees (*Pan troglodytes*). *J. Reprod. Fertil.*, 85: 43–50.

Marti, B., Tuomilehto, J., Saloman, V., et al. (1991). Body fat distribution in the Finnish population: Environmental determinants and predictive power for cardiovascular risk factor level. *J. Epidemiol. Community Health*, 45: 131–7.

Martin, R.D. (1983). *Human Brain Evolution in an Ecological Context (52nd James Arthur Lecture on the Evolution of the Human Brain)*. American Museum of Natural History: New York.

Martin, R.D. (1990). *Primate Origins and Evolution: A Phylogenetic Reconstruction*, Chapman Hall/Princeton University Press: London/New Jersey.

Martin, R.D. (2003). Human reproduction: A comparative background for medical hypotheses. *J. Reprod. Immunol.*, 59: 111–35.

Martin, R.D. and May, R.M. (1981). Outward signs of breeding. *Nature, Lond.*, 293: 7–9.

Martin, R.D., MacLarnon, A.M., Phillips, J.L., et al. (2006*a*). Comment on 'the brain of LB1, *Homo floresiensis*'. *Science*, 312: 999.

Martin, R.D., MacLarnon, A.M., Phillips, J.L., et al. (2006*b*). Flores hominid: New species or microcephalic dwarf? *Anat. Rec.*, 288A: 1123–45.

Martin, R.D., Willner, L.A., and Dettling, A. (1994). The evolution of sexual size dimorphism in primates. In *The Differences Between the Sexes*. (ed. R.V. Short and E. Balaban), pp. 159–200, Cambridge University Press: Cambridge, UK.

Marvan, R., Stevens, J.M.G., Roeder, A.D., et al. (2006). Male dominance rank, mating and reproductive success in captive bonobos (*Pan paniscus*). *Folia Primatol.*, 77: 364–76.

Masters, W.H. and Johnson, V.E. (1966). *Human Sexual Response*, J. and A. Churchill Ltd: London.

Masters, W.H. and Johnson, V.E. (1970). *Human Sexual Inadequacy*, Little Brown and Co: Boston, MA.

Mayr, E. (1963). *Animal Species and Evolution*, Belknap Press: Cambridge, MA.

Mazur, A., Halpern, C., and Udry, J.R. (1994). Dominant looking males copulate earlier. *Ethol. Sociobiol.*, 15: 87–94.

McBrearty, S. and Brooks, A. (2000). The revolution that wasn't: A new interpretation of the origin of modern human behavior. *J. Hum. Evol.*, 39: 453–63.

McCormack, S.A. (1971). Plasma testosterone concentration and binding in the chimpanzee: Effect of age. *Endocrinology*, 89: 1171–77.

McGinnis, P.R. (1979). Sexual behavior in free-living chimpanzees: Consort relationships. In *The Great Apes*. (eds. D.A. Hamburg and E.R. McCown), pp. 429–40. Benjamin Cummings: Menlo Park, CA.

McGrew, W.C. (1992). *Chimpanzee Material Culture: Implications for Human Evolution*, Cambridge University Press: Cambridge, UK.

Mead, M. (1967). *Male and Female: A Study of the Sexes in a Changing World*, William Morrow and Co.: New York.

Mealey, L. (2000). *Sex Differences: Developmental and Evolutionary Strategies*. Academic Press: New York.

Mellars, P. (2004). Neanderthals and the modern human colonization of Europe. *Nature, Lond.*, 432: 461–5.

Mellars, P. (2006). Why did human populations disperse from Africa ca. 60,000 years ago? A new model. *Proc. Nat. Acad. Sci. USA.*, 103: 9381–6.

Ménard, N., Scheffrahn, W., Vallet, D., et al. (1992). Application of blood protein electrophoresis and DNA fingerprinting to the analysis of paternity and social characteristics of wild Barbary macaques. In *Paternity in Primates: Genetic Tests and Theories.* (eds. R.D. Martin, A.F. Dixson, and E.J. Wickings), pp. 155–74, Karger: Basel.

Meston, C.M. and Buss, D.M. (2007). Why humans have sex? *Arch. Sex. Behav.*, 36: 477–507.

Miki, K., Qu, W., Goulding, E.H., et al. (2004). Glyceraldehyde 3-phosphate dehydrogenase-S, a sperm-specific glycolytic enzyme, is required for sperm motility and male fertility. *Proc. Nat. Acad. Sci. USA*, 101: 16501–6.

Miller, G. (2000). *The Mating Mind: How Sexual Choice Shaped the Evolution of Human Nature*, Doubleday: New York.

Miller, G., Tybur, J.M., and Jordan, B.D. (2007). Ovulatory cycle effects on tip earnings by lap dancers: Economic evidence for human estrus? *Evol. Hum. Behav.*, 28: 375–81.

Milton, K.M. (1985). Mating patterns of woolly spider monkeys, *Brachyteles arachnoides*: Implications for female choice. *Behav. Ecol. Sociobiol.*, 17: 53–9.

Mitani, J.C. (1985). Mating behaviour of male orang-utans in the Kutai Reserve, East Kalimantan, Indonesia. *Anim. Behav.*, 33: 392–402.

Mitra Setia, T. and van Schaik, C.P. (2007). The response of adult orang-utans to flanged male long calls: Inferences about their function. *Folia Primatol.*, 78: 215–26.

Møller, A.P. (1988). Ejaculate quality, testes size and sperm competition in primates. *J. Hum. Evol.*, 17: 479–88.

Møller, A.P. (1989). Ejaculate quality, testes size and sperm production in mammals. *Funct. Ecol.*, 3: 91–6.

Møller, A.P. (1991). Concordance of mammalian ejaculate features. *Proc. Roy. Soc. Lond. B.*, 246: 237–41.

Møller, A.P., Christe, P., and Lux, E. (1999). Parasitism, host immune function, and sexual selection. *Q. Rev. Biol.*, 74: 3–74.

Moore, H.D.M., Dvoráková, K., Jenkins, N., et al. (2002). Exceptional sperm cooperation in the wood mouse. *Nature, Lond.*, 418: 174–7.

Moore, H.D.M., Martin, M., and Birkhead, T.R. (1999). No evidence for killer sperm or other selective interactions between human spermatozoa in ejaculates of different males *in vitro*. *Proc. Roy. Soc. Lond. B.*, 266: 2343–50.

Moran, C., Hernandez, E., Ruiz, J.E., et al. (1999). Upper body obesity and hyperinsulinemia are associated with anovulation. *Gynecol. Obstet. Investig.*, 47: 1–5.

Morris, D. (1967). *The Naked Ape: A Zoologist's Study of the Human Animal*, Jonathan Cape: London.

Mortimer, D. (1983). Sperm transport in the human female reproductive tract. *Oxford Revs. Repro. Biol.*, 5: 30–61.

Morwood, M. and van Oosterzee, P. (2007). *A New Human: The Startling Discovery and Strange Story of the "Hobbits" of Flores, Indonesia*, Harper Collins, Smithsonian Books: New York.

Morwood, M.J., Soejono, R.P., Roberts, R.G., et al. (2004). Archaeology and age of a new hominin from Flores in eastern Indonesia. *Nature, Lond.*, 431: 1087–91.

Moynihan, M. (1964). Some behavior patterns of platyrrhine monkeys. 1. The night monkey (*Aotus trivirgatus*). *Smithson. Misc. Coll.*, 146: 1–84.

Mueller, U. and Mazur, A. (1997). Facial dominance in *Homo sapiens* as honest signalling of male quality. *Behav. Ecol.*, 8: 569–79.

Murdock, G.P. (1981). *Atlas of World Cultures*, University of Pittsburgh Press: Pittsburgh, PA.

Nadler, R.D. (1977). Sexual behavior of captive orangutans. *Arch. Sex. Behav.*, 6: 457–75.

Nadler, R.D. (1980). Reproductive physiology and behavior of gorillas. *J. Reprod. Fertil. Suppl.*, 28: 79–89.

Nadler, R.D. (1988). Sexual and reproductive behavior. In *Orang- Utan Biology*. (ed. J.H. Schwartz), pp. 105–16, Oxford University Press: New York.

Nakamura, R. (1961). Normal growth and maturation in the male genitalia of the Japanese. *J. Jpn. Soc. Urinary Syst.*, 52: 172–88.

Negus, V.E. (1949). *The Comparative Anatomy and Physiology of the Larynx*, Hafner Publishing Co: New York.

Nettle, D. (2002). Height and reproductive success in a cohort of British men. *Hum. Nat.*, 13: 473–91.

Nieschlag, E. (1979). The endocrine function of the human testis in regard to human sexuality. In *Sex, Hormones and Behaviour. CIBA Foundation Symposium (new series)*, Vol. 62, pp. 183–208. Excerpta Medica: Amsterdam.

Nishida, T. (ed.) (1990). *The Chimpanzees of the Mahale Mountains: Sexual and Life History Strategies*. University of Tokyo Press: Tokyo.

Nobile, P. (1982). Penis size: The difference between blacks and whites. *Forum: Int. J. Hum. Rel.*, 11: 21–8.

Nunn, C.L. (1999). The evolution of exaggerated sexual swellings in female primates and the graded-signal hypothesis. *Anim. Behav.*, 58: 229–46.

Nunn, C.L. and Altizer, S. (2006). *Infectious Diseases in Primates*, Oxford University Press: Oxford.

Oates, J.F., Bocian, C.M., and Terranova, C.J. (2000). The loud calls of black-and-white colobus monkeys: Their adaptive and taxonomic significance in light of new data. In *Old World Monkeys*. (eds. P.F. Whitehead and C.J. Jolly), pp. 431–52: Cambridge University Press: Cambridge, UK.

Obendorf, P.J., Oxnard, C.E., and Kefford, B.J. (2008). Are the small human-like fossils found on Flores human endocrine cretins? *Proc. Roy. Soc. Lond. B.*, 275: 1287–96.

O'Brien, S.J., Roelke, M.E., Marker, L., et al. (1985). Genetic basis for species vulnerability in the cheetah. *Science*, 227: 1428–34.

Okami, P. and Shackelford, T.K. (2001). Human sex differences in sexual psychology and behavior. *Ann. Revs. Sex Res.* 12: 186–241.

Olesen, H. (1948). *Morfologiske Sperma-og Testisundersogelser*, Munksgaard: Copenhagen.

Orgebin-Crist, M.C. and Olson, G.E. (1984). Epididymal sperm maturation. In *The Male in Farm Animal Reproduction.* (ed. M. Courot), pp. 80–102, Martinus Nijhoff Publishers: Dordrecht, The Netherlands.

Packer, C., Scheel, D., and Pusey, A.E. (1990). Why lions form groups- food is not enough. *Am. Nat.*, 136: 1–19.

Pagel, M. and Bodmer, W. (2003). A naked ape would have fewer parasites. *Proc. Roy. Soc. Lond. B.*, 270 (suppl. 1): S117-S119.

Palombit, R.A. (1994). Extra-pair copulations in a monogamous ape. *Anim. Behav.*, 47: 721–3.

Parker, G.A. (1970). Sperm competition and its evolutionary consequences in the insects. *Biol. Rev.*, 45: 525–67.

Parker, G.A. (1998). Sperm competition and the evolution of ejaculates: Towards a theory base. In *Sperm Competition and Sexual Selection.* (eds. T.R. Birkhead and A.P. Møller), pp. 3–54, Academic Press: San Diego, CA.

Pasquali, R., Gambineri, A., Anconetani, B., et al. (1999). The natural history of the metabolic syndrome in young women with the polycystic ovary syndrome and the effect of long-term oestrogen-progestagen treatment. *Clin. Endocr.*, 50: 517–27.

Pauerstein, C.J. and Eddy, C.A. (1979). Morphology of the Fallopian tube. In *The Biology of the Fluids of the Female Genital Tract.* (eds. F.K. Beller and G.F.B. Schumacher), pp. 299–317, Elsevier: North-Holland.

Paul, A., Kuester, J., Timme, A., et al. (1993). The association between rank, mating effort and reproductive success in male Barbary macaques (*Macaca sylvanus*). *Primates*, 34: 491–502.

Pawlowski, B. (1999*a*). Loss of oestrus and concealed ovulation in human evolution. The case against the sexual-selection hypothesis. *Curr. Anthropol.*, 40: 257–75.

Pawlowski, B. (1999*b*). Permanent breasts as a side effect of subcutaneous fat tissue increase in human evolution. *Homo*, 50: 149–62.

Pawlowski, B. (2003).Variable preferences for sexual dimorphism in height as a strategy for increasing the pool of potential partners in humans. *Proc. Roy. Soc. Lond. B.*, 270: 709–12.

Pawlowski, B. and Jasienska, G. (2005).Women's preferences for sexual dimorphism in height depend on menstrual cycle phase and expected duration of relationship. *Biol. Psychol.*, 70: 38–43.

Pawlowski, B. Dunbar, R.I.M., and Lipowicz, A. (2000). Tall men have more reproductive success. *Nature, Lond.*, 403: 156.

Peleg, G., Katzir, G., Peleg, O., et al. (2006). Hereditary family signature of facial expression. *Proc. Nat. Acad. Sci. USA.*, 103: 15921–6.

Pelligrini, R.J. (1973). Impressions of the male personality as a function of beardedness. *Psychology*, 10: 29–33.

Penton-Voak, I.S. and Perrett, D.I. (2000). Consistency and individual differences in facial attractiveness judgements: An evolutionary perspective. *Soc. Res.*, 67: 219–44.

Penton-Voak, I.S. and Perrett, D.I. (2001). Male facial attractiveness: perceived personality traits and shifting female preferences for male traits across the menstrual cycle. *Adv. Study Behav.*, 30: 219–59.

Penton-Voak, I.S., Little, A.C., Jones, B.C., et al. (2003). Female condition influences preferences for sexual dimorphism in faces of male humans (*Homo sapiens*). *J. Comp. Psychol.*, 117: 264–74.

Penton-Voak, I.S., Perrett, D.I., Castles, D.L., et al. (1999). Female preference for male faces changes cyclically. *Nature, Lond.*, 399: 741–2.

Peritz, E. and Rust, P.F. (1972). On the estimation of the non-paternity rate using more than one blood-group system. *Am. J. Hum. Genet.*, 24: 46–53.

Perrett, D.I., Lee, K.J., Penton-Voak, I.S., et al. (1998). Effects of sexual dimorphism on facial attractiveness. *Nature, Lond.*, 394: 884–7.

Perrett, D.I., May, K.A., and Yoshikawa, S. (1994). Facial shape and judgements of female attraciveness. *Nature, Lond.*, 368: 239–42.

Pillsworth, G.E. and Haselton, M.G. (2006). Women's sexual strategies: The evolution of long-term bonds and extrapair sex. *Ann. Revs. Sex Res.*, 17: 59–100.

Piñón, R. Jr. (2002). *Biology of Human Reproduction*, University Science Books: Sausalito, CA.

Pitnik, S., Jones, K.E., and Wilkinson, G.S. (2006). Mating system and brain size in bats. *Proc. Roy. Soc. Lond. B.*, 273: 719–24.

Pizzari, T., Cornwallis, C.K., Løvlie, H., et al. (2003). Sophisticated sperm allocation in male fowl. *Nature, Lond.*, 426: 70–4.

Platek, S.M. and Shackelford, T.K. (eds.) (2006). *Female Infidelity and Paternity Uncertainty.* Cambridge University Press: Cambridge, UK.

Plavcan, M.J. (2001). Sexual dimorphism in primate evolution. *Yrbk. Phys. Anthropol.* 44: 25–53.

Poiani, A. (2006). Complexity of seminal fluid: A review. *Behav. Ecol. Sociobiol.*, 60: 289–310.

Pond, C.M. (1998). *The Fats of Life*, Cambridge University Press: Cambridge, UK.

Pope, T.R. (1990). The reproductive consequences of male cooperation in the red howler monkey: Paternity exclusion in multi-male and single-male troops using genetic markers. *Behav. Ecol. Sociobiol.*, 27: 439–46.

Post, P.W., Daniels, F., and Binford, R.T. (1975). Cold injury and the evolution of 'white' skin. *Hum. Biol.*, 39: 131–43.

Potthoff, R.F. and Whittinghill, M. (1965). Maximum likelihood estimation of the proportion of nonpaternity. *Am. J. Hum. Genet.*, 17: 480–94.

Potts, M. and Short, R., (1999). *Ever Since Adam and Eve: The Evolution of Human Sexuality*. Cambridge University Press: Cambridge, UK.

Pound, N. (2002). Male interest in visual cues of sperm competition risk. *Evol. Hum. Behav.*, 23: 443–66.

Pound, N., Javed, M.H., Ruberto, C., et al. (2002). Duration of sexual arousal predicts semen parameters for masturbatory ejaculates. *Physiol. Behav.*, 76: 685–9.

Pound, N., Shackelford, T.K., and Goetz, A.T. (2006). Sperm competition in humans. In *Sperm Competition in Humans: Classic and Contemporary Readings*. (eds. T.K. Shackelford and N. Pound), pp. 3–31, Springer: New York.

Price, D. and Williams-Ashman, H.G. (1961). The accessory reproductive glands of mammals. In *Sex and Internal Secretions*, 3rd edn. (ed. W.C. Young), Vol. 1, pp. 366–448, Williams and Wilkins: Baltimore, MD.

Prins, G.S. and Zaneveld, L.J.D. (1980). Radiographic study of fluid transport in the rabbit *vas deferens* during sexual rest and after sexual activity. *J. Reprod. Fertil.*, 58: 311–9.

Putman, R. (1988). *The Natural History of Deer*, Christopher Helm: London.

Puts, D.A., Gaulin, S.J.C., and Verdolini, K. (2006). Dominance and the evolution of sexual dimorphism in human voice pitch. *Evol. Hum. Behav.*, 27: 283–96.

Ramachandran, S. et al. (2005). Support from the relationship of genetic and geographic distance in human populations for a serial founder effect originating in Africa. *Proc. Natl. Acad. Sci. USA.*, 102: 15942–7.

Ramm, S.A., Parker, G.A., and Stockley, P. (2005). Sperm competition and the evolution of male reproductive anatomy in rodents. *Proc. Roy. Soc. Lond. B.*, 272: 949–55.

Rantala, M.J. (1999). Human nakedness: Adaptation against ectoparasites? *Int. J. Parasitol.*, 29: 1987–9.

Rantala, M.J. (2007). Evolution of nakedness in *Homo sapiens*. *J. Zool. Lond.*, 273: 1–7.

Rebuffe-Scrive, M. (1991). Neuroregulation of adipose tissue: Molecular and hormonal mechanisms. *Int. J. Obes.*, 15: 83–6.

Reby, D. and McComb, K. (2003). Anatomical constraints generate honesty: Acoustical cues to age and weight in the roars of red deer stags. *Anim. Behav.*, 65: 519–30.

Reeder, D.M. (2003). The potential for cryptic female choice in primates: Behavioral, anatomical and physiological considerations. In *Sexual Selection and Reproductive Competition in Primates: New Perspectives and Directions*. (ed. C.B. Jones), pp. 255–303, American Society of Primatologists: Norman, OK.

Reichard, U. (1995). Extra-pair copulations in a monogamous gibbon (*Hylobates lar*). *Ethology*, 100: 99–112.

Reno, P.L., Meindl, R.S., McCollum, M.A., et al. (2003). Sexual dimorphism in *Australopithecus afarensis* was similar to that of modern humans. *Proc. Natl. Acad. Sci., USA*, 100: 9404–9.

Reynolds, V. (2005). *The Chimpanzees of the Budongo Forest: Ecology, Behaviour and Conservation*, Oxford University Press: Oxford.

Rhodes, G. (2006). The evolutionary psychology of facial beauty. *Ann. Rev. Psychol.*, 57: 199–226.

Rhodes, G., Chan, J., Zebrowitz, L.A., et al. (2003). Does sexual dimorphism in human faces signal health? *Proc. Roy. Soc. Lond. B. (Suppl).*, 270: S93–S95.

Richard, A. (1976). Patterns of mating in *Propithecus verreauxi verreauxi*. In *Prosimian Behaviour*. (eds. R.D. Martin, G.A. Doyle, and A.C. Walker), pp. 49–74, Duckworth: London.

Richards, G. (2006). Genetic, physiologic and ecogeographic factors contributing to variation in *Homo sapiens*: *Homo floresiensis* reconsidered. *J. Evol. Biol.*, 19: 1744–67.

Rightmire, G.P. (1990). *The Evolution of Homo Erectus*, Cambridge University Press: Cambridge, UK.

Rightmire, G.P. (1998). Human evolution in the middle Pleistocene: The Role of *Homo heidelbergensis*. *Evol. Anthropol.*, 6: 218–27.

Rijksen, H. (1978). *A Field Study on Sumatran Orang-Utans (Pongo pygmaeus abelii, Lesson, 1827): Ecology, Behaviour and Conservation*, Veenman and Zonen: Wageningen, The Netherlands.

Roberts, S.C., Havlicek, J., Flegr, J., et al. (2004). Female facial attractiveness increases during the fertile phase of the menstrual cycle. *Proc. Roy. Soc. Lond. B.*, 271: S270–S272.

Robillard, P.-Y., Dekker, G.A., and Hulsey, T.C. (2002). Evolutionary adaptations to pre-eclampsia/eclampsia in humans: low fecundability rate, loss of oestrus, prohibition of incest and systematic polyandry. *Am. J. Reprod. Immunol.*, 47: 104–11.

Rodger, J.C. and Bedford, J.M. (1982). Separation of sperm pairs and sperm-egg interaction in the opossum, *Didelphis virginiana*. *J. Reprod. Fertil.*, 64: 171–9.

Rodriguez, V. (1994). Function of the spermathecal muscle in *Chelymorpha alternans* Boheman (Coleoptera: Chrysomelidae: Cassidinae). *Physiol. Ent.*, 19: 198–202.

Rogers, A.R., Iltis, D., and Wooding, S. (2004). Genetic variation at the MC_1R locus and the time since loss of human body hair. *Curr. Anthropol.*, 45: 105–8.

Rolls, E.T. (2005). *Emotion Explained*, Oxford University Press: Oxford.

Roney, J.R., Hanson, K.N., Durante, K.M., et al. (2006). Reading men's faces: Women's mate attractiveness judgements track men's testosterone and interest in infants. *Proc. Roy. Soc. Lond. B.*, 273: 2169–75.

Rose, R.W., Nevison, C.M., and Dixson, A.F. (1997). Testes weight, body weight and mating systems in marsupials and monotremes. *J. Zool. Lond.*, 243: 523–31.

Rowe, N. (1996). *The Pictorial Guide to the Living Primates*, Pogonias Press: New York.

Rowell, T. (1972). Female reproduction cycles and social behavior in primates. *Adv. Stud. Behav.*, 4: 69–105.

Rowell, T. (1973). Social organization of wild talapoin monkeys. *Am. J. Phys. Anthropol.*, 38: 593–8.

Rowell, T.E. and Dixson, A.F. (1975). Changes in social organization during the breeding season of wild talapoin monkeys. *J. Reprod. Fertil.*, 43: 419–34.

Rushton, J.P. and Bogaert, A.F. (1987). Race differences in sexual behavior: Testing an evolutionary hypothesis. *J. Res. Person.*, 21: 529–51.

Russell, L.D. (1998). Sertoli cells, overview. In *Encyclopedia of Reproduction*, Vol. 4. (eds. E. Knobil and J.D. Neill), pp. 381–7, Academic Press: San Diego, CA.

Russell, L.D. and Griswold, M.D. (eds.) (1993). *The Sertoli Cell*. Cache River Press: Clearwater, FL.

Sakate, N., Elliott, R.M.A., Watson, P.F., and Holt, W.V. (2006). Sperm selection and competition in pigs may be mediated by the differential motility, activation and suppression of sperm subpopulations within the oviduct. *J. Exp. Biol.*, 209: 1560–1572.

Sanders, D. and Bancroft, J. (1982). Hormones and the sexuality of women—the menstrual cycle. *Clin. Endocr. Metab.*, 11: 639–59.

Sarich, V.M. and Wilson, A.C. (1967). Immunological time scale for hominid evolution. *Science*, 158: 1200–3.

Sasse, G., Müller, H., Chakraborty, R., et al. (1994). Estimating the frequency of non-paternity in Switzerland. *Hum. Hered.*, 44: 337–43.

Sawyer, G.J. and Deak, V. (2007). *The Last Human: A Guide to Twenty-two Species of Extinct Humans*, Yale University Press: New Haven, CT and London.

Sawyer, G.J. and Maley, B. (2005). Neanderthal reconstructed. *Anat. Rec.*, 283B: 23–31.

Schacht, L.E. and Gershowitz, H. (1963). Frequency of extra-marital children as determined by blood groups. In *Proceedings of the Second International Congress on Human Genetics*. (eds. F. Brockington, J. Francois, and L. Gedda), pp. 894–7, Instituto G. Mendel: Rome.

Schaller, G.B. (1963). *The Mountain Gorilla: Ecology and Behavior*, University of Chicago Press: Chicago, IL.

Schaller, G.B. (1972). *The Serengeti Lion: A Study of Predator-Prey Relations*, University of Chicago Press: Chicago, IL.

Scheib, J.E., Gangestad, S.W., and Thornhill, R. (1999). Facial attractiveness, symmetry and cues of good genes. *Proc. Roy. Soc. Lond. B.*, 266: 1913–7.

Schenck, A. and Kovacs, K.M. (1995). Multiple mating between black bears revealed by DNA fingerprinting. *Anim. Behav.*, 50: 1483–90.

Schön, M.A. (1971). The anatomy of the resonating mechanism in howling monkeys. *Folia Primatol.*, 15: 117–32.

Schreiner-Engel, P. (1980). Female sexual arousability: Its relation to gonadal hormones and the menstrual cycle, Unpublished PhD thesis, New York University, New York.

Schreiner-Engel, P., Schiavi, R.C., Smith, H., et al. (1981). Sexual arousability and the menstrual cycle. *Psychosom. Med.*, 43: 199–214.

Schultz, A.H. (1938). The relative weights of the testes in primates. *Anat. Rec.*, 72: 387–94.

Schulze, H. and Meier, B. (1995). Behavior of captive *Loris tardigradus nordicus*: A qualitative description, including some information about morphological bases of behavior. In *Creatures of the Dark: The Nocturnal Prosimians*. (eds. L. Alterman, G.A. Doyle, and M.K. Izard), pp. 221–49, Plenum Press: New York.

Schurmann, C. (1982). Mating behavior of wild orang-utans. In *The Orang-Utan: Its Ecology and Conservation*. (ed. L.E.M. de Boer), pp. 269–84, Dr. W. Junk Publishers: The Hague.

Scott, L. (1984). Reproductive behavior of adolescent female baboons (*Papio anubis*). In *Female Primates: Studies by Women Primatologists*. (ed. M. Small), pp. 77–102, Alan R. Liss: New York.

Scruton, D.M. and Herbert, J. (1970). The menstrual cycle and its effect upon behaviour in the talapoin monkey (*Miopithecus talapoin*). *J. Zool. Lond.*, 162: 419–36.

Sekulic, R. (1982). Daily and seasonal patterns of roaring and spacing in four red howler (*Alouatta seniculus*) troops. *Folia Primatol.*, 39: 22–48.

Setchell, J.M. and Dixson, A.F. (2001). Circannual changes in the secondary sexual adornments of semi-free ranging male and female mandrills (*Mandrillus sphinx*). *Am. J. Primatol.*, 53: 109–21.

Setchell, P.B. (1982). Spermatogenesis and spermatozoa. In *Reproduction in Mammals, Vol. 1. Germ Cells and Fertilization*. (eds. C.R. Austin and R.V. Short), pp. 63–101, Cambridge University Press: Cambridge, UK.

Settlage, D.S.F. and Hendrickx, A.G. (1974). Observations on coagulation characteristics of the rhesus monkey electroejaculate. *Biol. Reprod.*, 11: 619–23.

Seuánez, H.N. (1980). Chromosomes and spermatozoa of the African great apes. *J. Reprod. Fertil. Suppl.*, 28: 91–104.

Seuánez, H.N., Carothers, A.C., Martin, D.E., et al. (1977). Morphological abnormalities in the spermatozoa of the great apes and man. *Nature, Lond.*, 270: 345–7.

Shackelford, T.K. and Pound, N. (eds.) (2006). *Sperm Competition in Humans: Classic and Contemporary Readings.* Springer: New York.

Shackelford, T.K. and Pound, N. Goetz, A.T., et al. (2005). Female infidelity and sperm competition. In *The Handbook of Evolutionary Psychology*. (ed. D.M. Buss), pp. 372–93, Wiley & Sons: Hoboken, NJ.

Sharpe, R.M. (1994). Regulation of spermatogenesis. In *The Physiology of Reproduction*, Vol. 1., 2nd edn. (eds. E. Knobil and J.D. Neill), pp. 1363–434, Raven Press: New York.

Sheldon, W.H., Dupertuis, C.W., and McDermott, E. (1954). *Atlas of Men*, Harper and Brothers: New York.

Sheldon, W.H., Stevens, S.S., and Tucker, W.B. (1940). *The Varieties of Human Physique*, Harper and Brothers: New York.

Short, R.V. (1979). Sexual selection and its component parts, somatic and genital selection, as illustrated by man and the great apes. *Adv. Study Behav.*, 9: 131–58.

Short, R.V. (1980). The origins of human sexuality. In *Reproduction in Mammals, Vol. 8. Human Sexuality*. (eds. C.R. Austin and R.V. Short), pp. 1–33, Cambridge University Press: Cambridge, UK.

Short, R.V. (1984). Testis size, ovulation rate and breast cancer. In *One Medicine*. (eds. O.A. Ryder and M.L. Byrd), pp. 32–44, Springer-Verlag: Berlin.

Short, R.V. (1985). Species differences in reproductive mechanisms. In *Reproduction in Mammals, Vol. 4. Reproductive Fitness*. (eds. C.R. Austin and R.V. Short), pp. 24–61, Cambridge University Press: Cambridge, UK.

Short, R.V. (1997). The testis: The witness of the mating system, the site of mutation and the engine of desire. *Acta. Paediatr. Suppl.*, 422: 3–7.

Sillen-Tulberg, B. and Møller, A.P. (1993). The relationship between concealed ovulation and mating systems in primates: A phylogenetic analysis. *Am. Nat.*, 141: 1–25.

Sillero-Zubiri, C., Gottelli, D., and MacDonald, D.W. (1996). Male philopatry, extra pack copulations and inbreeding avoidance in Ethiopian wolves. *Behav. Ecol. Sociobiol.*, 38: 331–40.

Simmons, L.W. (2001). *Sperm Competition and its Evolutionary Consequences in the Insects*, Princeton University Press: Princeton, NJ.

Simmons, L.W., Firman, R.C., Rhodes, G., et al. (2004). Human sperm competition: Testis size, sperm production and rates of extrapair copulations. *Anim. Behav.*, 68: 297–302.

Singer, I. (1973). *The Goals of Human Sexuality*, W.W. Norton: New York.

Singh, D. (1994). Is thin really beautiful and good? Relationship between waist-to-hip (WHR) ratio and female attractiveness. *Pers. Ind. Diff.*, 16: 123–32.

Singh, D. (2002). Female mate choice at a glance: Relationship of waist-to-hip ratio to health, fecundity and attractiveness. *Neuroendocrinol. Lett.*, 23(suppl. 4): 81–91.

Singh, D. (2006). Universal allure of the hourglass figure: An evolutionary theory of female physical attractiveness. *Clin. Plastic. Surg.*, 33: 359–70.

Singh, D. Dixson, B.J., Jessop, T., et al. (in press). Cross cultural attractiveness of the female hourglass figure. *Evol. Hum. Behav.*

Singh, D. Renn, P., and Singh, A. (2007). Did the perils of abdominal obesity affect depiction of feminine beauty in sixteenth to eighteenth century British literature? Exploring the health and beauty link. *Proc. Roy. Soc. Lond. B.*, 274: 891–4.

Slob, A.K., Groeneveld, W.H., and Van der Werff Ten Bosch, J.J. (1986). Physiological changes during copulation in male and female stumptail macaques (*Macaca arctoides*). *Physiol. Behav.*, 38: 891–5.

Small, M.F. (1993). *Female Choices: Sexual Behavior of Female Primates*, Cornell University Press: Ithaca, NY.

Smith, R.L. (1984). Human sperm competition. In *Sperm Competition and the Evolution of Animal Mating Systems*. (ed. R.L. Smith), pp. 601–59, Academic Press, New York.

Smuts, B.B. (1985). *Sex and Friendship in Baboons*, Aldine: New York.

Smuts, B.B., Cheney, D., Seyfarth, R., et al. (ed.) (1987). *Primate Societies*. University of Chicago Press: Chicago, IL.

Snowdon, C.T. and Soini, P. (1988). The tamarins, genus *Saguinus*. In *Ecology and Behavior of Neotropical Primates*, Vol. 2. (eds. R.A. Mittermeier, A.B. Rylands, A. Coimbra-Filho, and G.A.B. Fonseca), pp. 223–98, World Wildlife Fund: Washington, DC.

Sodhi, V.K. and Sausker, W.F. (1988). Dermatoses of pregnancy. *Am. Fam. Physician*, 37: 131–8.

Soini, P. (1988). The pigmy marmoset, genus *Cebuella*. In *Ecology and Behavior of Neotropical Primates*, Vol. 2. (eds. R.A. Mittermeier, A.B. Rylands, A. Coimbra-Filho, and G.A.B. Fonseca), pp. 79–129, World Wildlife Fund: Washington, DC.

Soler, C., Núñez, M., Gutiérrez, R., et al. (2003). Facial attractiveness in men provides clues to semen quality. *Evol. Hum. Behav.*, 24: 199–207.

Sorokowski, P. and Pawlowski, B. (2008). Adaptive preferences for leg length in a potential partner. *Evol. Hum. Behav.*, 29: 86–91.

Spoor, F., Leakey, M.G., Gathogo, P.N., et al. (2007). Implications of new early *Homo* fossils from Ileret, east of Lake Turkana, Kenya. *Nature, Lond.*, 448: 688–91.

Spyropoulos, E., Borousas, D., Mavrikos, S., et al. (2002). Size of external genital organs and somatometric parameters among physically normal men younger than 40 years old. *Urology*, 60: 485–90.

Stammbach, E. (1987). Desert, forest and montane baboons: Multilevel societies. In *Primate Societies*. (eds. B.B. Smuts, D.L. Cheney., R.M. Seyfarth, R.W. Wrangham, and T.T. Struhsaker), pp. 112–20, University of Chicago Press: Chicago, IL.

Stanislaw, H. and Rice, F.J. (1988). Correlation between sexual desire and menstrual cycle characteristics. *Arch. Sex Behav.*, 17: 499–508.

Stanyon, R., Consigliere, S., and Moresalchi, M.A. (1993). Cranial capacity in hominid evolution. *Hum. Evol.*, 8: 205–16.

Starck, D. and Schneider, R. (1960). Respirationsorgane. In *Primatologia*, Vol. 3. (eds. H. Hofer, A.H. Schultz, and D. Starck), pp. 423–587, Karger: Basel.

Stockley, P. (2002). Sperm competition risk and male genital anatomy: Comparative evidence for reduced duration of female sexual receptivity in primates with penile spines. *Evol. Ecol.*, 16: 123–37.

Stockley, P. Searle, J.B. MacDonald, D.W., et al. (1993). Female multiple mating behaviour in the common shrew as a strategy to reduce inbreeding. *Proc. Roy. Soc. Lond. B.*, 254: 173–9.

Strait, D.S., Grine, F.E., and Moniz, M.A. (1997). A reappraisal of early hominid phylogeny. *J. Human. Evol.*, 32: 17–82.

Stringer, C. (2002). Modern human origins: Progress and prospects. *Phil. Trans. Roy. Soc. Lond. B.*, 357: 563–79.

Stringer, C. and Andrews, P. (2005). *The Complete World of Human Evolution*, Thames and Hudson: London.

Stringer, C.B. and Gamble, C.S. (1993). *In Search of the Neanderthals*, Thames and Hudson: London.

Štulhofer, A. (2006). How (un)important is penis size for women with heterosexual experience? *Arch. Sex. Behav.*, 35: 5–6.

Suggs, R.C. (1966). *Marquesan Sexual Behavior*, Harcourt Brace: New York.

Susman, R.L. (1988). New postcranial remains from Swartkrans and their bearing on the functional morphology and behavior of *Paranthropus robustus*. In *Evolutionary History of the Robust Australopithecines*. (ed. F.E. Grine), pp. 144–72, Aldine, Hawthorne, New York.

Susman, R.L. (1994). Fossil evidence of early hominid tool use. *Science* 265: 1570–3.

Suwa, G., Kono, R.T., Katoh, S., et al. (2007). A new species of great ape from the late Miocene epoch in Ethiopia. *Nature, Lond.*, 448: 921–4.

Suzuki, F. and Racey, P. (1984). Light and electron microscopical observations on the male excurrent duct system of the common shrew (*Sorex araneus*). *J. Reprod. Fertil.*, 70: 419–28.

Swaddle, J.P. and Reierson, G.W. (2002). Testosterone increases perceived dominance but not attractiveness in human faces. *Proc. Roy. Soc. Lond. B.*, 269: 2285–9.

Sykes, B. and Irven, C. (2000). Surnames and the Y chromosome. *Am. J. Hum. Genet.*, 66: 1417–9.

Symons, D. (1979). *The Evoluton of Human Sexuality*, Oxford University Press: Oxford.

Symons, D. (1995). Beauty is in the adaptations of the beholder. In *Sexual Nature, Sexual Culture*, (eds. P.R. Abramson and S.D. Pinkerson), pp. 80–118, University of Chicago Press: Chicago, IL.

Taggart, D.A., Breed, W.G., Temple-Smith, P.D., et al. (1998). Reproduction, mating strategies and sperm competition in marsupials and monotremes. In *Sperm Competition and Sexual Selection*. (eds. T.R. Birkhead and A.P. Møller), pp. 623–66, Academic Press: San Diego, CA.

Taggart, D.A., Johnson, J.L., O'Brien, H.P., et al. (1993). Why do spermatozoa of American marsupials form pairs? A clue from the analysis of sperm-pairing in the epididymis of the grey short-tailed possum, *Monodelphis domestica*. *Anat. Rec.*, 236: 465–78.

Tanner, J.M. (1978). *Foetus into Man: Physical Growth from Conception to Maturity*, Harvard University Press: Cambridge, MA.

Tanner, N. and Zihlman, A. (1976). Women in evolution: Innovation and selection in human origins. *Signs: J. Women and Culture and Society*, 1(3): 586–608.

Tassinary, L.G. and Hansen, K.A. (1998). A critical test of the WHR hypothesis of female sexual attractiveness. *Psychol. Sci.*, 9: 150–5.

Tattersall, I. and Schwartz, J.H. (2001). *Extinct Humans*, Westview Press: Boulder, CO.

Taub, D.M. (1980). Female choice and mating strategies among wild Barbary macaques (*Macaca sylvanus*). In *The Macaques: Studies in Ecology, Behavior and Evolution*. (ed. D.G. Lindburg), pp. 287–344, Van Nostrand Reinhold: New York.

Telford, S.R. and Jennions, M.D. (1998). Establishing cryptic female choice in animals. *TREE*, 13: 216–8.

Thompson-Handler, N., Malenky, R.K., and Badrian, N. (1984). Sexual behavior of *Pan Paniscus* under natural conditions in the Lomako Forest, Equateur, Zaire. In *The Pigmy Chimpanzee: Evolutionary Biology and Behavior*. (ed. R.L. Susman), pp. 347–68, Plenum: New York.

Thorne, A.G. and Wolpoff, M.H. (1981). Regional continuity in Australasian Pleistocene hominid evolution. *Am. J. Phys. Anthropol.*, 55: 337–49.

Thornhill, R. (1983). Cryptic female choice and its implications in the scorpionfly *Harpobittacus nigriceps*. *Am. Nat.*, 122: 765–88.

Thornhill, R. (2006). Human sperm competition and woman's dual sexuality. In *Sperm Competition in Humans:*

Classic and Contemporary Readings. (eds. T.K. Shackelford and N. Pound), pp. V–XVII. Springer: New York.

Thornhill, R., Gangestad, S.W., and Comer, R. (1996). Human female orgasm and mate fluctuating asymmetry. *Anim. Behav.*, 50: 1601–1615.

Titze, I.R. (2000). *Principles of Voice Production*, National Center for Voice and Speech: Iowa City, IA.

Tokuda, K., Simms, R.C., and Jensen, J.D. (1968). Sexual behavior in a captive group of pigtail macaques (*Macaca nemestrina*). *Primates*, 9: 283–94.

Tovée, M.J. and Cornelissen, P.L. (1999). The mystery of female beauty. *Nature, Lond.*, 399: 215–6.

Tovée, M.J., Maisey, D.S., Emery, J.L., et al. (1999). Visual cues to female sexual attractiveness. *Proc. Roy. Soc. Lond. B.*, 266: 211–8.

Troisi, A. and Carosi, M. (1998). Female orgasm rate increases with male dominance in Japanese macaques. *Anim. Behav.*, 56: 1261–6.

Tudge, C. (2000). *The Variety of Life*, Oxford University Press: Oxford.

Tutin, C.E.G. (1979). Mating patterns and reproductive strategies in a community of wild chimpanzees (*Pan troglodytes schweinfurthii*). *Behav. Ecol. Sociobiol.*, 6: 29–38.

Tutin, C.E.G. (1980). Reproductive behaviour of wild chimpanzees in the Gombe National Park, Tanzania. *J. Reprod. Fertil. Suppl.*, 28: 43–57.

Utami Atmoko, S. and Van Hooff, J.A.R.A.M. (2004). Alternative male reproductive tactics: male bimaturism in orang-utans. In *Sexual Selection in Primates: New and Comparative Perspectives*. (eds. P. Kappeler and C. Van Schaik), pp. 196–207, Cambridge University Press: Cambridge, UK.

Utami, S.S., Goosens, B., Bruford, M.W., et al. (2002). Male bimaturism and reproductive success in Sumatran orang-utans. *Behav. Ecol.*, 13: 643–52.

Van den Berghe, P.L. and Frost, P. (1986). Skin color preference, sexual dimorphism and sexual selection: A case of gene culture co-evolution? *Ethnic Rac. Stud.*, 9: 87–113.

Van Hooff, J.A.R.A.M. (1967). The facial displays of the catarrhine monkeys and apes. In *Primate Ethology*. (ed. D. Morris), pp. 7–68, Weidenfeld and Nicholson: London.

Van Hooff, M.H., Voorhorst, F.J., Kaptein, M.B., et al. (2000). Insulin, androgen, and gonadotropin concentration, body mass index, and waist-to-hip ratio in the first years after menarche in girls with regular menstrual cycles, irregular menstrual cycles or oligomenorrhea. *J. Clin. Endocr. Metab.*, 85: 1394–400.

Van Noordwijk, M.A. (1985). Sexual behaviour of Sumatran long-tailed macaques (*Macaca fascicularis*). *Z. Tierpsychol.*, 70: 277–296.

Van Roosmalen, M.G.M. and Klein, L.L. (1988). The spider monkeys, genus *Ateles*. In *Ecology and Behavior of Neotropical Primates*, Vol. 2. (eds. R.A. Mittermeier, A.B. Rylands, A. Coimbra-Filho, and G.A.B. Fonseca), pp. 455–537, World Wildlife Fund: Washington, D.C.

Van Wagenen, G. (1936). The coagulating function of the cranial lobe of the prostate gland in the monkey. *Anat. Rec.*, 118: 231–51.

Velasquez, H.E., Bellabarba, G.A., Mendoza, S., et al. (2000). Post prandial triglyceride response in patients with polycystic ovary syndrome: Relationship with waist-to-hip ratio and insulin. *Fertil. Steril.*, 74: 1159–63.

Waage, J.K. (1979). Dual function of the damselfly penis: Sperm removal and transfer. *Science*, 203: 916–8.

Wagner, G. and Green, R. (1981). *Impotence: Physiological, Psychological, Surgical Diagnosis and Treatment*, Plenum: New York.

Walker, A. and Leakey, R. (ed.) (1993). *The Nariokotome Homo erectus Skeleton*. Harvard University Press: Cambridge, MA.

Walker, A. and Shipman, P. (1996). *The Wisdom of the Bones*, Knopf: New York.

Wallach, S.J.R. and Hart, B.L. (1983). The role of the striated penile muscles of the male rat in seminal plug dislodgement and deposition. *Physiol. Behav.*, 31: 815–21.

Wallen, K. (1982). Influence of female hormonal state upon rhesus monkey sexual behavior varies with space for social interactions. *Science*, 217: 375–7.

Wallen, K. (1995). The evolution of female sexual desire. In: *Sexual Nature Sexual Culture*. (eds. P.R. Abramson and S.D. Pinkerton), pp. 57–79, University of Chicago Press: Chicago, IL.

Wallen, K. and Lloyd, E.A. (2008). Clitoral variability compared with penile variability supports nonadaptation of female orgasm. *Evol. Devel.*, 10: 1–2.

Wallen, K. and Winston, L.A. (1984). Social complexity and hormonal influences on sexual behavior in rhesus monkeys (*Macaca mulatta*). *Physiol. Behav.*, 32: 629–37.

Wallis, S.J. (1983). Sexual behaviour and reproduction of *Cercocebus albigena johnstonii* in Kibale Forest, Western Uganda. *Int. J. Primatol.*, 4: 153–66.

Ward, P.I. (2000). Cryptic female choice in the yellow dung fly *Scathophaga stercoraria* (L.). *Evolution*, 54: 1680–86.

Wasser, P.M. (1982). The evolution of male loud calls among mangabeys and baboons. In *Primate Communication*. (eds. C.T. Snowdon and C.H. Brown), pp. 117–43, Cambridge University Press: Cambridge, UK.

Wasser, S.K. and Waterhouse, M.L. (1983). The establishment and maintenance of sex biases. In *The Social Behavior of Female Vertebrates*. (ed. S.K. Wasser), pp. 19–35, Academic Press: New York.

Weber, J., Czarnetzki, A., and Pusch, C.M. (2005). Comment on the brain of LB1, *Homo floresiensis*. *Science*, 310: 236.

Wedekind, C. (2007). Body odours and body odour preferences in humans. In *The Oxford Handbook of Evolutionary Psychology*. (eds. R.I.M. Dunbar and L. Barrett), pp. 315–20, Oxford University Press: Oxford.

Weir, B.J. (1974). Reproductive characteristics of hystricomorph rodents. *Symp. Zool. Soc. Lond.*, 34: 265–301.

Weir, B.J. and Rowlands, I.W. (1973). Reproductive strategies of mammals. *Ann. Rev. Ecol. System.*, 4: 139–63.

Wellings, K., Field, J., Johnson, A.M., et al. (1994). *Sexual Behaviour in Britain: The National Survey of Sexual Attitudes and Lifestyles*, Penguin Books: London.

Wessells, H., Lue, T.F., and McAninch, J.W. (1996). Penile length in the flaccid and erect states: Guidelines for penile augmentation. *J. Urol.*, 156: 995–7.

Wetsman, A. and Marlowe, F. (1998). How universal are preferences for female waist-to-hip ratios? Evidence from the Hadza from Tanzania. *Evol. Hum. Behav.*, 20: 228–9.

Wheeler, P.E. (1984). The evolution of bipedality and loss of functional body hair in hominids. *J. Hum. Evol.*, 13: 91–8.

Whitcome, K.K., Shapiro, L.J., and Lieberman, D.E. (2007). Fetal load and the evolution of lumbar lordosis in bipedal hominins. *Nature, Lond.*, 450: 1075–8.

White, T.D., Asfaw, B., DeGusta, D., et al. (2003). Pleistocene *Homo sapiens* from middle Awash, Ethiopia. *Nature, Lond.*, 423: 742–7.

Wickings, E.J. and Dixson, A.F. (1992*a*). Testicular function, secondary sexual development, and social status in male mandrills (*Mandrillus sphinx*). *Physiol. Behav.*, 52: 909–16.

Wickings, E.J. and Dixson, A.F. (1992*b*). Development from birth to sexual maturity in a semi-free-ranging colony of mandrills (*Mandrillus sphinx*) in Gabon. *J. Reprod. Fertil.*, 95: 129–38.

Wilcox, A.J., Baird, D.D., Dunson, D.B., et al. (2004). On the frequency of intercourse around ovulation: Evidence for biological influences. *Human Repro.*, DOI: 10.1093/humrep/deh 305.

Wilcox, A.J., Weinberg, C.R., and Baird, D.D. (1995). Timing of sexual intercourse in relation to ovulation: Effects on the probability of conception, survival of the pregnancy and sex of the baby. *New Engl. J. Med.*, 333: 1517–21.

Wildt, D.E., Doyle, U., Stone, S.C., et al. (1977). Correlation of perineal swelling with serum ovarian hormone levels, vaginal cytology and ovarian follicular development during the baboon reproductive cycle. *Primates*, 18: 261–70.

Wildt, L., Kissler, S., Licht, P., et al. (1998). Sperm transport in the human female genital tract and its modulation by oxytocin as assessed by hysterosalpingoscintigraphy, hysterotonography. *Hum. Reprod. Update*, 4: 655–66.

Wilkinson, G.S. and McCracken, G.F. (2003). Bats and balls: Sexual selection and sperm competition in the Chiroptera. In *Bat Ecology*. (eds. T.H. Kunz and M.B. Fenton), pp. 128–55, University of Chicago Press: Chicago, IL.

Wilson, A.C. and Sarich, V.M. (1969). A molecular time scale for human evolution. *Proc. Nat. Acad. Sci.USA*, 63: 1088–93.

Wilson, C.G. (2008). Male genital mutilation: An adaptation to sexual conflict. *Evol. Hum. Behav.*, 29: 149–64.

Wilson, N., Tubman, S., Eady, P., et al. (1997). Female genotype affects male success in sperm competition. *Proc. Roy. Soc. Lond. B.*, 264: 1491–5.

Wolfe, L.D. (1984). Japanese macaque female sexual behavior: A comparison of Arashiyama East and West. In *Female Primates: Studies by Women Primatologists* (ed. M.F. Small), pp. 141–57, Alan R. Liss: New York.

Wolfe, L.D. (1991). Human Evolution and the sexual behavior of female primates. In *Understanding Behavior: What Primate Studies Tell Us About Human Behavior*. (eds. J. Loy, and C.B. Peters), pp. 121–51, Oxford University Press: New York.

Wong, R. and Ellis, C.N. (1984). Physiologic skin changes in pregnancy. *J. Am. Acad. Dermatol.*, 10: 929–43.

Wood, B. (1992). Origin and evolution of the genus *Homo*. *Nature, Lond.*, 355: 783–90.

Wood, B.A. (2001). Human Evolution: Overview. *Encyclopedia of Life Sciences*. DOI: 10.1038/npg.els.0001573, John Wiley and Sons.

Wood, B. and Collard, M. (1999). The human genus. *Science*, 284: 65–71.

Wrangham, R. (1979). On the evolution of ape social systems. *Soc. Sci. Inform.*, 18: 335–68.

Wrangham, R. (1987). The significance of African apes for reconstructing human social evolution. In *The Evolution of Human Behavior: Primate Models*. (ed. W.G. Kinzey), pp. 51–71, State University of New York Press: Albany, NY.

Wright, P.C. (1994). The behavior and ecology of the owl monkey. In *Aotus: The Owl Monkey*. (eds. J.F. Baer, R.E. Weller, and I. Kakoma), pp. 97–112, Academic Press: San Diego, IL.

Yamagiwa, J. (2006). Playful encounters: the development of homosexual behaviour in male mountain gorillas. In *Homosexual Behaviour in Animals: An Evolutionary Perspective*. (eds. V. Sommer and P.L. Vasey), pp. 273–93, Cambridge University Press: Cambridge, UK.

Yanagimachi, R. (1994). Mammalian fertilization. In *The Physiology of Reproduction*, Vol. 1., 2nd edn. (eds. E. Knobil and J.D. Neill), pp. 189–317, Raven Press: New York.

Yu, D.W. and Shepard, G.H. (1998). Is beauty in the eye of the beholder? *Nature, Lond.*, 396: 321–2.

Zachmann, M., Prader, A., Kind, H.P., et al. (1974). Testicular volume during adolescence. Cross-sectional and longitudinal studies. *Hel. Paediat. Acta.*, 29: 61–72.

Zavos, P.M. (1985). Seminal parameters of ejaculates collected from oligospermic and normospermic patients via masturbation and at intercourse with the use of a silastic seminal fluid collection device. *Fertil. Steril.*, 44: 517–20.

Zavos, P.M. (1988). Seminal parameters of ejaculates collected at intercourse with the use of a seminal collection device with different levels of precoital stimulation. *J. Androl.*, 9: 36.

Zollikofer, C.P.E., Ponce de León, M.S., Lieberman, D.E., et al. (2005). Virtual cranial reconstruction of *Sahelanthropus tchadensis*. *Nature, Lond.*, 434: 755–9.

Zuckerman, S. (1932). *The Social Life of Monkeys and Apes*, Harcourt, Brace and Co.: New York.

Zuckerman, S. (1981). *The Social Life of Monkeys and Apes*, 2nd edn. Routledge and Kegan Paul: London.

Zuckerman, S. Ashton, E.H., Flinn, R.M., et al. (1973). Some locomotor features of the pelvic girdle in primates. *Symp. Zool. Soc. Lond.*, 33: 71–165.

INDEX

Note: page numbers in *italics* refer to Figures and Tables.

Printed in the United States
By Bookmasters